Simplified Engineering for Architects and Builders

Simplified Engineering for Architects and Builders

||||||||||||||||||||||||||||||||||||||||||||||||||||||||||||||||||||||||||||||||||||||||

The Late Harry Parker, M.S.
Formerly Professor of Architectural Construction
University of Pennsylvania

SIXTH EDITION

prepared by

JAMES AMBROSE, M.S.
Professor of Architecture
University of Southern California

A Wiley-Interscience Publication
JOHN WILEY & SONS
New York   Chichester   Brisbane   Toronto   Singapore

Library of Congress Cataloging in Publication Data:

Parker, Harry, 1887–
    Simplified engineering for architects and builders.

    "A Wiley-Interscience publication."
    Includes index.
    1. Structural engineering.   I. Ambrose, James E.
II. Title.
TA633.P3   1983   690′.21   83-14512
ISBN 0-471-86611-3

Printed in the United States of America

10 9 8 7 6

Preface to the Sixth Edition

Publication of this edition of Professor Parker's book presents the opportunity for yet another generation of designers and builders to utilize this highly popular work. The purpose and intentions of the work remain essentially the same as those stated by Professor Parker in his preface to the first edition, which follows. If anything, the need for a simplified presentation of structural design for buildings is even more acute now with the increased complexity and diversity of the technology and the sophistication of design theory and procedures.

In developing this new edition, I began with the basic premise that no book could have endured in popularity for so long if there was something fundamentally wrong with it. I have therefore sought to retain the original character of the work as much as possible while giving it the lightest of trims required to bring the material into conformance with current technology, construction practices, and design procedures. Some new material has been added, including treatments of structural steel bolts, formed steel decks, and nailed joints in wood. Data and criteria have been brought into conformance with the eighth edition of the *Manual of Steel Construction,* the 1982 edition of the *National Design Specification for Wood Construction,* the 1977 edition of the ACI *Building Code Requirements for Reinforced Concrete,* and the 1982 edition of the *Uniform Building Code.*

A major revision in this edition is the inclusion of data and computations in the metric-based SI (Systeme Internationale) units. While work in the building field continues to utilize the old English (now more properly called U.S.) units, it is quite likely that those now beginning to become involved in this field will need to develop the facility to work in both unit systems. Work in this edition is developed primarily in the traditional units, but wherever feasible, the corresponding material has been included in SI units. This presents the opportunity for the reader to learn whichever of the two systems he or she is least familiar with, utilizing the work as a running translation.

As in previous editions the level of mathematical work has been kept well below that requiring facility with anything more than simple algebra, geometry, and elementary trigonometry. In general, the work is intended to be understandable to persons lacking any formal training in engineering. Readers having some such background preparation may find it possible to spend less time on the material in Part 1, although a review of this fundamental material may serve some purpose.

I am grateful to the American Institute of Steel Construction, American Concrete Institute, National Forest Products Association, Steel Deck Institute, Steel Joist Institute, and International Conference of Building Officials for permission to reprint and abstract materials from their publications. I am also, as ever, highly appreciative of the excellent work and the support given to me by the editors and production staff at John Wiley and Sons. I have recently had the opportunity to teach a course using the previous edition of this book and am grateful to the School of Architecture at the University of Southern California and to the students for the highly illuminating feedback I received.

Finally, I am indebted once more to my family for their patience and endurance and for the direct assistance given to me by my wife, Peggy, and my daughter, Julie.

JAMES AMBROSE

Westlake Village, California
August 1983

Preface to the First Edition

IIIIIIIIIIIIIIIIIIIIIIIIIIIIIIIIIIIIIIIIIIIIIIIIIIIIIIIIIIII

To the average young architectural draftsman or builder, the problem of selecting the proper structural member for given conditions appears to be a difficult task. Most of the numerous books on engineering which are available assume that the reader has previously acquired a knowledge of fundamental principles and, thus, are almost useless to the beginner. It is true that some engineering problems are exceedingly difficult, but it is also true that many of the problems that occur so frequently are surprisingly simple in their solution. With this in mind, and with a consciousness of the seeming difficulties in solving structural problems, this book has been written.

In order to understand the discussions of engineering problems, it is essential that the student have a thorough knowledge of the various terms which are employed. In addition, basic principles of forces in equilibrium must be understood. The first section of this book, "Principles of Mechanics," is presented for those who have not had such technical preparation, as well as for those who wish a brief review of the subject. Following this section are structural problems involving the most commonly used building materials, wood, steel, reinforced concrete, and roof trusses. A major portion of the book is devoted to numerous problems and their solution, the purpose of which is to explain practical procedure in the design of structural members. Similar examples are given to be solved by the student. Although handbooks published by the manufacturers are necessities to the more advanced student, a great number of appropriate tables are presented herewith so that sufficient data are directly at hand to those using this book.

Care has been taken to avoid the use of advanced mathematics, a knowledge of arithmetic and high-school algebra being all that is required to follow the discussions presented. The usual formulas employed in the solution of structural problems are given with explanations of the terms involved and their application, but only the most elementary of these formulas are derived. These derivations are given to show how simple they are and how the underlying principle involved is used in building up a formula that has a practical application.

No attempt has been made to introduce new methods of calculation, nor have all the various methods been included. It has been the desire of the author to present to those having little or no knowledge of the subject simple solutions of everyday structural problems. Whereas thorough technical training is to be desired, it is hoped that this presentation of fundamentals will provide valuable working knowledge and, perhaps, open the doors to more advanced study.

HARRY PARKER

Philadelphia, Pennsylvania
March 1938

Contents

‖‖

Introduction

IIIIIIIIIIIIIIIIIIIIIIIIIIIIIIIIIIIIIIIIIIIIIIIIIIIIII

Those unfamiliar with the terms and basic principles of structural mechanics should study Part 1 thoroughly before attempting the succeeding parts. Others who have had previous work in mechanics may wish to review this material, since it is fundamental to the technical discussions in all other parts. The following suggestions are offered as aids to study:

1. Take up each item in the sequence presented and be certain that each is thoroughly understood before continuing with the next.

2. Since each problem to be solved is prepared to illustrate some basic principle or procedure, read it carefully and make sure you understand exactly what is wanted before starting to solve it.

3. Whenever possible make a sketch showing the conditions and record the data given. Such diagrams frequently show at a glance the problem to be solved and the procedure necessary for its solution.

4. Make a habit of checking your answers to problems. Confidence in the accuracy of one's computations is best gained by self-checking. However, in order to provide an occasional outside check, answers to some of the exercise problems are given at the end of the book. Such problems

TABLE 1. Units of Measurement: U.S. System

Name of Unit	Abbreviation	Use
Length		
Feet	ft	large dimensions, building plans, beam spans
Inches	in.	small dimensions, size of member cross sections
Area		
Square feet	ft^2	large areas
Square inches	in^2	small areas, properties of cross sections
Volume		
Cubic feet	ft^3	large volumes, quantities of materials
Cubic inches	in^3	small volumes
Force, mass		
Pounds	lb	specific weight, force, load
Kips	k	1000 pounds
Pounds per foot	lb/ft	linear load (as on a beam)
Kips per foot	k/ft	linear load (as on a beam)
Pounds per square foot	lb/ft^2, psf	distributed load on a surface
Kips per square foot	k/ft^2, ksf	distributed load on a surface
Pounds per cubic foot	lb/ft^3, pcf	relative density, weight
Moment		
Foot-pounds	ft-lb	rotational or bending moment
Inch-pounds	in-lb	rotational or bending moment
Kip-feet	k-ft	rotational or bending moment
Kip-inches	k-in	rotational or bending moment
Stress		
Pounds per square foot	lb/ft^2, psf	soil pressure
Pounds per square inch	lb/in^2, psi	stresses in structures

TABLE 1. (*Continued*)

Name of Unit	Abbreviation	Use
Kips per square foot	k/ft², ksf	soil pressure
Kips per square inch	k/in², ksi	stresses in structures
Temperature		
Degrees Fahrenheit	°F	temperature

are indicated by an asterisk (*) following the problem number where it occurs in the text.

5. If you do not own a pocket calculator, get one at the first opportunity. The ability to use this computational aid is readily acquired, and it will shortly become indispensable.

6. In solving problems, form the habit of writing the denomination of each quantity. The solution of an equation will be a number. It may be so many pounds, or is it pounds per square inch? Are the units foot-pounds or inch-pounds? Adding the names of the quantities signifies an exact knowledge of the quantity and frequently prevents subsequent errors. Abbreviations are commonly used for this purpose, and those employed in this book are identified in the following discussion of units of measurement.

Units of Measurement

At the time of preparation of this edition, the building industry in the United States is still in a state of confused transition from the use of English units (feet, pounds, etc.) to the new metric-based system referred to as the SI units (for Système International). Although a complete phase-over to SI units seems inevitable, at the time of this writing the construction-materials and products suppliers in the United States are still resisting it. Consequently, the AISC Manual and most building codes and other widely used references are still in the old units. (The old system is now more appropriately called the U.S. system because England no longer uses it!) Although it results in some degree of clumsiness in the work, we have chosen to give the data and computations in this book in both units as much as is practicable. The technique is generally to perform the work in U.S. units and immediately follow it with the equivalent work in SI units enclosed in brackets [thus] for separation and identity.

Table 1 lists the standard units of measurement in the U.S. system with the abbreviations used in this work and a description of the type of use in structural work. In similar form Table 2 gives the corresponding units in the SI system. For more ready access the conversion units used in shifting from one system to the other appear on the inside back cover of this book.

Symbols

The following "shorthand" symbols are frequently used:

Symbol	Reading
$>$	is greater than
$<$	is less than
$\geqq$	equal to or greater than
$\leqq$	equal to or less than
$6'$	six feet
$6''$	six inches
Σ	the sum of
ΔL	change in L

Computations

In most professional design firms structural computations are now commonly done with computers, particularly when the work is complex or repetitive. Anyone aspiring to participation in professional design work is advised to acquire the background and experience necessary to the application of computer-aided techniques. The computational work in this book is simple and can be performed easily with a pocket calculator. The reader who has

TABLE 2. Units of Measurement: SI System

Name of Unit	Abbreviation	Use
Length		
Meters	m	large dimensions, building plans, beam spans
Millimeters	mm	small dimensions, size of member cross sections
Area		
Square meters	m^2	large areas
Square millimeters	mm^2	small areas, properties of cross sections
Volume		
Cubic meters	m^3	large volumes
Cubic millimeters	mm^3	small volumes
Mass		
Kilograms	kg	mass of materials (equivalent to weight in U.S. system)
Kilograms per cubic meter	kg/m^3	density
Force (*load on structures*)		
Newtons	N	force or load
Kilonewtons	kN	1000 Newtons
Stress		
Pascals	Pa	stress or pressure (one pascal = one N/m^2)
Kilopascals	kPa	1000 Pascals
Megapascals	MPa	1,000,000 Pascals
Gigapascals	GPa	1,000,000,000 Pascals
Temperature		
Degrees Celcius	°C	temperature

not already done so is advised to obtain one. The "pocket slide rule" type with eight-digit capacity is quite sufficient.

Structural computations can for the most part be rounded off. Accuracy beyond the third place is seldom significant, and this is the level used in this work. In some examples more accuracy is carried in early stages of the computation to ensure the desired degree in the final answer. All the work in this book, however, was performed on an eight-digit pocket calculator.

I

PRINCIPLES OF STRUCTURAL MECHANICS

1

Forces and Stresses

||||||||||||||||||||||||||||||||||||||||||||||||||||||||||||

1-1 Introduction

Mechanics is the science that treats of the action of forces on
material bodies, and *statics* is that branch of mechanics which
treats of motionless bodies held in equilibrium by the balanced
external forces acting on them.

Strength of materials considers the behavior of material bodies
in resisting the action of external forces, the stresses developed
within the bodies, and the deformations that result from the external forces.

Taken together, these two subjects constitute the field of *structural mechanics*, and it is the purpose of the chapters in this
section to present the key principles of structural mechanics that
form the basis of structural design.

1-2 Forces

A force is that which tends to change the state of rest or motion of
a body. It may be considered as pushing or pulling a body at a
definite point and in a definite direction. Such a force tends to
give motion to a body at rest, but this tendency may be neutral-
ized by the action of another force or set of forces. A force is
completely determined when its magnitude, direction, sense (or
sign), line of action, and point of application are known. In build-

ing construction we are concerned primarily with forces in equilibrium, that is, with bodies at rest. For example, a steel column supports a given load which, owing to gravity, is downward. The column transfers the load to the footing below. The resulting earth pressure on the footing equals the load in magnitude; its sense is upward, and it is called the *reaction.* The two forces are both vertical in direction, have the same line of action, and are equal in magnitude, but they are opposite in sense (or sign). The result is equilibrium, that is, no motion. The units of force in English units are pounds, tons, and so on. In engineering work it is common to use a unit called a kip, meaning 1000 pounds. In SI (Systeme International) units, mass is measured in kilograms (kg), but force is measured in Newtons (N) or kilonewtons (1000 Newtons) (kN).

1-3   Direct Stress

Stresses are the means by which a body develops internal resistance to external forces. The hanger bar shown in Fig. 1-1a supports a suspended load P acting along the vertical axis of the bar. The load constitutes an external tensile force that tends to stretch the bar, and the bar resists the tendency to elongate by developing an internal tensile force equal in magnitude to the external force. This internal force is developed by stresses in the material

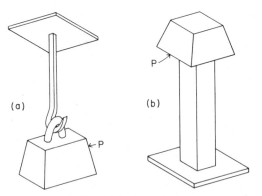

FIGURE 1-1.   Direct stress.

of the bar. Under this condition of axial loading, the tensile stress produced is called a *direct stress*.

A characteristic of direct stress is that the internal force may be assumed to be evenly distributed over the cross-sectional area of the body under stress. Thus if force P in Fig. 1-1a is 30 kips [133.44 kN] and the cross-sectional area of the hanger bar is 2 square inches (written in^2) [1290.4 mm^2], each unit of area is stressed to 30 ÷ 2 = 15 kips per square inch (written ksi) [103,425 kPa], or 15,000 pounds per square inch (psi). This tensile stress per unit of area is called a *unit stress* to distinguish it from the total internal force of 30,000 pounds (lb). By calling the load or external force P, the area of the bar's cross section A, and the unit stress f, the fundamental relationship governing direct stress may be stated

$$f = \frac{P}{A} \quad \text{or} \quad P = fA \quad \text{or} \quad A = \frac{P}{f}$$

When using this relationship, remember the assumptions on which it is based: the loading is axial and the stresses are evenly (uniformly) distributed over the cross section. Note also that if any two of the quantities are known, the third may be found.

Example. Suppose a wrought iron rod with a diameter of 1.5 in. [38.1 mm] is used as a hanger in an arrangement similar to that shown in Fig. 1-1a. If the allowable unit tensile stress for the wrought iron is 12,000 psi [82,740 kPa], determine the load that the rod will safely support.

Solution: (1) To find the area of the rod cross section

$$A = \pi r^2 = 3.1416 \left(\frac{1.5}{2}\right)^2 = 1.767 \text{ in}^2$$

$$\left[A = 3.1416 \left(\frac{38.1}{2}\right)^2 = 1140 \text{ mm}^2\right]$$

(2) Since the allowable unit stress is 12,000 psi [82,740 kPa], the load the rod may carry is

$$P = fA = (12,000)(1.767) = 21,204 \text{ lb}$$
$$[P = (82,740)(1140)(10)^{-6} = 94,324 \text{ N}]$$

1-4 Kinds of Stress

The three basic kinds of stress that concern us are *tension, compression,* and *shear.* As we observed in Section 1-3, tensile forces tend to stretch a structural member. Compressive forces tend to shorten members, and shearing forces tend to make parts of a structure slide past each other.

In addition to occurring under conditions of direct stress, tension and compression are also developed in structural members subjected to bending or flexure (Fig. 1-2e). This is explained in detail later.

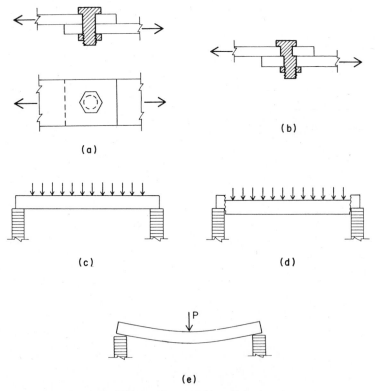

FIGURE 1-2.   Shear and bending stresses.

1-5 Compression

The load P on the short, square block B shown in Fig. 1-1b exerts an axial force on the block that tends to shorten it. This external compressive force is resisted by an internal compressive force equal to P. The *unit compressive stress* is given by the direct stress formula $f = P/A$. However, this relationship holds for short compression members only. As the ratio of length to least width of compression members increases, other factors enter the problem; these are considered under the design of steel, wood, and reinforced concrete columns in Parts 2, 3, and 4, respectively.

Example. A short timber post with nominal cross-sectional dimensions of 8 × 8 in. [203.2 mm] supports an axial load of 50,000 lb [222.4 kN], as indicated in Fig. 1-1b. Find the unit compressive stress developed in the post.

Solution: (1) The nominal size of the post is 8 × 8 in., but the actual size is $7\frac{1}{2} \times 7\frac{1}{2}$ in. [190.5 mm]; therefore the area of the cross section is $(7.5)^2 = 56.25$ in^2 [36,290 mm^2] (see Table 4-7).

(2) With $P = 50,000$ lb the unit stress is thus

$$f = \frac{P}{A} = \frac{50,000}{56.25} = 888.9 \text{ psi}$$

$$\left[f = \frac{222.4 \times 10^6}{36,290} = 6128 \text{ kPa} \right]$$

1-6 Shear

A shearing stress occurs when two forces act on a body in opposite directions but not in the same plane. This condition is illustrated in Fig. 1-2a, which shows two plates held together by a bolt. Under the action of the forces P, the plates tend to shear the bolt at their plane of contact as indicated in Fig. 1-2b. Another illustration of shear is given in Fig. 1-2c, where a load W rests on a beam which in turn is supported on walls at its ends. It is evident from the sketch that the beam might fail by dropping between the walls (Fig. 1-2d). This type of shear is discussed in Chapter 3.

Whereas the direct tensile and compressive stresses discussed in Sections 1-3 and 1-5 act at right angles to the cross sections of the members considered, shearing stress acts transversely, or parallel, to the cross section. With respect to the bolt in Figs. 1-2a and b, the unit shearing stress f_v is given by the formula $f_v = P/A$, where P is the shearing force and A is the cross-sectional area of the bolt. It will be noted that this expression is similar to the direct stress formula $f = P/A$. However, it must be understood clearly that the *physical situations represented by the two formulas are quite different.*

Example.   The forces P in the plates illustrated in Fig. 1-2a are each 5000 lb [22.24 kN], and the bolt has a diameter of $\frac{3}{4}$ in. [19.05 mm]. What is the unit shearing stress?
Solution: (1) The area of the bolt may be computed from the formula $A = \pi r^2$ or found directly from Table 7-1, which gives a value of 0.4418 in² [285.0 mm²].
(2) Since $P = 5000$ lb, the unit shearing stress is

$$f_v = \frac{P}{A} = \frac{5000}{0.4418} = 11{,}317 \text{ psi}$$

$$\left[f_v = \frac{22.4 \times 10^6}{285.0} = 78{,}596 \text{ kPa} \right]$$

1-7   Bending

Figure 1-2e illustrates a simple beam with a concentrated load P at the center of the span. This is an example of *bending, or flexure*. The fibers in the upper part of the beam are in compression, and those in the lower part are in tension. Although steel and concrete are not fibrous materials in the sense that wood is, the concept of infinitely small fibers is useful in the study of stress relationships within any material. These stresses are not uniformly distributed over the cross section of the beam and cannot be computed by the direct stress formula. The expression used to compute the value of the bending stress in either tension or compression is known as the *beam formula*, or the *flexure formula*, and is considered in Chapter 4.

Problem 1-7-A*. A wrought iron bar sustains a tensile force of 40 kips [177.92 kN]. If the allowable unit tensile stress is 12 ksi [82,740 kPa], what is the required cross-sectional area of the bar?

Problem 1-7-B. What axial load may be placed on a short timber post, whose actual cross-sectional dimensions are $9\frac{1}{2} \times 9\frac{1}{2}$ in. [241.3 mm], if the allowable unit compressive stress is 1100 psi [7585 kPa]?

Problem 1-7-C*. What should be the diameter of the bolt shown in Fig. 1-2a if the shearing force is 9000 lb [40.03 kN] and the allowable unit shearing stress is 15 ksi [103,425 kPa]?

Problem 1-7-D. The allowable bearing capacity of a soil is 8000 psf [383 kPa]. What should be the length of the side of a square footing if the total load (including the weight of the footing) is 240 kips [1067.5 kN]?

Problem 1-7-E. If a steel bolt with a diameter of $1\frac{1}{4}$ in. [31.75 mm] is used for the fastener shown in Fig. 1-2a, find the shearing force that can be transmitted across the joint if the allowable unit shearing stress in the bolt is 15 ksi [103,425 kPa].

Problem 1-7-F*. A short, hollow, cast iron column is circular in cross section, the outside diameter being 10 in. [254 mm] and the thickness of the shell $\frac{3}{4}$ in. [19.05 mm]. If the allowable unit compressive stress is 9 ksi [62,055 kPa], what load will the column support?

Problem 1-7-G. Determine the minimal cross-sectional area of a steel bar required to support a tensile force of 50 kips [222.4 kN] if the allowable unit tensile stress is 20 ksi [137,900 kPa].

Problem 1-7-H. A short, square timber post supports a load of 115 kips [511.5 kN]. If the allowable unit compressive stress is 1000 psi [6895 kPa], what nominal size square timber should be used? (See Table 4-7).

1-8 Deformation

Whenever a force acts on a body, there is an accompanying change in shape or size of the body. In structural mechanics this is called *deformation*. Regardless of the magnitude of the force, some deformation is always present, although often it is so small that it is difficult to measure even with the most sensitive instruments. In the design of structures it is often necessary that we know what the deformation in certain members will be. A floor joist, for instance, may be large enough to support a given load safely but may *deflect* (the term for deformation that occurs with bending) to such an extent that the plaster ceiling below will crack, or the floor may feel excessively springy to persons walking on it. For the usual cases we can readily determine what the deformation will be. This is considered in more detail later.

1-9 Hooke's Law

As a result of experiments with clock springs, Robert Hooke, a mathematician and physicist working in the seventeenth century, developed the theory that "deformations are directly proportional to stresses." In other words, if a force produces a certain deformation, twice the force will produce twice the amount of deformation. This law of physics is of utmost importance in structural engineering although, as we shall find, Hooke's Law holds true only up to a certain limit. (UP TO ELASTIC LIMIT)

OR YIELD STRESS
36 K

1-10 Elastic Limit and Yield Point

Suppose that we place a bar of structural steel with a cross-sectional area of 1 sq in. [645.2 mm^2] into a machine for making tension tests. We measure its length accurately and then apply a tensile force of 5000 lb [22.24 kN], which, of course, produces a unit tensile stress of 5000 psi [34,475 kPa] in the bar. We measure the length again and find that the bar has lengthened a definite amount, which we will call x inches. On applying 5000 lb more, we note that the amount of lengthening is now $2(x)$, or twice the amount noted after the first 5000 lb. If the test is continued, we will find that for each 5000-lb increment of additional load, the length of the bar will increase the same amount as noted when the initial 5000 lb was applied; that is, the deformations (length changes) are directly proportional to the stresses. So far Hooke's Law has held true, but after we reach a unit stress of about 36,000 psi [248,220 kPa], the length increases more than x for each additional 5000 lb of load. This unit stress is called the *elastic limit,* or the *yield stress,* and it varies for different grades of steel. Beyond this stress limit, Hooke's Law will no longer apply.

Another phenomenon may be noted in this connection. If we make the test again, we will discover that when any applied load which produces a unit stress *less* than the elastic limit is removed, the bar returns to its original length. If a load producing a unit stress *greater* than the elastic limit is removed, we will find that the bar has permanently increased its length. This permanent deformation is called the *permanent set*. This fact permits another way of defining the elastic limit: it is that unit stress beyond which the material does not return to its original length when the load is removed.

If our test is continued beyond the elastic limit, we quickly reach a point where the deformation increases without any increase in the load. The unit stress at which this deformation occurs is called the *yield point;* it has a value only slightly higher than the elastic limit. Since the yield point, or yield stress, as it is sometimes called, can be determined more accurately by test than the elastic limit, it is a particularly important unit stress. Nonductile materials such as wood and cast iron have poorly defined elastic limits and no yield point.

1-11 Ultimate Strength

After passing the yield point, the steel bar of the test described in the preceding section again develops resistance to the increasing load. When the load reaches a sufficient magnitude, rupture occurs. The unit stress in the bar just before it breaks is called the *ultimate strength.* For the grade of steel assumed in our test, the ultimate strength occurs at about 70,000 psi.

Structural members are designed so that stresses under normal service conditions will not exceed the elastic limit, even though there is considerable reserve strength between this value and the ultimate strength. This procedure is followed because deformations produced by stresses above the elastic limit are permanent and hence change the shape of the structure.

1-12 Factor of Safety

The degree of uncertainty that exists, with respect to both actual loading of a structure and uniformity in the quality of materials, requires that some reserve strength be built into the design. This degree of reserve strength is the *factor of safety*. Although there is no general agreement on the definition of this term, the following discussion will serve to fix the concept in mind.

Consider a structural steel that has an ultimate tensile unit stress of 58,000 psi [399,910 kPa], a yield-point stress of 36,000 psi [248,220 kPa], and an allowable stress of 22,000 psi [151,690 kPa]. If the factor of safety is defined as the ratio of the ultimate strength to the allowable stress, its value is 58,000 ÷ 22,000, or 2.64. On the other hand, if it is defined as the ratio of the yield-point stress to the allowable stress, its value is 36,000 ÷ 22,000, or 1.64. This is a considerable variation, and since failure of a structural member begins when it is stressed beyond the elastic limit, the higher value may be misleading. Consequently, the term *factor of safety* is not employed extensively today. Building codes generally specify the allowable unit stresses that are to be used in design for the grades of structural steel to be employed.

If one should be required to pass judgment on the safety of a structure, the problem resolves itself into considering each structural element, finding its actual unit stress under the existing loading conditions, and comparing this stress with the allowable stress prescribed by the local building regulations. This procedure is called *investigation*.

1-13   Modulus of Elasticity

We have seen that, within the elastic limit of a material, deformations are directly proportional to the stresses. Now we shall compute the magnitude of these deformations by use of a number (ratio), called the *modulus of elasticity*, that indicates the degree of *stiffness* of a material.

A material is said to be stiff if its deformation is relatively small when the unit stress is high. As an example, a steel rod 1 sq in. [645.2 mm²] in cross-sectional area and 10 ft [3.05 m] long will elongate about 0.008 in. [0.203 mm] under a tensile load of 2000 lb [8.90 kN]. But a piece of wood of the same dimensions will stretch about 0.24 in. [6.096 mm] with the same tensile load. We say that the steel is stiffer than the wood because, for the same unit stress, the deformation is not so great.

Modulus of elasticity is defined as the unit stress divided by the unit deformation. Unit deformation refers to the percent of deformation and is usually called strain. It is dimensionless, since it is expressed as a ratio, as follows:

$$\text{strain} = s = \frac{e}{l}$$

in which s = the unit strain,
$\quad e$ = the actual dimensional change,
$\quad l$ = the original length of the member.

The modulus of elasticity is represented by the letter E, expressed in pounds per square inch, and has the same value in compression and tension for most structural materials. Letting f represent the unit stress and s the unit deformation, we have, by definition,

$$E = \frac{f}{s}$$

From Section 1-3 we remember that $f = P/A$. It is obvious that, if l represents the length of the member and e the total deformation, then s, the deformation per unit of length, must equal the total deformation divided by the length, or $s = e/l$. Now, if we substitute these values in the equation determined by definition,

$$E = \frac{f}{s} = \frac{P/A}{e/l} = \frac{P}{A} \times \frac{l}{e}$$

This can also be written

$$e = \frac{Pl}{AE}$$

in which e = total deformation in inches,
$\quad P$ = force in pounds,
$\quad l$ = length in inches,
$\quad A$ = cross-sectional area in square inches,
$\quad E$ = modulus of elasticity in pounds per square inch.

Note that E is expressed in the same units as f (pounds per square inch) because in the equation $E = f/s$, s is a dimensionless number. For steel E = 29,000,000 psi [200,000,000 kPa], and for wood, depending on the species and grade, it varies from something less than 1,000,000 psi [6,895,000 kPa] to about 1,900,000 psi [13,100,000 kPa]. For concrete E ranges from about 2,000,000 psi [13,790,000 kPa] to about 5,000,000 psi [34,475,000 kPa] for common structural grades. The important thing to remember is that the foregoing formula is valid only when the unit stress lies within the elastic limit of the material.

Example.   A 2-in.[50.8mm] diameter round steel rod 10 ft [3.05 m] long is subjected to a tensile force of 60 kips [266.88 kN]. How much will it elongate under the load?

Solution: (1) The area of the 2-in. rod is 3.1416 sq in. [2027 mm²].

(2) Checking to determine whether the stress in the bar is within the elastic limit, we find that

$$f = \frac{P}{A} = \frac{60}{3.1416} = 19.1 \text{ ksi}$$

$$\left[f = \frac{266.88 \times 10^6}{2027} = 131,663 \text{ kPa} \right]$$

which is within the elastic limit of structural steel, so the formula for finding the deformation is applicable.

(3) From data, $P = 60$ kips, $l = 120$ (length in inches), $A = 3.1416$, and $E = 29,000,000$. Substituting these values, we calculate that the total lengthening of the rod is

$$e = \frac{Pl}{AE} = \frac{(60,000)(120)}{(3.1416)(29,000,000)} = 0.079 \text{ in.}$$

$$\left[e = \frac{(266.88 \times 10^6)(3050)}{(2027)(200,000,000)} = 2.0 \text{ mm} \right]$$

1-14 Allowable Unit Stresses

In the examples and problems dealing with the direct stress equation, we have differentiated between the unit stress developed in a member sustaining a given load ($f = P/A$) and the *allowable unit stress* used when determining the size of a member required to carry a given load ($A = P/f$). The latter form of the equation is, of course, the one used in design.

From the discussion in Sections 1-8 through 1-12, we can see that the allowable unit stresses should be set within the elastic limit of the structural material being used. The procedures for establishing allowable unit stresses in tension, compression, shear, and bending are different for different materials and are prescribed in specifications promulgated by the American Society for Testing and Materials. In general, allowable stresses for structural steel are expressed as fractions of the yield stress, those for wood involve an adjustment of clear wood strength as modified by lumber grading rules and conditions of use, and allowable stresses for concrete are given as fractions of the specified compressive strength of concrete. Tables 5-3, 10-1, and 13-1 give allowable stresses for steel, wood, and reinforced concrete construction, respectively, as recommended by the industry associations concerned. These are the American Institute of Steel Construction, the National Forest Products Association, and the American Concrete Institute. When scanning these tables, you will notice that they contain several terms that have not been introduced in this book thus far. These will be identified in subsequent sections dealing with the design of members to which they apply. However, in order to provide information for convenient reference when solving the problems at the end of this chapter, selected data from the more complete allowable stress tables are presented in Table 1-1.

In actual design work, the building code governing the construction of buildings in the particular locality must be consulted for specific requirements. Many municipal codes are revised infrequently and, consequently, may not be in agreement with current editions of the industry-recommended allowable stresses. Unless otherwise noted, the allowable stresses used in this book are those given in the three tables referenced above.

1-15 Use of Direct Stress Formula

Except for shear the stresses we have discussed so far have been direct or axial stresses. This, we recall, means they are assumed to be uniformly distributed over the cross section. The examples and problems presented fall under three general types: first, the

TABLE 1-1.   Selected Stress Values for Common Structural Materials

Material and Property	Common Values	
	(in pounds per sq in.)	(in kPa)
Structural steel		
Yield strength	36,000	248,220
Allowable tension	22,000	151,690
Allowable shear (on rivets)	15,000	103,425
E	29,000,000	200,000,000
Concrete		
f'_c (specified compressive strength)	3,000	20,685
Usable compression (in bearing)	900	6,206
Shear on concrete, in beams	60	414
E	3,100,000	21,374,500
Structural lumber		
1. Douglas fir, select structural grade, posts & timbers Compression parallel to grain	1,150	7,929
E	1,600,000	11,032,000
2. Southern pine, No. 1 dense SR grade, 5 in. & thicker compression parallel to grain	925	6,378
E	1,600,000	11,032,000

Source: Taken from Tables 5-3, 10-1, and 13-1.

design of structural members ($A = P/f$); second, the determination of safe loads ($P = fA$); third, the investigation of members for safety ($f = P/A$). The following examples will serve to fix in mind each of these types.

Example 1.   Design (determine the size of) a short, square post of Southern pine, No. 1 dense SR grade, to carry an axial compressive load of 30,000 lb [133,440 N].

Solution: (1) Referring to Table 1-1, we find that the allowable unit compressive stress for this wood parallel to the grain is 925 psi [6378 kPa].

(2) The required area of the post is

$$A = \frac{P}{f} = \frac{30,000}{925} = 32.43 \text{ in}^2$$

$$\left[A = \frac{133,440 \times 10^3}{6378} = 20,922 \text{ mm}^2 \right]$$

(3) From Table 4-7 an area of 30.25 in² [19,517 mm²] is provided by a 6 × 6-in. post with a dressed size of $5\frac{1}{2}$ × $5\frac{1}{2}$ in. [139.7 mm].

Example 2. Determine the safe axial compressive load for a short, square concrete pier with a side dimension of 2 ft [0.6096 m].

Solution: (1) The area of the pier is 4 ft² or 576 in² [0.3716 m²].

(2) Table 1-1 gives the allowable unit compressive stress for concrete as 900 psi [6206 kPa].

(3) Therefore the safe load on the pier is

$$P = (f)(A) = (900)(576) = 518,400 \text{ lb}$$

$$\left[P = (6206)(0.3716) = 2306 \text{ kN} \right]$$

Example 3. A running track in a gymnasium is hung from the roof trusses by steel rods, each of which supports a tensile load of 11,200 lb [49,818 N]. The round rods have a diameter of $\frac{7}{8}$ in. [22.23 mm] with the ends *upset,* that is, made larger by forging. This upset is necessary if the full cross-sectional area of the rod (0.601 in²) [388 mm²] is to be utilized; otherwise the cutting of the threads will reduce the cross section of the rod. Investigate this design to determine whether it is safe.

Solution: (1) Since the gross area of the hanger rod is effective, the unit stress developed is

$$f = \frac{P}{A} = \frac{11,200}{0.601} = 18,636 \text{ psi}$$

$$\left[f = \frac{49,818 \times 10^3}{388} = 128,397 \text{ kPa} \right]$$

(2) Table 1-1 gives the allowable unit tensile stress for steel as 22,000 psi [151,690 kPa], which is greater than that developed by the loading. Therefore the design is safe.

Shearing Stress Formula. The foregoing manipulations of the direct stress formula can, of course, be carried out also with the shearing stress formula $f_v = P/A$. However, as pointed out in Section 1-6, it must be borne in mind that the shearing stress acts transversely to the cross section—not at right angles to it. Furthermore, while the shearing stress equation applies directly to the situation illustrated by Fig. 1-2a and b, it requires modification for application to beams (Fig. 1-2c and d). The latter situation will be considered in more detail later.

Review Problems

Problem 1-15-A*. What force must be applied to a steel bar, 1 in. [25.4 mm] square and 2 ft [610 mm] long, to produce an elongation of 0.016 in. [0.4064 mm]?

Problem 1-15-B. How much will a nominal 8 × 8-in. [actually 190.5-mm] Douglas fir post, 12 ft [3.658 m] long, shorten under an axial load of 45 kips [200 kN]?

Problem 1-15-C*. A routine quality control test is made on a structural steel bar 1 in. [25.4 mm] square and 16 in. [406 mm] long. The data developed during the test show that the bar elongated 0.0111 in. [0.282 mm] when subjected to a tensile force of 20.5 kips [91.184 kN]. Compute the modulus of elasticity of the steel.

Problem 1-15-D. A ½-in.[12.7mm] diameter steel cable 100 ft [30.48 m] long supports a load of 4 kips [17.79 kN]. How much will it elongate?

Problem 1-15-E. What should be the minimum cross-sectional area of a steel rod to support a tensile load of 26 kips [115.648 kN]?

Problem 1-15-F. A short, square post of Douglas fir, select structural grade, is to support an axial load of 61 kips [271.3 kN]. What should its nominal dimensions be?

Problem 1-15-G. A steel rod has a diameter of 1.25 in. [31.75 mm]. What safe tensile load will it support if its ends are upset?

Problem 1-15-H. What safe load will a short, 12 × 12-in. [actually 292.1-mm] Southern pine post support if the grade of the wood is No. 1 dense SR?

Problem 1-15-I.   A short post of Douglas fir, select structural grade, with nominal dimensions of 6 × 8 in. [actually 139.7 × 190.5 mm] supports an axial load of 50 kips [222.4 kN]. Investigate this design to determine whether it is safe.

Problem 1-15-J.   A short concrete pier, 1 ft 6 in. [457.2 mm] square, supports an axial load of 150 kips [667.2 kN]. Is the construction safe?

Problem 1-15-K.   The shearing load on a $\frac{7}{8}$-in.[22.23mm] diameter bolt in a lap joint (Fig. 1-2a) is 8.5 kips [37.808 kN]. Is this a safe condition?

2

Moments and Reactions

||||||||||||||||||||||||||||||||||||||||||||||||||||||||||||||

In the explanations of the relationships that are illustrated in this chapter, the units used for forces and dimensions are of less significance than their numeric values. For this reason, and for sake of brevity and simplicity, the numeric calculations in the text have been done using only English units. For those readers who wish to use metric units, however, the examples and exercise problems have been provided with dual unit work.

2-1 Moment of a Force

The term *moment of a force* is commonly used in engineering problems; it is of utmost importance that you understand exactly what the term means. It is fairly easy to visualize a length of 3 ft, an area of 16 sq in., or a force of 100 lb. A moment, however, is less readily comprehended; it is a force multiplied by a distance. *A moment is the tendency of a force to cause rotation about a given point or axis.* The magnitude of the moment of a force about a given point is the magnitude of the force (pounds, kips, etc.) multiplied by the distance (feet, inches, etc.) to the point. The point is called the *center of moments,* and the distance, which is called the *lever arm* or *moment arm,* is measured by a line drawn through the center of moments *perpendicular to the line of action* of the force. Moments are expressed in compound units such as

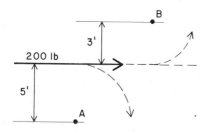

FIGURE 2-1. Moment of a force.

foot-pounds and inch-pounds or kip-feet and kip-inches. In summary,

> moment of force = magnitude of force × moment arm

Consider the horizontal force of 200 lb shown in Fig. 2-1. If point A is the center of moments, the lever arm of the force is 5 ft 0 in. Then the moment of the 200-lb force with respect to point A is $200 \times 5 = 1000$ ft-lb. In this illustration the force tends to cause a *clockwise* rotation (shown by the dotted arrow) about point A and is called a positive moment. If point B is the center of moments, the moment arm of the force is 3 ft 0 in. Therefore, the moment of the 200-lb force about point B is $200 \times 3 = 600$ ft-lb. With respect to point B, the force tends to cause *counterclockwise* rotation; it is called a negative moment. It is important to remember that we can never consider the moment of a force without having in mind the particular point or axis about which it tends to cause rotation.

Figure 2-2 represents two forces acting on a bar which is supported at point A. The moment of force P_1 about point A is

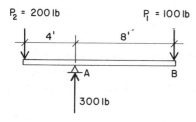

FIGURE 2-2.

100 × 8 = 800 ft-lb, and it is clockwise or positive. The moment of force P_2 about point A is 200 × 4 = 800 ft-lb. The two moment values are the same, but P_2 tends to produce a counterclockwise, or negative, moment about point A. In other words, the positive and negative moments are equal in magnitude and are in equilibrium; that is, there is no motion. Another way of stating this is to say that the sum of the positive and negative moments about point A is zero, or

$$(P_1 \times 8) - (P_2 \times 4) = 0$$

Stated more generally, *if a system of forces is in equilibrium, the algebraic sum of the moments is zero.* This is one of the laws of equilibrium.

In Fig. 2-2 point A was taken as the center of moments, but the fundamental law holds for any point that might be selected. For example, if we take point B as the center of moments, the moment of the upward supporting force of 300 lb acting at A is clockwise (positive) and that of P_2 is counterclockwise (negative). Then

$$(300 \times 8) - (200 \times 12) = 2400 \text{ ft-lb} - 2400 \text{ ft-lb} = 0$$

Note that the moment of force P_1 about point B is 100 × 0 = 0; it is therefore omitted in writing the equation. The reader should be satisfied that the sum of the moments is zero also when the center of moments is taken at the left end of the bar under the point of application of P_2.

2-2 Laws of Equilibrium

When a body is acted on by a number of forces, each force tends to move the body. If the forces are of such magnitude and position that their combined effect produces no motion of the body, the forces are said to be in *equilibrium*. The three fundamental laws of equilibrium are:

1. The algebraic sum of all the vertical forces equals zero.
2. The algebraic sum of all the horizontal forces equals zero.
3. The algebraic sum of the moments of all the forces about any point equals zero.

These laws, sometimes called the conditions for equilibrium, may be expressed as follows (the symbol Σ indicates a summation, i.e., an algebraic addition of all similar terms involved in the problem):

$$\Sigma V = 0 \qquad \Sigma H = 0 \qquad \Sigma M = 0$$

The law of moments, $\Sigma M = 0$, was discussed in the preceding section.

We shall defer consideration of $\Sigma H = 0$ for the time being. Our immediate concern is with vertical loads acting on beams where the expression $\Sigma V = 0$ is another way of saying that *the sum of the downward forces equals the sum of the upward forces.* Thus the bar of Fig. 2-2 satisfies $\Sigma V = 0$ because the upward supporting force of 300 lb equals the sum of P_1 and P_2, the downward forces.

2-3 Moment of Forces on a Beam

Figure 2-3a shows two downward forces of 100 lb and 200 lb acting on a beam. The beam has a length of 8 ft between the supports; the supporting forces, which are called *reactions*, are 175 lb and 125 lb. The four forces are in equilibrium, and therefore the two laws, $\Sigma V = 0$ and $\Sigma M = 0$, apply. Let us see if this is true.

First, because the forces are in equilibrium, the sum of the downward forces must equal the sum of the upward forces. The sum of the downward forces, the loads, is $100 + 200 = 300$ lb; and the sum of the upward forces, the reactions, is $175 + 125 = 300$ lb. We can write $100 + 200 = 175 + 125$; this is a true statement.

Second, because the forces are in equilibrium, the sum of the moments of the forces tending to cause clockwise rotation (positive moments) must equal the sum of the moments of the forces tending to produce counterclockwise rotation (negative moments) about any center of moments. Let us first write an equation of moments about point A at the right-hand support. The force tending to cause clockwise rotation (shown by the curved arrow) about this point is 175 lb; its moment is $175 \times 8 = 1400$ ft-lb. The forces tending to cause counterclockwise rotation *about*

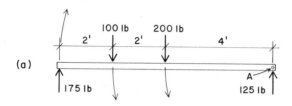

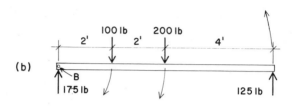

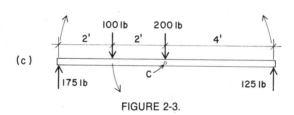

FIGURE 2-3.

the same point are 100 lb and 200 lb, and their moments are (100 × 6) and (200 × 4) ft-lb. Therefore we can write

$$(175 \times 8) = (100 \times 6) + (200 \times 4)$$

$$1400 = 600 + 800$$

$$1400 \text{ ft-lb} = 1400 \text{ ft-lb}$$

which is true.

The upward force of 125 lb is omitted from the above equation because its lever arm about point *A* is 0 ft, and consequently its moment is zero. Thus we see that a force passing through the center of moments does not cause rotation about that point.

Let us try again. This time we select point B at the left support as the center of moments. (See Fig. 2-3b.) By the same reasoning we can write

$$(100 \times 2) + (200 \times 4) = (125 \times 8)$$

$$200 + 800 = 1000$$

$$1000 \text{ ft-lb} = 1000 \text{ ft-lb}$$

Again the law holds. In this case the force 175 lb has a lever arm of 0 ft about the center of moments and its moment is zero.

Suppose we select any point, such as point C in Fig. 2-3c, as the center of moments; then

$$(175 \times 4) = (100 \times 2) + (125 \times 4)$$

$$700 = 200 + 500$$

$$700 \text{ ft-lb} = 700 \text{ ft-lb}$$

We have seen that the law of moments holds in each case. It is of great importance that we understand this principle thoroughly before going on. Remember that the loads and reactions are usually in units of pounds or kips and that the moments are compound quantities, usually foot-pounds or kip-feet, the result of multiplying a force by a distance. When loads are given in kips, there is no intrinsic reason why moments could not be stated as "foot-kips," which would be consistent with "foot-pounds." However, foot-pounds and kip-feet (or inch-pounds and kip-inches) are the terms commonly used in practice for the compound units in which moments are expressed.

Problem 2-3-A.   Figure 2-4 represents a beam in equilibrium with three loads and two reactions. Select five different centers of moments and write the equation of moments for each, showing that the sum of the clockwise moments equals the sum of the counterclockwise moments.

2-4   Types of Beams

A beam is a structural member that resists transverse loads. The supports for beams are usually at or near the ends, and the supporting upward forces are called reactions. As noted in Section

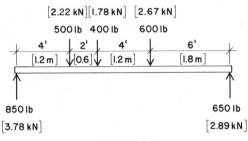

FIGURE 2-4.

1-7, the loads acting on a beam tend to *bend* it rather than shorten or lengthen it. *Girder* is the name given to a beam that supports smaller beams; all girders are beams insofar as their structural action is concerned. There are, in general, five types of beams which are identified by the number, kind, and position of the supports. Figure 2-5 shows diagrammatically the different types and also the shape each beam tends to assume as it bends (deforms) under the loading. In ordinary steel or reinforced concrete beams, these deformations are not usually visible to the eye, but as noted in Section 1-8, some deformation is always present.

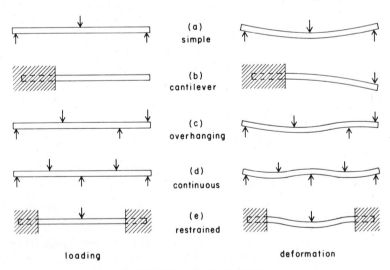

FIGURE 2-5. Types of beams.

A *simple beam* rests on a support at each end, the ends of the beam being free to rotate (Fig. 2-5a).

A *cantilever beam* is supported at one end only. A beam embedded in a wall and projecting beyond the face of the wall is a typical example (Fig. 2-5b).

An *overhanging beam* is a beam whose end or ends project beyond its supports. Fig. 2-5c indicates a beam overhanging one support only.

A *continuous beam* rests on more than two supports (Fig. 2-5d). Continuous beams are commonly used in reinforced concrete and welded steel construction.

A *restrained beam* has one or both ends restrained or *fixed* against rotation (Fig. 2-5e).

2-5 Kinds of Loads

The two types of loads that commonly occur on beams are called *concentrated* and *distributed*. A concentrated load is assumed to act at a definite point, such as a column resting on a beam. A distributed load is one that acts over a considerable length of the beam. A concrete floor slab supported by a beam is an example of a distributed load. If the distributed load exerts a force of equal magnitude for each unit of length of the beam, it is known as a *uniformly distributed load*. Obviously, a distributed load need not extend over the entire length of the beam.

2-6 Reactions

We have already defined reactions as the upward forces acting at the supports which hold in equilibrium the downward forces or loads. The left and right reactions are usually called R_1 and R_2, respectively.

If we have a beam 18 ft in length with a concentrated load of 9000 lb located 9 ft from the supports, it is readily seen that each upward force at the supports will be equal and will be one-half the load in magnitude, or 4500 lb. But consider, for instance, the 9000-lb load placed 10 ft from one end, as shown in Fig. 2-6. What will the upward supporting forces be? Certainly they will not be equal.

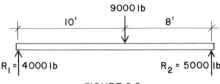

FIGURE 2-6.

Now this is where the principle of moments applies. (See Fig. 2-6.) Let us write an equation of moments, taking the center of moments about the right-hand support R_2.

$$18R_1 = 9000 \times 8$$

$$R_1 = \frac{72,000}{18}$$

$$R_1 = 4000 \text{ lb}$$

Because we know that the sum of the loads is equal to the sum of the reactions, we can easily compute R_2, for

$$R_1 + R_2 = 9000$$

$$4000 + R_2 = 9000$$

$$R_2 = 5000 \text{ lb}$$

To check this value of R_2, write an equation of moments about the left-hand support R_1.

$$18R_2 = 9000 \times 10$$

$$R_2 = \frac{90,000}{18}$$

$$R_2 = 5000 \text{ lb}$$

Example.   A simple beam 20 ft [6.096 m] in length has three concentrated loads, as indicated in Fig. 2-7. Find the magnitudes of the reactions.

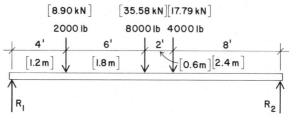

FIGURE 2-7.

Solution: (1) With the right-hand support as the center of moments, write the equation of moments.

$$20R_1 = (2000 \times 16) + (8000 \times 10) + (4000 \times 8)$$

$$20R_1 = \quad 32{,}000 \quad + \quad 80{,}000 \quad + \quad 32{,}000$$

$$20R_1 = 144{,}000$$

$$R_1 = \frac{144{,}000}{20} = 7200 \text{ lb } [32.026 \text{ kN}]$$

(2) The sum of the reactions equals the sum of the loads.

$$R_1 + R_2 = 2000 + 8000 + 4000$$

Thus

$$7200 + R_2 = 14{,}000$$

$$R_2 = 14{,}000 - 7200 = 6800 \text{ lb}$$

$$[R_2 = 62.272 - 32.026 = 30.246 \text{ kN}]$$

(3) To check R_2, write the equation of moments about R_1.

$$20(6800) = (2000 \times 4) + (8000 \times 10) + (4000 \times 12)$$

$$136{,}000 = 8000 \quad + \quad 80{,}000 \quad + \quad 48{,}000$$
$$136{,}000 = 136{,}000$$

Problems 2-6-A-B*-C-D-E*-F-G-H. Eight simple beams with concentrated loads are shown in Fig. 2-8. Find the reactions.

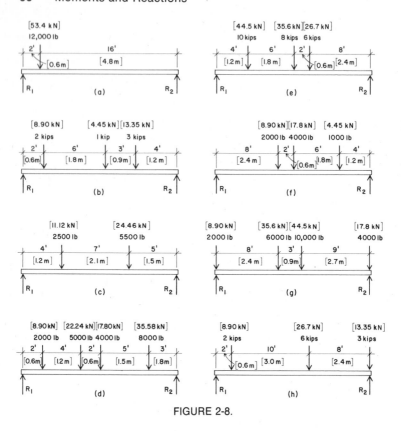

FIGURE 2-8.

2-7 Distributed Loads

So far we have considered only concentrated loads in computing the magnitudes of reactions. The method of dealing with distributed loads is quite similar but there is one key point to remember: *a distributed load on a beam produces the same reactions as a concentrated load of the same magnitude acting through the center of gravity of the distributed load.* If we bear in mind that the center of gravity of a uniformly distributed load lies at the middle of its length, the problem becomes a very simple one.

Example 1. A simple beam 16 ft [4.8768 m] long carries a concentrated load of 8000 lb [35.584 kN] and a uniformly distributed load of 14,000 lb [62.272 kN] arranged as shown in Fig. 2-9a. Find the reactions.

Solution: (1) Note that the uniformly distributed load extends over a length of 10 ft [3.048 m]. Let us write an equation of moments about R_2, considering that the uniformly distributed load acts at the middle of its length, or 5 ft from R_2. (See Fig. 2-9b.)

$$16R_1 = (8000 \times 12) + (14,000 \times 5)$$

$$16R_1 = \quad 96,000 \quad + \quad 70,000$$

$$16R_1 = 166,000$$

$$R_1 = \frac{166,000}{16} = 10,375 \text{ lb } [46.148 \text{ kN}]$$

(2) To find R_2, take the center of moments at R_1. The lever arm of the 14,000-lb load is 11 ft because we consider that the uniformly distributed load acts at its midpoint, or 11 ft from R_1.

$$16R_2 = (8000 \times 4) + (14,000 \times 11)$$

$$16R_2 = \quad 32,000 \quad + \quad 154,000$$

$$16R_2 = 186,000$$

$$R_2 = \frac{186,000}{16} = 11,625 \text{ lb } [51.708 \text{ kN}]$$

(3) When we check these results, the sum of the loads should equal the sum of the reactions, or

$$8000 + 14,000 = 10,375 + 11,625$$

$$22,000 = 22,000$$

$$[35.584 + 62.272 = 46.148 + 51.708]$$

$$[97.856 = 97.856]$$

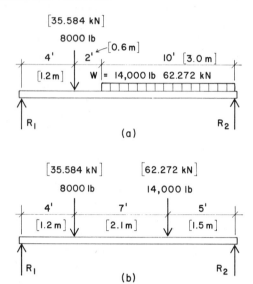

FIGURE 2-9.

Example 2. A beam has a uniformly distributed load of 20 kips [89.0 kN] extending over a length of 8 ft [2.4 m] and a concentrated load of 8 kips [35.6 kN] arranged as shown in Fig. 2-10. Find the reactions.

Solution: (1) The center of gravity of the 20-kip load is 6 ft from R_1 and 8 ft from R_2. The equation for moments about R_2 is

$$14R_1 = (20 \times 8) + (8 \times 6)$$

$$14R_1 = 160 + 48 = 208$$

$$R_1 = \frac{208}{14} = 14.857 \text{ kips } [66.114 \text{ kN}]$$

(2) With R_1 as the center of moments,

$$14R_2 = (20 \times 6) + (8 \times 8)$$

$$14R_2 = 120 + 64 = 184$$

$$R_2 = \frac{184}{14} = 13.143 \text{ kips } [58.486 \text{ kN}]$$

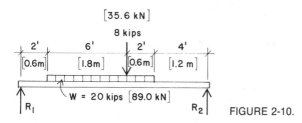

FIGURE 2-10.

(3) Checking results, we find

$$20 + 8 = 14.857 + 13.143$$

$$28 = 28$$

$$[89.0 + 35.6 = 66.114 + 58.486]$$

$$[124.6 = 124.6]$$

Problems 2-7-A-B*-C-D*-E-F-G-H*-I-J.   Compute the reactions for the 10 beams shown in Fig. 2-11, and check the results in each case.

2-8   Overhanging Beams

The method of computing reactions for overhanging beams is the same as that employed in the preceding examples. Select one of the reactions as the center of moments. On one side of the equation place the sum of the moments tending to cause clockwise rotation, and on the other side place the sum of the moments tending to cause rotation in the opposite direction. When writing a moment equation, bear two points in mind: (1) *be consistent and take the same center of moments for each force,* and (2) *consider uniformly distributed loads to act at their midpoints.*

Example 1.   Find the reactions for the overhanging beam shown in Fig. 2-12*a* and check the results.
Solution: (1) Select R_1 as the center of moments. The forces tending to cause clockwise rotation about R_1 are the three loads,

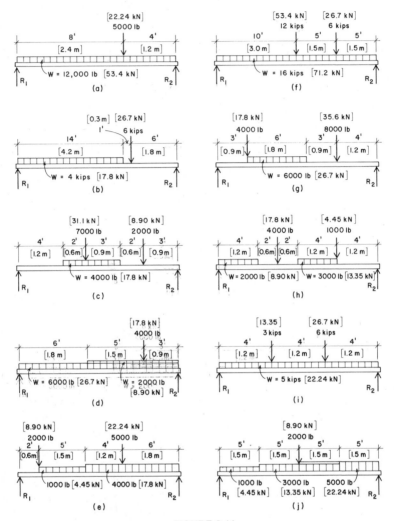

FIGURE 2-11.

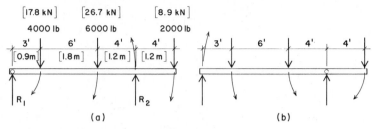

FIGURE 2-12.

and the only force tending to cause counterclockwise rotation is R_2. Note the directions of the arrows. Therefore

$$13R_2 = (4000 \times 3) + (6000 \times 9) + (2000 \times 17)$$

$$13R_2 = 12{,}000 + 54{,}000 + 34{,}000 = 100{,}000$$

$$R_2 = \frac{100{,}000}{13} = 7692 \text{ lb } [34.214 \text{ kN}]$$

(2) To find R_1, take the center of moments at R_2. The forces tending to cause clockwise rotation about this point are R_1 and the 2000-lb load; those tending to cause counterclockwise rotation about the same point are the 4000-lb and 6000-lb loads. Note the directions of the arrows in Fig. 2-12b. Then

$$13R_1 + (2000 \times 4) = (4000 \times 10) + (6000 \times 4)$$

$$13R_1 = 40{,}000 + 24{,}000 - 8000 = 56{,}000$$

$$R_1 = \frac{56{,}000}{13} = 4308 \text{ lb } [19.162 \text{ kN}]$$

(3) If these are correct results, the sum of the loads should equal the sum of the reactions; therefore

$$4000 + 6000 + 2000 = 4308 + 7692$$

$$12{,}000 = 12{,}000$$

$$[17.8 + 26.7 + 8.9 = 34.214 + 19.162]$$

$$[53.4 = 53.4]$$

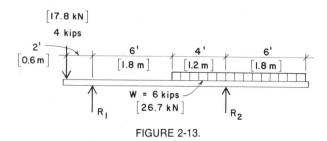

FIGURE 2-13.

Example 2. The overhanging beam shown in Fig. 2-13 supports a concentrated load of 4 kips [17.8 kN] and a uniformly distributed load of 6 kips [26.7 kN], arranged as shown. Find the reactions and check the results.

Solution: (1) Note that this beam overhangs both of its reactions. The uniformly distributed load extends over a length of 10 ft, and its midpoint lies 1 ft to the right of R_2. With R_2 as the center of moments,

$$10R_2 + (4 \times 2) = (6 \times 11)$$

$$10R_2 = 66 - 8 = 58$$

$$R_2 = \frac{58}{10} = 5.8 \text{ kips } [25.81 \text{ kN}]$$

(2) With R_2 as the center of moments,

$$10R_1 + (6 \times 1) = (4 \times 12)$$

$$10R_1 = 48 - 6 = 42$$

$$R_1 = 42 \div 10 = 4.2 \text{ kips } [18.69 \text{ kN}]$$

(3) Checking the results, we find

$$4 + 6 = 5.8 + 4.2$$

$$10 = 10$$

$$[17.8 + 26.7 = 25.81 + 18.69]$$

$$[44.5 = 44.5]$$

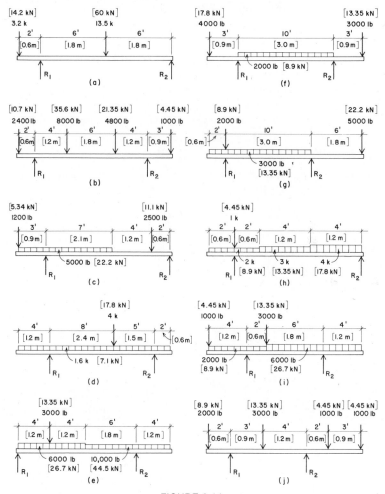

FIGURE 2-14.

Note: The importance of your ability to compute the magnitudes of the reactions of beams cannot be stressed too highly. The principle involved occurs time after time in engineering problems; therefore it is necessary to understand thoroughly the foregoing discussion before proceeding. In addition to working the problems below, make up some of your own to test your knowledge. As the above examples show, it is a simple matter to check your results, for after you have computed the reactions by the principle of moments, the sum of the reactions must equal the sum of the loads.

Problems 2-8-A-B-C*-D*-E-F-G-H-I*-J. Compute the reactions and check the results for the beams shown in Fig. 2-14.

3

Shear and Bending Moment

||||||||||||||||||||||||||||||||||||||||||||||||||||||||||||||

3-1   Introduction

Figure 3-1*a* represents a simple beam with a uniformly distributed
load *W* over its entire length. Examination of an actual beam so
loaded probably would not reveal any effects of the loading on the
beam. However, there are three distinct major tendencies for the
beam to fail. Figures 3-1*b*, *c*, and *d* illustrate the three phenom-
ena, with deformations greatly exaggerated.

First, there is a tendency for the beam to fail by dropping
between the supports (Fig. 3-1*b*). This is called *vertical shear*.
Second, the beam may fail by bending (Fig. 3-1*c*). Third, there is a
tendency for the fibers of the beam to slide past each other in a
horizontal direction (Fig. 3-1*d*). The name given to this action is
horizontal shear, and it will be discussed further under steel,
wood, and reinforced concrete construction. Naturally, a beam
properly designed does not fail in any of the ways just mentioned,
but these tendencies to fail are always present and must be considered in structural design.

The forces that prevent failure are supplied by the resisting
stresses developed within the beam. Our problem in design is to
select beams with dimensions that will provide adequate material

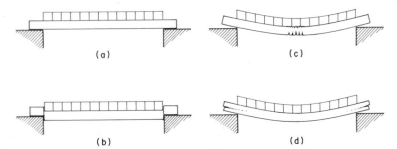

FIGURE 3-1. Shear and bending failure in beams.

to develop these resisting stresses. In this chapter we shall be concerned with methods of measuring the magnitude of shearing and bending forces caused by the beam loading or, as sometimes stated, by the external forces on the beam.

3-2 Vertical Shear

We can define vertical shear as the tendency for one part of a beam to move vertically with respect to an adjacent part. *The magnitude of the shear at any section in the length of a beam is equal to the algebraic sum of the vertical forces on either side of the section.* Vertical shear is usually represented by the letter V. In computing its values in our examples and problems, we consider the forces to the left of the section, but keep in mind that the same result will be obtained if we work with the forces on the right. We may say then that *the vertical shear at any section of a beam is equal to the reactions minus the loads to the left of the section.* Fix this definition firmly in mind. Now, if we wish to find the magnitude of the vertical shear at any section in the length of a beam, we simply repeat the foregoing statement and write an equation accordingly. It follows from this procedure that the maximum value of the shear for simple beams is equal to the greater reaction.

If the loads and reactions are in units of pounds or kips, the magnitude of the vertical shear will be in units of pounds or kips also. Form the habit of writing the denomination of the units after the numerical values; it will prevent many errors. As in Chapter 2 the work in this chapter is done exclusively in English units, since the units themselves are of less significance than their numerical values. For a metric "translation" simply read Newtons for pounds, kilonewtons for kips, and meters for feet. Corresponding metric data are provided for the problems, however, for the readers who may prefer to use them.

Example 1.   Figure 3-2*a* illustrates a simple beam with two concentrated loads of 600 lb and 1000 lb. Our problem is to find the value of the vertical shear at various points along the length of the beam. Although the weight of the beam constitutes a uniformly distributed load, it is neglected in this example.

Solution: (1) The reactions are computed by the principle of moments previously described, and we find that they are R_1 = 1000 lb and R_2 = 600 lb.

(2) Consider first the value of the vertical shear V at an infinitely short distance to the right of R_1. Applying the rule that the shear is equal to the reactions minus the loads to the left of the section, we write $V = R_1 - 0$, or $V = 1000$ lb. The zero represents the value of the loads to the left of the section, which, of course, is zero. Now take a section 1 ft to the right of R_1; again $V_{(x=1)}$ = $R_1 - 0$, or $V_{(x=1)} = 1000$ lb. The subscript $(x = 1)$ indicates the position of the section at which the shear is taken, the distance of the section from R_1. We find that the shear is still 1000 lb and has the same magnitude up to the 600-lb load. The next section to consider is a very short distance to the right of the load 600 lb. Then $V_{(x=2+)} = 1000 - 600 = 400$ lb. Because there are no loads intervening, the shear continues to be the same magnitude up to

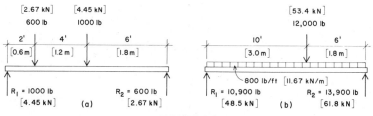

FIGURE 3-2.

the load 1000 lb. At a section a short distance to the right of the 1000-lb load, $V_{(x=6+)} = 1000 - (600 + 1000) = -600$ lb. This magnitude continues up to the right-hand reaction R_2.

The preceding example dealt only with concentrated loads. Let us see if we can apply the same procedure to a beam having a uniformly distributed load in addition to a concentrated load.

Example 2. The beam shown in Fig. 3-2b supports a concentrated load of 12,000 lb located 6 ft from R_2 and a uniformly distributed load of 800 pounds per linear foot (lb per lin ft) over its entire length. Compute the value of the vertical shear at various sections along the span.

Solution: (1) Note that the uniform load is given in lb per lin ft (frequently abbreviated to lb/ft). The symbol for uniform load per foot is w, the capital letter W being used to represent the *total* uniformly distributed load. In this instance $w = 800$ lb and $W = 800 \times 16 = 12,800$ lb. The equation of moments about R_2 as the center is

$$16R_1 = (800 \times 16 \times 8) + (12,000 \times 6)$$
$$16R_1 = 102,400 + 72,000$$
$$R_1 = \frac{174,400}{16} = 10,900 \text{ lb}$$

In a similar manner R_2 is found to be 13,900 lb. In the quantity $(800 \times 16 \times 8)$, the load (800×16) lb has a lever arm of 8 ft, the distance of its center of gravity to the reaction.

(2) Following the rule used in the preceding example, write the value of V at various sections along the beam.

$$V_{(x=0)} = 10,900 - 0 = 10,900 \text{ lb}$$
$$V_{(x=1)} = 10,900 - 800 = 10,100 \text{ lb}$$
$$V_{(x=5)} = 10,900 - (800 \times 5) = 6900 \text{ lb}$$
$$V_{(x=10-)} = 10,900 - (800 \times 10) = 2900 \text{ lb}$$
$$V_{(x=10+)} = 10,900 - [(800 \times 10) + 12,000] = -9100 \text{ lb}$$
$$V_{(x=16)} = 10,900 - [(800 \times 16) + 12,000] = -13,900 \text{ lb}$$

Note that the value of the vertical shear at the supports has the same magnitude as the reactions.

3-3 Shear Diagrams

In the two preceding examples we computed the value of the shear at several sections along the length of the beams. In order to visualize the results we have obtained, we may make diagrams to plot these values. They are called *shear diagrams* and are constructed as explained below.

To make such a diagram, first draw the beam to scale and locate the loads. This has been done in Fig. 3-3*a* and *b* by repeating the load diagrams of Fig. 3-2*a* and *b*, respectively. Beneath the beam draw a horizontal base line representing zero shear. Above and below this line, plot at any convenient scale the values of the shear at the various sections; the positive, or plus, values are placed above the line and the negative, or minus, values below. In Fig. 3-3*a*, for instance, the value of the shear at R_1 is +1000 lb. The shear continues to have the same value up to the load of 600 lb, at which point it drops to +400 lb. The same value continues up to the next load, 1000 lb, where it drops to −600 lb and continues to the right-hand reaction. Obviously, to draw a shear diagram it is necessary to compute the values at significant points only. Having made the diagram, we may readily find the value of the shear at any section of the beam by scaling the vertical distance in the diagram. The shear diagram for the beam in Fig. 3-3*b* is made in the same manner.

There are two important facts to note concerning the vertical shear. The first is the maximum value. We see that the diagrams in each case confirm our earlier observation that the maximum shear is at the reaction having the greater value, and its magnitude is equal to that of the greater reaction. In Fig. 3-3*a* the maximum shear is 1000 lb, and in Fig. 3-3*b* it is 13,900 lb. We disregard the positive or negative signs in reading the maximum values of the shear, for the diagrams are merely conventional methods of representing the absolute numerical values.

The other important fact is to note the point at which the shear changes from a plus to a minus quantity. We call it *the point a*

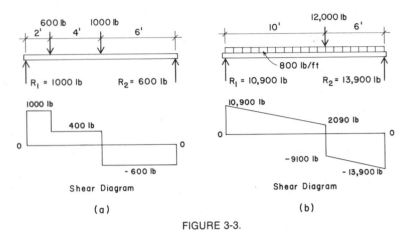

FIGURE 3-3.

which the shear passes through zero. In Fig. 3-3a it is under the 1000-lb load, 6 ft from R_1. In Fig. 3-3b the shear passes through zero 10 ft from R_1. The reason for finding this point is that *the greatest tendency for the beam to fail by bending is at the point at which the shear passes through zero.*

Problems 3-3-A-B*-C*-D-E-F-G-*H-I-J. For the beams indicated in Fig. 3-4, draw the shear diagrams and note in each instance the value of the maximum shear and the point at which the shear passes through zero.

3-4   Bending Moment

The forces that tend to cause bending in a beam are the reactions and the loads. Consider the section X-X, 6 ft from R_1. (See Fig. 3-5.) The force R_1, or 2000 lb, tends to cause a clockwise rotation about this point. Because the force is 2000 lb and the lever arm is 6 ft, the moment of the force is $2000 \times 6 = 12,000$ ft-lb. This same value may be found by considering the forces to the right of the section X-X. Let us see. There are two forces to the right of section X-X: R_2, which is 6000 lb, and the load 8000 lb, with lever arms of 10 and 6 ft, respectively. The moment of the reaction is $6000 \times 10 = 60,000$ ft-lb, and its direction is counterclockwise

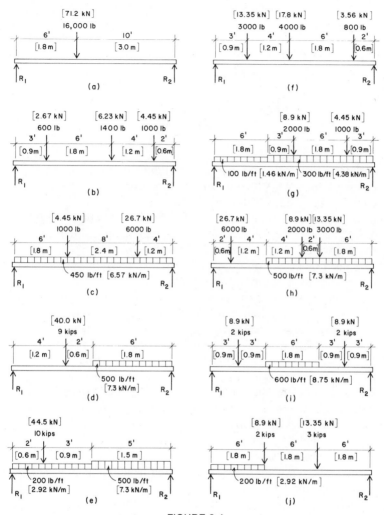

FIGURE 3-4.

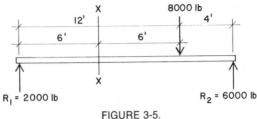

FIGURE 3-5.

with respect to the section X-X. The moment of the force 8000 lb is $8000 \times 6 = 48{,}000$ ft-lb, and its direction is clockwise. Then $60{,}000$ ft-lb $- 48{,}000$ ft-lb $= 12{,}000$ ft-lb, the resultant moment tending to cause counterclockwise rotation about the section X-X. This is the same magnitude as the moment of the forces on the left which tends to cause a clockwise rotation.

Thus it makes no difference whether we consider the forces to the right of the section or the left; the magnitude of the moment is the same. It is called the *bending moment* because it is the moment of the forces that cause bending stresses in the beam. Its magnitude varies throughout the length of the beam. For instance, at 4 ft from R_1 it is only 2000×4, or 8000 ft-lb. *The bending moment is the algebraic sum of the moments of the forces on either side of the section.* For simplicity, let us take the forces on the left; then we may say *the bending moment at any section of a beam is equal to the moments of the reactions minus the moments of the loads to the left of the section.* Because the bending moment is the result of multiplying forces by distances, the denominations are foot-pounds or kip-feet.

Almost everyone confuses shear and bending moment at first. Remember that the shear is the result of subtracting loads from reactions, the units being pounds or kips; the bending moment is the result of subtracting *moments* of loads from *moments* of reactions, with units of foot-pounds or kip-feet.

3-5   Bending Moment Diagrams

The construction of bending moment diagrams follows the procedure used for shear diagrams. The beam span is drawn to scale showing the locations of the loads. Below this, and usually below the shear diagram, a horizontal base line is drawn representing zero bending moment. Then the bending moments are computed at various sections along the beam span, and the values are plotted vertically to any convenient scale. In simple beams all bending moments are positive and therefore are plotted above the base line. In overhanging or continuous beams we shall find negative moments, and these are plotted below the base line.

Example. The load diagram in Fig. 3-6 shows a simple beam with two concentrated loads. Draw the shear and bending moment diagrams.

Solution: (1) R_1 and R_2 are first computed and are found to be 16,000 lb and 14,000 lb, respectively. These values are recorded on the load diagram.

(2) The shear diagram is drawn as described in Section 3-3. Note that in this instance it is necessary to compute the shear at only one section (between the concentrated loads) because there

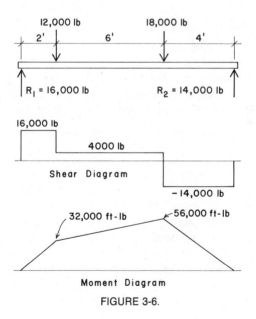

FIGURE 3-6.

is no distributed load, and we know that the shear at the reactions is equal in magnitude to the reactions.

(3) Because the value of the bending moment at any section of the beam is equal to the moments of the reactions minus the moments of the loads to the left of the section, the moment at R_1 must be zero, for there are no forces to the left. We might also say that R_1 has a lever arm of zero and $R_1 \times 0 = 0$. Other values in the length of the beam are computed as follows. The subscripts ($x = 1$, etc.) show the distance from R_1 at which the bending moment is computed.

$$M_{(x=1)} = (16,000 \times 1) = 16,000 \text{ ft-lb}$$

$$M_{(x=2)} = (16,000 \times 2) = 32,000 \text{ ft-lb}$$

$$M_{(x=5)} = (16,000 \times 5) - (12,000 \times 3) = 44,000 \text{ ft-lb}$$

$$M_{(x=8)} = (16,000 \times 8) - (12,000 \times 6) = 56,000 \text{ ft-lb}$$

$$M_{(x=10)} = (16,000 \times 10) - [(12,000 \times 8) + (18,000 \times 2)]$$
$$= 28,000 \text{ ft-lb}$$

$$M_{(x=12)} = (16,000 \times 12) - [(12,000 \times 10) + (18,000 \times 4)] = 0$$

The result of plotting these values is shown in the bending moment diagram of Fig. 3-6. More moments were computed than were actually necessary. We know that the bending moments at the supports of simple beams are zero, and in this instance only the bending moments directly under the loads were needed.

3-6   Relation between Shear and Bending Moments

In simple beams the shear diagram passes through zero at some point between the supports. As stated earlier, an important principle in this respect is that *the bending moment has a maximum magnitude wherever the shear passes through zero*. In Fig. 3-6 the shear passes through zero under the 18,000-lb load, that is, at ($x = 8$). Note that the bending moment has its greatest value at this same point, 56,000 ft-lb. In order to design beams, we must know the value of the maximum bending moment. Frequently we draw only enough of the shear diagram to find the section at which the shear passes through zero and then compute the bending moment at this point.

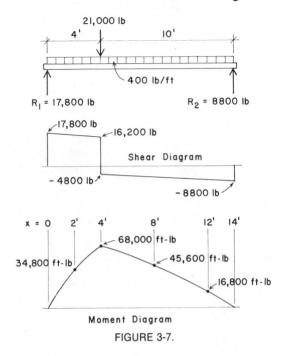

FIGURE 3-7.

Example 1. Draw the shear and bending moment diagrams for the beam shown in Fig. 3-7, which carries a uniformly distributed load of 400 lb per lin ft and a concentrated load of 21,000 lb located 4 ft from R_1.

Solution: (1) Computing the reactions, we find

$$14R_1 = (21,000 \times 10) + (400 \times 14 \times 7)$$

$$R_1 = 17,800 \text{ lb}$$

$$\text{total load} = 21,000 + (400 \times 14) = 26,600 \text{ lb}$$

Therefore

$$R_1 + R_2 = 26,600$$

$$17,800 + R_2 = 26,600$$

$$R_2 = 26,600 - 17,800 = 8800 \text{ lb}$$

(2) Computing the value of the shear at essential points, we find

$V_{(at\ R_1)} = 17,800$ lb

$V_{(x=4-)} = 17,800 - (400 \times 4) = 16,200$ lb

$V_{(x=4+)} = 17,800 - [(400 \times 4) + 21,000] = -4800$ lb

$V_{(at\ R_2)} = -8800$ lb

Note that the shear passes through zero under the 21,000-lb load; therefore at this point we shall expect to find that the bending moment is a maximum.

(3) Computing the value of the bending moment at selected points, we get

$M_{(x=0)} = 0$

$M_{(x=2)} = (17,800 \times 2) - (400 \times 2 \times 1) = 34,800$ ft-lb

In the foregoing equation the reaction to the left of the section is 17,800 lb and its lever arm is 2 ft. The load to the left of the section is (400 × 2) lb and its lever arm is 1 ft, the distance from the center of the load, (400 × 2) lb, to the center of moments.

$M_{(x=4)} = (17,800 \times 4) - (400 \times 4 \times 2) = 68,000$ ft-lb

$M_{(x=8)} = (17,800 \times 8) - [(400 \times 8 \times 4) + (21,000 \times 4)]$
$\qquad = 45,600$ ft-lb

$M_{(x=12)} = (17,800 \times 12) - [(400 \times 12 \times 6) + (21,000 \times 8)]$
$\qquad = 16,800$ ft-lb

$M_{(x=14)} = (17,800 \times 14) - [(400 \times 14 \times 7) + (21,000 \times 10)] = 0$

From the two preceding examples (Figs. 3-6 and 3-7), it will be observed that the shear diagram for the parts of the beam on which no loads occur is represented by horizontal lines. For the parts of the beam on which a uniformly distributed load occurs, the shear diagram consists of straight inclined lines. The bending moment diagram is represented by straight inclined lines when only concentrated loads occur and by a curved line if the load is distributed.

Occasionally, when a beam has both concentrated and uni-

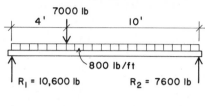

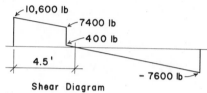

Shear Diagram   FIGURE 3-8.

formly distributed loads, the shear does not pass through zero under one of the concentrated loads. This frequently occurs when the distributed load is relatively large compared with the concentrated loads. Since it is necessary in designing beams to find the maximum bending moment, we must know the point at which it occurs. This, of course, is the point where the shear passes through zero, and its location is readily determined by the procedure illustrated in the following example.

Example 2. The load diagram in Fig. 3-8 shows a beam with a concentrated load of 7000 lb, applied 4 ft from the left reaction, and a uniformly distributed load of 800 lb per lin ft extending over the full span. Compute the maximum bending moment on the beam.

Solution: (1) Compute the values of the reactions. These are found to be $R_1 = 10,600$ lb and $R_2 = 7600$ lb and are recorded on the load diagram.

(2) Constructing the shear diagram in Fig. 3-8, we see that the shear passes through zero at some point between the concentrated load of 7000 lb and the right reaction. Call this distance x ft from R_1. Now, we know that the value of the shear at this section is zero; therefore we write an expression for the shear for this point, using the terms of the reaction and loads, and equate the quantity to zero. This equation contains the distance x:

$$V_{(at\,x)} = 10,600 - [7000 + (800x)] = 0$$

$$800x = 3600$$

$$x = 4.5 \text{ ft}$$

(3) The value of the bending moment at this section is given by the expression

$$M_{(x=4.5)} = (10,600 \times 4.5) - \left[(7000 \times 0.5) + \left(800 \times 4.5 \times \frac{4.5}{2}\right)\right]$$

$$M = 36,100 \text{ ft-lb}$$

Under certain conditions the value of the shear is zero for the entire distance between concentrated loads. This occurs when a simple beam has two equal loads at equal distances from the supports. The value of the bending moment is the same at any section between the loads.

Example 3.   The load diagram of Fig. 3-9 shows a simple beam with a span of 18 ft, supporting concentrated loads of 10,000 lb

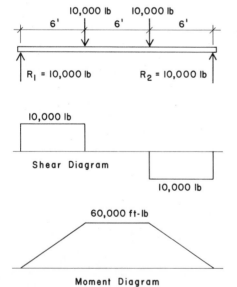

FIGURE 3-9.

located 6 ft from each reaction. Compute the maximum bending moment, and draw the moment diagram.

Solution: (1) Because the beam is symmetrically loaded, each reaction is equal to half the total load, or $\frac{1}{2} \times$ 20,000 = 10,000 lb.

(2) The shear diagram is readily constructed as shown. Note that the shear has a zero value at all points between the loads. According to the rule, the bending moment must be a maximum over this portion of the beam.

(3) Compute M at various sections along the span and plot the diagram.

$$M_{(x=6)} = 10,000 \times 6 = 60,000 \text{ ft-lb}$$

$$M_{(x=9)} = (10,000 \times 9) - (10,000 \times 3) = 60,000 \text{ ft-lb}$$

When such a problem occurs, therefore, we know that the maximum bending moment is equal to one of the loads multiplied by its distance to the nearest reaction. In this example only two concentrated loads are considered. Actually, the weight of the beam constitutes a uniformly distributed load; hence the shear passes through zero at the center of the span. Because the weight of the beam is often quite small when compared with the concentrated loads, it is sometimes neglected in the computations. In this chapter and the preceding one, consideration of beam weight has been omitted in order to focus on the separate effects of concentrated and uniform loads. It will be considered, however, in the sections on steel, wood, and reinforced concrete design.

Problems 3-6-A*-B*-C-D*-E-F-G-H-I-J*. Draw the shear and bending moment diagrams for the beams shown in Fig. 3-10. In each case, note the magnitude of the maximum shear and the maximum bending moment.

3-7 Negative Bending Moment: Overhanging Beams

When a simple beam bends, it has a tendency to assume the shape shown in Fig. 3-11*a*. In this case the fibers in the upper part of the beam are in compression. For this condition we say the bending moment is positive (+). Another way to describe a positive bending moment is to say that it is positive when the curve assumed by the bent beam is concave upward. When a beam projects beyond

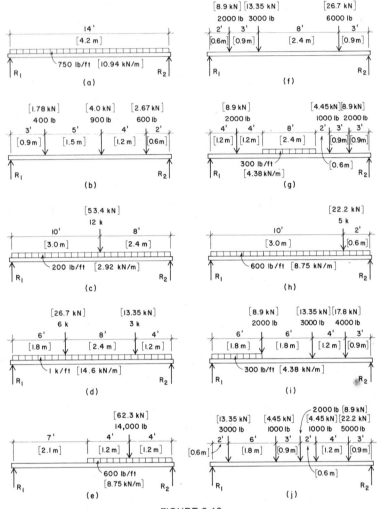

FIGURE 3-10.

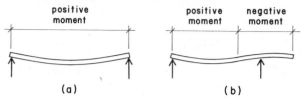

FIGURE 3-11. Positive and negative moment.

a support (Fig. 3-11*b*), this portion of the beam has tensile stresses in its upper part. The bending moment for this condition is called negative (−); the beam is bent concave downward. If we construct moment diagrams, following the method previously described, the positive and negative moments are shown graphically.

Example 1. Draw the shear and bending moment diagrams for the overhanging beam shown in Fig. 3-12.
Solution: (1) Computing the reactions, we find

$$12R_2 = 600 \times 16 \times 8 \qquad R_2 = 6400 \text{ lb}$$

$$12R_1 = 600 \times 16 \times 4 \qquad R_1 = 3200 \text{ lb}$$

(2) Computing the values of the shear, we find

$$V_{(at\ R_1)} = +3200 \text{ lb}$$

$$V_{(x=12-)} = 3200 - (600 \times 12) = -4000 \text{ lb}$$

$$V_{(x=12+)} = (3200 + 6400) - (600 \times 12) = +2400 \text{ lb}$$

$$V_{(x=16)} = (3200 + 6400) - (600 \times 16) = 0$$

To find the point at which the shear passes through zero betwen the supports (see Example 2, Section 3-6), calculate

$$3200 - 600x = 0$$

$$x = 5.33 \text{ ft}$$

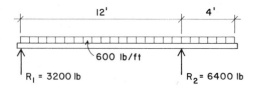

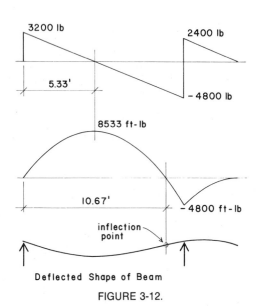

Deflected Shape of Beam

FIGURE 3-12.

(3) Computing the values of the bending moment, we find

$M_{(at\ R_1)} = 0$

$M_{(x=5.33)} = (3200 \times 5.33) - \left(600 \times 5.33 \times \dfrac{5.33}{2}\right) = 8533$ ft-lb

$M_{(x=12)} = (3200 \times 12) - (600 \times 12 \times 6) = -4800$ ft-lb

$M_{(x=16)} = 0$

To draw the bending moment diagram accurately, we may compute the magnitudes at other points; it will be a curved line.

It is seen in plotting the shear values that there are two points at which the shear passes through zero: at $x = 5.33$ ft and at $x = 12$ ft. The bending moment diagram shows that maximum values, one for a positive moment and the other a negative moment, are found at each point. When we design beams, we are concerned only with the maximum value, regardless of whether it is positive or negative. The shear diagram does not indicate which of the two points gives the greater value of the bending moment, and often it is necessary to compute the value at each point at which the shear passes through zero, to determine which one is greater numerically. For this beam it is 8533 ft-lb.

Inflection Point. The bending moment diagram in Fig. 3-12 indicates a point between the supports at which the value of $M = 0$. This is called the *inflection point*; it is the point at which the curvature reverses as it changes from concave to convex. It is important to know the position of the inflection point in the study of reinforced concrete beams, for this is the position at which the tensile steel reinforcement is bent upward.

In this problem call x the distance from the left support to the point at which $M = 0$. Then writing an expression for the value of the bending moment and equating it to zero,

$$(3200 \times x) - \left(600 \times x \times \frac{x}{2}\right) = 0$$

$$3200x - 300x^2 = 0$$

$$x = 10.67 \text{ ft}$$

Examine the curve of the bending moment. You will see that the curve is symmetrical between the left reaction and the inflection point. Now, since the curve reaches its highest point at 5.33 ft from the left support, the inflection point occurs at 2×5.33, or 10.67 ft from R_1. This is the same result found by solving for x in the foregoing equation.

For the beam and load shown in Fig. 3-12, note that the value of the maximum vertical shear is -4000 lb. It occurs immediately to the left of the right-hand support.

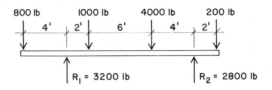

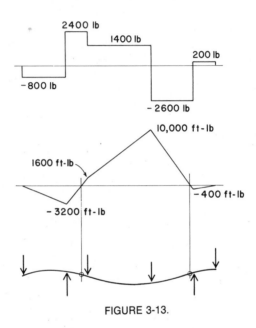

FIGURE 3-13.

Example 2. Compute the maximum bending moment for the overhanging beam shown in Fig. 3-13.

Solution: (1) Computing the reactions, we find

$$12R_1 + (200 \times 2) = (800 \times 16) + (1000 \times 10) + (4000 \times 4)$$

$$R_1 = 3200 \text{ lb}$$

$$\text{total load} = 800 + 1000 + 4000 + 200 = 6000 \text{ lb}$$

Therefore

$$R_2 = 6000 - 3200 = 2800 \text{ lb}$$

(2) Computing the values of the shear, we find

$V_{(x=1)} = -800$ lb

$V_{(x=4+)} = 3200 - 800 = +2400$ lb

$V_{(x=6+)} = 3200 - (800 + 1000) = +1400$ lb

$V_{(x=12+)} = 3200 - (800 + 1000 + 4000) = -2600$ lb

$V_{(x=16+)} = (3200 + 2800) - (800 + 1000 + 4000) = +200$ lb

With these values we now plot the shear diagram. We note that the shear passes through zero at three points, R_1, R_2, and under the 4000-lb load. We expect the bending moment to reach maximum values at these points.

(3) Computing the values of the bending moment, we find

$M_{(x=0)} = 0$

$M_{(x=4)} = -(800 \times 4) = -3200$ ft-lb

$M_{(x=6)} = (3200 \times 2) - (800 \times 6) = +1600$ ft-lb

$M_{(x=12)} = (3200 \times 8) - [(800 \times 12) + (1000 \times 6)]$

$= +10,000$ ft-lb

$M_{(x=16)} = (3200 \times 12) - [(800 \times 16) + (1000 \times 10)$

$+ (4000 \times 4)] = -400$ ft-lb

$M_{(x=18)} = 0$

The maximum value of the bending moment (10,000 ft-lb) occurs under the 4000-lb load.

The value of the maximum vertical shear is −2600 lb.

Note that the bending moment diagram changes from a plus value to a minus value at two points. There are two inflection points. If it were possible to scale the moment diagram of Fig. 3-13 with sufficient accuracy, we would find that one of these inflection points is located approximately 1.3 ft to the right of R_1 and the other approximately 0.15 ft to the left of R_2. The reader should check the accuracy of these locations by writing moment equations about each inflection point. If the locations are "exactly'' correct, the value of the bending moment at each point will, by definition, equal zero.

Problems 3-7-A*-B-C*-D-E-F-G*-H-I-J.   Draw the shear and bending moment diagrams for the beams shown in Fig. 3-14. Determine the maximum bending moment in each case.

3-8   Cantilever Beams

When we compute the shear and bending moment for cantilever beams, it is convenient to draw the fixed end at the right and to compute the shear and bending moment values to the left, as in the preceding examples.

Example 1.   The cantilever beam shown in Fig. 3-15a projects 12 ft from the face of the wall and has a concentrated load of 800 lb at the unsupported end. Draw the shear and moment diagrams. What are the values of the maximum shear and maximum bending moment?

Solution: (1) The value of the shear is -800 lb throughout the entire length of the beam.

(2) The bending moment is maximum at the wall; its value is -9600 ft-lb.

$$M_{(x=0)} = 0$$

$$M_{(x=1)} = -(800 \times 1) = -800 \text{ ft-lb}$$

$$M_{(x=2)} = -(800 \times 2) = -1600 \text{ ft-lb}$$

$$M_{(x=12)} = -(800 \times 12) = -9600 \text{ ft-lb}$$

Example 2.   Draw the shear and bending moment diagrams for the cantilever beam, shown in Fig. 3-15b, which carries a uniformly distributed load of 500 lb per lin ft over its full length.

Solution: (1) Computing the values of the shear, we find

$$V_{(x=1)} = -(500 \times 1) = -500 \text{ lb}$$

$$V_{(x=2)} = -(500 \times 2) = -1000 \text{ lb}$$

$$V_{(x=10)} = -(500 \times 10) = -5000 \text{ lb, the maximum value}$$

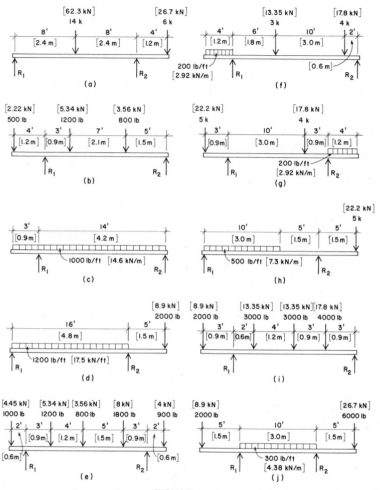

FIGURE 3-14.

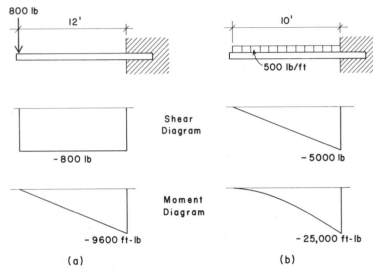

800 lb

12'

10'

500 lb/ft

Shear
Diagram

−800 lb

−5000 lb

Moment
Diagram

−9600 ft-lb

−25,000 ft-lb

(a)

(b)

FIGURE 3-15.

(2) Computing the moments, we find

$M_{(x=0)} = 0$

$M_{(x=2)} = -(500 \times 2 \times 1) = -1000$ ft-lb

$M_{(x=4)} = -(500 \times 4 \times 2) = -4000$ ft-lb

$M_{(x=10)} = -(500 \times 10 \times 5) = -25,000$ ft-lb, the maximum value

Example 3.   The cantilever beam indicated in Fig. 3-16 has a concentrated load of 2000 lb and a uniformly distributed load of 600 lb per lin ft at the positions shown. Draw the shear and bending moment diagrams. What are the magnitudes of the maximum shear and maximum bending moment?

Solution: (1) Computing the values of the shear, we find

$$V_{(x=1)} = -2000 \text{ lb}$$

$$V_{(x=8)} = -2000 \text{ lb}$$

$$V_{(x=10)} = -[2000 + (600 \times 2)] = -3200 \text{ lb}$$

$$V_{(x=14)} = -[2000 + (600 \times 6)] = -5600 \text{ lb}$$

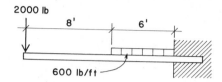

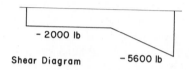

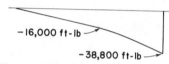

FIGURE 3-16.

(2) Computing the moments, we find

$$M_{(x=0)} = 0$$

$$M_{(x=4)} = -(2000 \times 4) = -8000 \text{ ft-lb}$$

$$M_{(x=8)} = -(2000 \times 8) = -16,000 \text{ ft-lb}$$

$$M_{(x=10)} = -[(2000 \times 10) + (600 \times 2 \times 1)] = -21,200 \text{ ft-lb}$$

$$M_{(x=14)} = -[(2000 \times 14) + (600 \times 6 \times 3)] = -38,800 \text{ ft-lb}$$

The maximum shear is 5600 lb, and the maximum bending moment is 38,800 ft-lb. In cantilever beams the maximum values of both shear and moment occur at the support; the beams are concave downward and the bending moment is negative throughout the length. The reader should be satisfied that this statement holds when the fixed end of a cantilever beam is taken to the left. It is suggested that Example 3 be reworked with Fig. 3-16 reversed, left for right. All numerical results will be the same, but the shear diagram will be positive over its full length.

Problems 3-8-A-B*-C-D*.   Draw the shear and bending moment diagrams for the cantilever beams shown in Fig. 3-17. State the maximum shear and maximum bending moment values in each beam.

3-9   Bending Moment Formulas

The method of computing bending moments presented thus far in this chapter enables us to find the maximum value under a wide variety of loading conditions. However, certain conditions occur so frequently that it is convenient to use formulas that give the maximum values directly. Structural design handbooks contain many such formulas; two of the most commonly used formulas are derived in the following sections, and a few others are given in Section 3-12.

3-10   Concentrated Load at Center of Span

A simple beam with a concentrated load at the center of the span occurs very frequently in practice. Call the load P and the span length between supports L, as indicated in the load diagram of Fig. 3-18a. For this symmetrical loading each reaction is $P/2$, and it is readily apparent that the shear will pass through zero at distance $x = L/2$ from R_1. Therefore the maximum bending moment occurs at the center of the span, under the load. Now let us compute the value of the bending moment at this section.

$$M_{(x=L/2)} = \frac{P}{2} \times \frac{L}{2}$$

or

$$M = \frac{PL}{4}$$

Note that this value is given in Case 1, Fig. 3-19. This formula is well worth remembering. Observe how quickly bending moments are computed by its use.

Example.   A simple beam 20 ft in length has a concentrated load of 8000 lb at the center of the span. Compute the maximum bending moment.

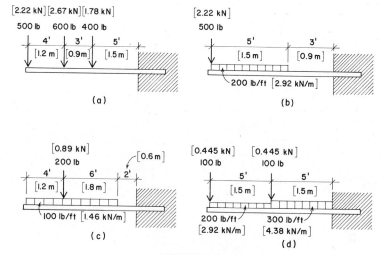

(a)

(b)

(c)

(d)

FIGURE 3-17.

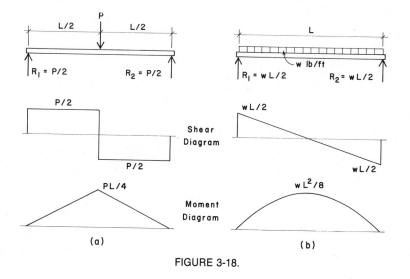

Shear Diagram

Moment Diagram

(a)

(b)

FIGURE 3-18.

71

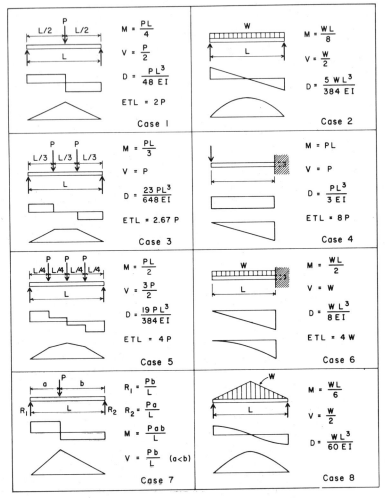

Case 1

$$M = \frac{PL}{4}$$

$$V = \frac{P}{2}$$

$$D = \frac{PL^3}{48\,EI}$$

ETL = 2 P

Case 2

$$M = \frac{WL}{8}$$

$$V = \frac{W}{2}$$

$$D = \frac{5\,WL^3}{384\,EI}$$

Case 3

$$M = \frac{PL}{3}$$

$$V = P$$

$$D = \frac{23\,PL^3}{648\,EI}$$

ETL = 2.67 P

Case 4

$$M = PL$$

$$V = P$$

$$D = \frac{PL^3}{3\,EI}$$

ETL = 8 P

Case 5

$$M = \frac{PL}{2}$$

$$V = \frac{3P}{2}$$

$$D = \frac{19\,PL^3}{384\,EI}$$

ETL = 4 P

Case 6

$$M = \frac{WL}{2}$$

$$V = W$$

$$D = \frac{WL^3}{8\,EI}$$

ETL = 4 W

Case 7

$$R_1 = \frac{Pb}{L}$$

$$R_2 = \frac{Pa}{L}$$

$$M = \frac{Pab}{L}$$

$$V = \frac{Pb}{L} \quad (a<b)$$

Case 8

$$M = \frac{WL}{6}$$

$$V = \frac{W}{2}$$

$$D = \frac{WL^3}{60\,EI}$$

FIGURE 3-19.   Values for typical beam loadings.

Solution: The formula giving the value of the maximum bending moment for this condition is $M = PL/4$. Therefore

$$M = \frac{8000 \times 20}{4} = 40,000 \text{ ft-lb}$$

If we want the results in inch-pounds, as we frequently do in design, we simply multiply by 12; that is

$$M = 40,000 \text{ ft-lb} = (40,000 \times 12)$$

or

$$480,000 \text{ in-lb}$$

3-11 Simple Beam with Uniform Load

This is probably the most common beam loading. It occurs time and again. Let us call the span L and the load w lb per lin ft, as indicated in Fig. 3-18*b*. The total load on the beam is wL; hence each reaction is $wL/2$. The maximum bending moment occurs at the center of the span at distance $L/2$ from R_1. Writing the value of M for this section, we have

$$M_{(x=L/2)} = \left(\frac{wL}{2} \times \frac{L}{2}\right) - \left(\frac{wL}{2} \times \frac{L}{4}\right)$$

$$M = \frac{wL^2}{4} - \frac{wL^2}{8}$$

$$M = \frac{wL^2}{8}$$

If, instead of being given the load per linear foot, we are given the total uniformly distributed load, we call the total load W. Because $wL = W$, the value of the maximum bending moment can be written $M = wL^2/8$, or $WL/8$. (See Case 2, Fig. 3-19.) Remember this formula. You will use it many times. Its convenience is demonstrated in the practical example below.

Example. A simple beam 14 ft long has a uniformly distributed load of 800 lb per lin ft. Compute the maximum bending moment.

Solution: The formula that gives the maximum bending moment for a simple beam with a uniformly distributed load is $M = wL^2/8$. Substituting these values, we find

$$M = \frac{800 \times 14 \times 14}{8}$$

$$M = 19,600 \text{ ft-lb}$$

or

$$M = 19,600 \times 12 = 235,200 \text{ in-lb}$$

Suppose in this problem we had been given the total load of 11,200 lb instead of 800 lb per ft. Then

$$M = \frac{WL}{8}$$

$$M = \frac{11,200 \times 14}{8} = 19,600 \text{ ft-lb}$$

$$19,600 \text{ ft-lb} = 235,200 \text{ in-lb}$$

The result, of course, is the same.

3-12   Typical Loadings: Formulas for Beam Behavior

Some of the most common beam loadings are shown in Fig. 3-19. In addition to the formulas for maximum shear V and maximum bending moment M, expressions for maximum deflection D are given also. (Discussion of deflection formulas will be deferred for the time being but will be considered under beam design in subsequent sections.)

In Fig. 3-19, if the loads P and W are in pounds or kips, the vertical shear V will also be in units of pounds or kips. When the loads are given in pounds or kips and the span in feet, the bending moment M will be in units of foot-pounds or kip-feet.

An extensive series of beam diagrams and formulas is contained in Part 2 of the *Manual of Steel Construction,* 8th ed., published by the American Institute of Steel Construction.

Problem 3-12-A*.   A simple beam with a span of L ft has three concentrated loads of P lb each, located at the quarter points of the span.

Compute the values of the maximum shear and maximum bending moment, using the terms P and L.

Problem 3-12-B*.   Four concentrated loads of P lb each are placed at the fifth points of the span of a simple beam. If the span length is L ft, compute the magnitudes of the maximum shear and maximum bending moment in terms of P and L.

Problem 3-12-C.   Draw the shear and moment diagrams for the beam and loading described in Problem 3-12-A, noting significant values in terms of P and L (similar to Fig. 3-19).

Problem 3-12-D.   For the beam and loading described in Problem 3-12-B, draw and label the shear and bending moment diagrams in terms of P and L.

4

Theory of Bending and Properties of Sections

||||||||||||||||||||||||||||||||||||||||||||||||||||||

4-1   Resisting Moment

We learned in the preceding chapter that bending moment is a measure of the tendency of the external forces on a beam to deform it by bending. We will now consider the action within the beam that resists bending and is called the *resisting moment*.

Figure 4-1a shows a simple beam, rectangular in cross section, supporting a single concentrated load *P*. Figure 4-1b is an enlarged sketch of the left-hand portion of the beam between the reaction and section *X-X*. From the preceding discussions we know that the reaction R_1 tends to cause a clockwise rotation about point *A* in the section under consideration; this we have defined as the bending moment at the section. In this type of beam the fibers in the upper part are in compression, and those in the lower part are in tension. There is a horizontal plane separating the compressive and tensile stresses; it is called the *neutral surface,* and at this plane there are neither compressive nor tensile stresses with respect to bending. The line in which the neutral surface intersects the beam cross section (Fig. 4-1c) is called the *neutral axis,* NA.

Call *C* the sum of all the compressive stresses acting on the upper part of the cross section, and call *T* the sum of all the tensile

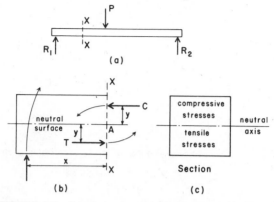

FIGURE 4-1. Development of bending stress in a beam.

stresses acting on the lower part. It is the sum of the moments of these stresses at the section that holds the beam in equilibrium; this is called the *resisting moment* and is equal to the bending moment in magnitude. The bending moment about point A is $R_1 \times x$, and the resisting moment about the same point is $(C \times y) + (T \times y)$. The bending moment tends to cause a clockwise rotation, and the resisting moment tends to cause a counterclockwise rotation. If the beam is in equilibrium, these moments are equal, or

$$R_1 \times x = (C \times y) + (T \times y)$$

that is, the bending moment equals the resisting moment. This is the theory of flexure (bending) in beams. For any type of beam, we can compute the bending moment; and if we wish to design a beam to withstand this tendency to bend, we must select a member with a cross section of such shape, area, and material that it is capable of developing a resisting moment equal to the bending moment.

4-2 The Flexure Formula

The flexure formula, $M = fS$, is an expression for resisting moment that involves the size and shape of the beam cross section

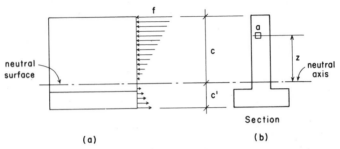

FIGURE 4-2.   Distribution of bending stress on a beam section.

(represented by S in the formula) and the material of which the beam is made (represented by f). It is used in the design of all homogeneous beams, that is, beams made of one material only, such as steel or wood. You will never need to derive this formula, but you will use it many times. The following brief derivation is presented to show the principles on which the formula is based.

Figure 4-2 represents a partial side elevation and the cross section of a homogeneous beam subjected to bending stresses. The cross section shown is unsymmetrical about the neutral axis, but this discussion applies to a cross section of any shape. In Fig. 4-2a let c be the distance of the fiber farthest from the neutral axis, and let f be the unit stress on the fiber at distance c. If f, the extreme fiber stress, does not exceed the elastic limit of the material, the stresses in the other fibers are directly proportional to their distances from the neutral axis. That is to say, if one fiber is twice the distance from the neutral axis than another fiber, the fiber at the greater distance will have twice the stress. The stresses are indicated in the figure by the small lines with arrows, which represent the compressive and tensile stresses acting toward and away from the section, respectively. If c is in inches, the unit stress on a fiber at 1 in. distance is f/c. Now imagine an infinitely small area a at z distance from the neutral axis. The unit stress on this fiber is $(f/c) \times z$, and because this small area contains a square inches, the total stress on fiber a is $(f/c) \times z \times a$. The *moment* of the stress on fiber a at z distance is

$$\frac{f}{c} \times z \times a \times z \qquad \text{or} \qquad \frac{f}{c} \times a \times z^2$$

We know, however, that there is an extremely large number of these minute areas, and if we use the symbol Σ to represent the sum of this very large number, we can write

$$\Sigma \frac{f}{c} \times a \times z^2$$

which means the sum of the moments of all the stresses in the cross section with respect to the neutral axis. This we know is the *resisting moment,* and it is equal to the bending moment.

Therefore

$$M = \frac{f}{c} \Sigma \, a \times z^2$$

The quantity $\Sigma \, a \times z^2$ may be read "the sum of the products of all the elementary areas times the square of their distances from the neutral axis." We call this the *moment of inertia* and represent it by the letter I. Therefore, substituting in the above, we have

$$M = \frac{f}{c} \times I \quad \text{or} \quad M = \frac{fI}{c}$$

This is known as the *flexure formula* or *beam formula,* and by its use we may design any beam that is composed of a single material. The expression may be simplified further by substituting S for I/c, called the *section modulus,* a term that is described more fully in Section 4-6. Making this substitution, the formula becomes

$$M = fS$$

Use of the flexure formula in the design and investigation of beams is discussed in Section 4-7.

4-3   Properties of Sections: Structural Shapes

Each of the terms *moment of inertia* and *section modulus* used in the preceding article represents a *property* of a particular beam cross section. Other properties are the area of the section (A), the radius of gyration (r), and the position of the centroid of the cross-sectional area. These properties and other useful data are tabu-

TABLE 4-1. Properties of Wide Flange (WF) Shapes

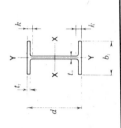

Designation	Area A in²	Depth d in.	Web Thickness t_w in.	Flange Width b_f in.	Flange Thickness t_f in.	k in.	Elastic Properties Axis X-X I in⁴	S in³	r in.	Axis Y-Y I in⁴	S in³	r in.
W 36 × 300	88.3	36.74	0.945	16.655	1.680	2.81	20,300	1110	15.2	1300	156	3.83
× 260	76.5	36.26	0.840	16.550	1.440	2.56	17,300	953	15.0	1090	132	3.78
× 230	67.6	35.90	0.760	16.470	1.260	2.38	15,000	837	14.9	940	114	3.73
× 194	57.0	36.49	0.765	12.115	1.260	2.19	12,100	664	14.6	375	61.9	2.56
× 170	50.0	36.17	0.680	12.030	1.100	2.00	10,500	580	14.5	320	53.2	2.53
× 150	44.2	35.85	0.625	11.975	0.940	1.88	9,040	504	14.3	270	45.1	2.47
× 135	39.7	35.55	0.600	11.950	0.790	1.69	7,800	439	14.0	225	37.7	2.38
W 33 × 241	70.9	34.18	0.830	15.860	1.400	2.19	14,200	829	14.1	932	118	3.63
× 201	59.1	33.68	0.715	15.745	1.150	1.94	11,500	684	14.0	749	95.2	3.56
× 152	44.7	33.49	0.635	11.565	1.055	1.88	8,160	487	13.5	273	47.2	2.47

× 130	38.3	33.09	0.580	11.510	0.855	6,710	1.69	406	13.2	218	37.9	2.39
× 118	34.7	32.86	0.550	11.480	0.740	5,900	1.56	359	13.0	187	32.6	2.32
W 30 × 211	62.0	30.94	0.775	15.105	1.315	10,300	2.13	663	12.9	757	100	3.49
× 173	50.8	30.44	0.655	14.985	1.065	8,200	1.88	539	12.7	598	79.8	3.43
× 124	36.5	30.17	0.585	10.515	0.930	5,360	1.69	355	12.1	181	34.4	2.23
× 108	31.7	29.83	0.545	10.475	0.760	4,470	1.56	299	11.9	146	27.9	2.15
× 99	29.1	29.65	0.520	10.450	0.670	3,990	1.44	269	11.7	128	24.5	2.10
W 27 × 178	52.3	27.81	0.725	14.085	1.190	6,990	1.88	502	11.6	555	78.8	3.26
× 146	42.9	27.38	0.605	13.965	0.975	5,630	1.69	411	11.4	443	63.5	3.21
× 102	30.0	27.09	0.515	10.015	0.830	3,620	1.56	267	11.0	139	27.8	2.15
× 84	24.8	26.71	0.460	9.960	0.640	2,850	1.38	213	10.7	106	21.2	2.07
W 24 × 162	47.7	25.00	0.705	12.955	1.220	5,170	2.00	414	10.4	443	68.4	3.05
× 131	38.5	24.48	0.605	12.855	0.960	4,020	1.75	329	10.2	340	53.0	2.97
× 104	30.6	24.06	0.500	12.750	0.750	3,100	1.50	258	10.1	259	40.7	2.91
× 84	24.7	24.10	0.470	9.020	0.770	2,370	1.56	196	9.79	94.4	20.9	1.95
× 69	20.1	23.73	0.415	8.965	0.585	1,830	1.38	154	9.55	70.4	15.7	1.87
× 55	16.2	23.57	0.395	7.005	0.505	1,350	1.94	114	9.11	29.1	8.30	1.34
W 21 × 147	43.2	22.06	0.720	12.510	1.150	3,630	1.88	329	9.17	376	60.1	2.95
× 122	35.9	21.68	0.600	12.390	0.960	2,960	1.69	273	9.09	305	49.2	2.92
× 101	29.8	21.36	0.500	12.290	0.800	2,420	1.56	227	9.02	248	40.3	2.89
× 83	24.3	21.43	0.515	8.355	0.835	1,830	1.56	171	8.67	81.4	19.5	1.83
× 68	20.0	21.13	0.430	8.270	0.685	1,480	1.44	140	8.60	64.7	15.7	1.80
× 57	16.7	21.06	0.405	6.555	0.650	1,170	1.38	111	8.36	30.6	9.35	1.35

TABLE 4-1. (Continued)

Designation	Area A	Depth d	Web Thickness t_w	Flange Width b_f	Flange Thickness t_f	k	Elastic Properties Axis X-X			Axis Y-Y		
							I	S	r	I	S	r
	in²	in.	in.	in.	in.	in.	in⁴	in³	in.	in⁴	in³	in.
× 44	13.0	20.66	0.350	6.500	0.450	1.19	843	81.6	8.06	20.7	6.36	1.26
W 18 × 119	35.1	18.97	0.655	11.265	1.060	1.75	2,190	231	7.90	253	44.9	2.69
× 97	28.5	18.59	0.535	11.145	0.870	1.56	1,750	188	7.82	201	36.1	2.65
× 76	22.3	18.21	0.425	11.035	0.680	1.38	1,330	146	7.73	152	27.6	2.61
× 65	19.1	18.35	0.450	7.590	0.750	1.44	1,070	117	7.49	54.8	14.4	1.69
× 55	16.2	18.11	0.390	7.530	0.630	1.31	890	98.3	7.41	44.9	11.9	1.67
× 46	13.5	18.06	0.360	6.060	0.605	1.25	712	78.8	7.25	22.5	7.43	1.29
× 35	10.3	17.70	0.300	6.000	0.425	1.13	510	57.6	7.04	15.3	5.12	1.22
W 16 × 100	29.4	16.97	0.585	10.425	0.985	1.69	1,490	175	7.10	186	35.7	2.51
× 77	22.6	16.52	0.455	10.295	0.760	1.44	1,110	134	7.00	138	26.9	2.47
× 57	16.8	16.43	0.430	7.120	0.715	1.38	758	92.2	6.72	43.1	12.1	1.60
× 45	13.3	16.13	0.345	7.035	0.565	1.25	586	72.7	6.65	32.8	9.34	1.57
× 36	10.6	15.86	0.295	6.985	0.430	1.13	448	56.5	6.51	24.5	7.00	1.52
× 26	7.68	15.69	0.250	5.500	0.345	1.06	301	38.4	6.26	9.59	3.49	1.12

W 14 × 730	215.0	22.42	3.070	17.890	4.910	5.56	14,300	1280	8.17	4720	527	4.69
× 605	178.0	20.92	2.595	17.415	4.160	4.81	10,800	1040	7.80	3680	423	4.55
× 455	134.0	19.02	2.015	16.835	3.210	3.88	7,190	756	7.33	2560	304	4.38
× 370	109.0	17.92	1.655	16.475	2.660	3.31	5,440	607	7.07	1990	241	4.27
× 283	83.3	16.74	1.290	16.110	2.070	2.75	3,840	459	6.79	1440	179	4.17
× 211	62.0	15.72	0.980	15.800	1.560	2.25	2,660	338	6.55	1030	130	4.07
× 159	46.7	14.98	0.745	15.565	1.190	1.88	1,900	254	6.38	748	96.2	4.00
× 120	35.3	14.48	0.590	14.670	0.940	1.63	1,380	190	6.24	495	67.5	3.74
× 90	26.5	14.02	0.440	14.520	0.710	1.38	999	143	6.14	362	49.9	3.70
× 68	20.0	14.04	0.415	10.035	0.720	1.50	723	103	6.01	121	24.2	2.46
× 53	15.6	13.92	0.370	8.060	0.660	1.44	541	77.8	5.89	57.7	14.3	1.92
× 43	12.6	13.66	0.305	7.995	0.530	1.31	428	62.7	5.82	45.2	11.3	1.89
× 34	10.0	13.98	0.285	6.745	0.455	1.00	340	48.6	5.83	23.3	6.91	1.53
× 30	8.85	13.84	0.270	6.730	0.385	0.94	291	42.0	5.73	19.6	5.82	1.49
× 26	7.69	13.91	0.255	5.025	0.420	0.94	245	35.3	5.65	8.91	3.54	1.08
× 22	6.49	13.74	0.230	5.000	0.335	0.88	199	29.0	5.54	7.00	2.80	1.04
W 12 × 336	98.8	16.82	1.775	13.385	2.955	3.69	4,060	483	6.41	1190	177	3.47
× 279	81.9	15.85	1.530	13.140	2.470	3.19	3,110	393	6.16	937	143	3.38
× 210	61.8	14.71	1.180	12.790	1.900	2.63	2,140	292	5.89	664	104	3.28
× 152	44.7	13.71	0.870	12.480	1.400	2.13	1,430	209	5.66	454	72.8	3.19
× 106	31.2	12.89	0.610	12.220	0.990	1.69	933	145	5.47	301	49.3	3.11
× 79	23.2	12.38	0.470	12.080	0.735	1.44	662	107	5.34	216	35.8	3.05
× 58	17.0	12.19	0.360	10.010	0.640	1.38	475	78.0	5.28	107	21.4	2.51

TABLE 4-1. (Continued)

Designation	Area A in²	Depth d in.	Web Thickness t_w in.	Flange Width b_f in.	Flange Thickness t_f in.	k in.	Axis X-X I in⁴	Axis X-X S in³	Axis X-X r in.	Axis Y-Y I in⁴	Axis Y-Y S in³	Axis Y-Y r in.
× 50	14.7	12.19	0.370	8.080	0.640	1.38	394	64.7	5.18	56.3	13.9	1.96
× 40	11.8	11.94	0.295	8.005	0.515	1.25	310	51.9	5.13	44.1	11.0	1.93
× 30	8.79	12.34	0.260	6.520	0.440	0.94	238	38.6	5.21	20.3	6.24	1.52
× 22	6.48	12.31	0.260	4.030	0.425	0.88	156	25.4	4.91	4.66	2.31	0.847
× 19	5.57	12.16	0.235	4.005	0.350	0.81	130	21.3	4.82	3.76	1.88	0.822
× 16	4.71	11.99	0.220	3.990	0.265	0.75	103	17.1	4.67	2.82	1.41	0.773
× 14	4.16	11.91	0.200	3.970	0.225	0.69	88.6	14.9	4.62	2.36	1.19	0.753
W 10 × 112	32.9	11.36	0.755	10.415	1.250	1.88	716	126	4.66	236	45.3	2.68
× 88	25.9	10.84	0.605	10.265	0.990	1.63	534	98.5	4.54	179	34.8	2.63
× 60	17.6	10.22	0.420	10.080	0.680	1.31	341	66.7	4.39	116	23.0	2.57
× 45	13.3	10.10	0.350	8.020	0.620	1.25	248	49.1	4.32	53.4	13.3	2.01
× 33	9.71	9.73	0.290	7.960	0.435	1.06	170	35.0	4.19	36.6	9.20	1.94
× 26	7.61	10.33	0.260	5.770	0.440	0.88	144	27.9	4.35	14.1	4.89	1.36
× 19	5.62	10.24	0.250	4.020	0.395	0.81	96.3	18.8	4.14	4.29	2.14	0.874
× 15	4.41	9.99	0.230	4.000	0.270	0.69	68.9	13.8	3.95	2.89	1.45	0.810
× 12	3.54	9.87	0.190	3.960	0.210	0.63	53.8	10.9	3.90	2.18	1.10	0.785

Elastic Properties

W 8 × 67	19.7	9.00	0.570	8.280	0.935	1.44	272	60.4	3.72	88.6	21.4	2.12
× 48	14.1	8.50	0.400	8.110	0.685	1.19	184	43.3	3.61	60.9	15.0	2.08
× 40	11.7	8.25	0.360	8.070	0.560	1.06	146	35.5	3.53	49.1	12.2	2.04
× 35	10.3	8.12	0.310	8.020	0.495	1.00	127	31.2	3.51	42.6	10.6	2.03
× 31	9.13	8.00	0.285	7.995	0.435	0.94	110	27.5	3.47	37.1	9.27	2.02
× 28	8.25	8.06	0.285	6.535	0.465	0.94	98.0	24.3	3.45	21.7	6.63	1.62
× 24	7.08	7.93	0.245	6.495	0.400	0.88	82.8	20.9	3.42	18.3	5.63	1.61
× 21	6.16	8.28	0.250	5.270	0.400	0.81	75.3	18.2	3.49	9.77	3.71	1.26
× 18	5.26	8.14	0.230	5.250	0.330	0.75	61.9	15.2	3.43	7.97	3.04	1.23
× 15	4.44	8.11	0.245	4.015	0.315	0.75	48.0	11.8	3.29	3.41	1.70	0.876
× 13	3.84	7.99	0.230	4.000	0.255	0.69	39.6	9.91	3.21	2.73	1.37	0.843
× 10	2.96	7.89	0.170	3.940	0.205	0.63	30.8	7.81	3.22	2.09	1.06	0.841
W 6 × 25	7.34	6.38	0.320	6.080	0.455	0.81	53.4	16.7	2.70	17.1	5.61	1.52
× 20	5.87	6.20	0.260	6.020	0.365	0.75	41.4	13.4	2.66	13.3	4.41	1.50
× 15	4.43	5.99	0.230	5.990	0.260	0.63	29.1	9.72	2.56	9.32	3.11	1.46
× 16	4.74	6.28	0.260	4.030	0.405	0.75	32.1	10.2	2.60	4.43	2.20	0.966
× 12	3.55	6.03	0.230	4.000	0.280	0.63	22.1	7.31	2.49	2.99	1.50	0.918
× 9	2.68	5.90	0.170	3.940	0.215	0.56	16.4	5.56	2.47	2.19	1.11	0.905
W 5 × 19	5.54	5.15	0.270	5.030	0.430	0.81	26.2	10.2	2.17	9.13	3.63	1.28
× 16	4.68	5.01	0.240	5.000	0.360	0.75	21.3	8.51	2.13	7.51	3.00	1.27
W 4 × 13	3.83	4.16	0.280	4.060	0.345	0.69	11.3	5.46	1.72	3.86	1.90	1.00

Source: Adapted from data in the *Manual of Steel Construction*, 8th ed., with permission of the publishers, American Institute of Steel Construction.

TABLE 4-2. Properties of American Standard (S) Shapes (I-Beams)

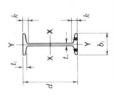

			Web	Flange Width	Flange Thickness		Elastic Properties					
							Axis X-X			Axis Y-Y		
Designation	Area A	Depth d	Thickness t_w	b_f	t_f	k	I	S	r	I	S	r
	in²	in.	in.	in.	in.	in.	in⁴	in³	in.	in⁴	in³	in.
S 24 × 121	35.6	24.50	0.800	8.050	1.090	2.00	3160	258	9.43	83.3	20.7	1.53
× 106	31.2	24.50	0.620	7.870	1.090	2.00	2940	240	9.71	77.1	19.6	1.57
× 100	29.3	24.00	0.745	7.245	0.870	1.75	2390	199	9.02	47.7	13.2	1.27
× 90	26.5	24.00	0.625	7.125	0.870	1.75	2250	187	9.21	44.9	12.6	1.30
× 80	23.5	24.00	0.500	7.000	0.870	1.75	2100	175	9.47	42.2	12.1	1.34
S 20 × 96	28.2	20.30	0.8000	7.200	0.920	1.75	1670	165	7.71	50.2	13.9	1.33
× 86	25.3	20.30	0.660	7.060	0.920	1.75	1580	155	7.89	46.8	13.3	1.36
× 75	22.0	20.00	0.635	6.385	0.795	1.63	1280	128	7.62	29.8	9.32	1.16
× 66	19.4	20.00	0.505	6.255	0.795	1.63	1190	119	7.83	27.7	8.85	1.19

S 18 × 70	20.6	18.00	0.711	6.251	0.691	1.50	926	103	6.71	24.1	7.72	1.08
× 54.7	16.1	18.00	0.461	6.001	0.691	1.50	804	89.4	7.07	20.8	6.94	1.14
S 15 × 50	14.7	15.00	0.550	5.640	0.622	1.38	486	64.8	5.75	15.7	5.57	1.03
× 42.9	12.6	15.00	0.411	5.501	0.622	1.38	447	59.6	5.95	14.4	5.23	1.07
S 12 × 50	14.7	12.00	0.687	5.477	0.659	1.44	305	50.8	4.55	15.7	5.74	1.03
× 40.8	12.0	12.00	0.462	5.252	0.659	1.44	272	45.4	4.77	13.6	5.16	1.06
× 35	10.3	12.00	0.428	5.078	0.544	1.19	229	38.2	4.72	9.87	3.89	0.980
× 31.8	9.35	12.00	0.350	5.000	0.544	1.19	218	36.4	4.83	9.36	3.74	1.00
S 10 × 35	10.3	10.00	0.594	4.944	0.491	1.13	147	29.4	3.78	8.36	3.38	0.901
× 25.4	7.46	10.00	0.311	4.661	0.491	1.13	124	24.7	4.07	6.79	2.91	0.954
S 8 × 23	6.77	8.00	0.441	4.171	0.426	1.00	64.9	16.2	3.10	4.31	2.07	0.798
× 18.4	5.41	8.00	0.271	4.001	0.426	1.00	57.6	14.4	3.26	3.73	1.86	0.831
S 7 × 20	5.88	7.00	0.450	3.860	0.392	0.94	42.4	12.1	2.69	3.17	1.64	0.734
× 15.3	4.50	7.00	0.252	3.662	0.392	0.94	36.7	10.5	2.86	2.64	1.44	0.766
S 6 × 17.25	5.07	6.00	0.465	3.565	0.359	0.88	26.3	8.77	2.28	2.31	1.30	0.675
× 12.5	3.67	6.00	0.232	3.332	0.359	0.88	22.1	7.37	2.45	1.82	1.09	0.705
S 5 × 14.75	4.34	5.00	0.494	3.284	0.326	0.81	15.2	6.09	1.87	1.67	1.01	0.620
× 10	2.94	5.00	0.214	3.004	0.326	0.81	12.3	4.92	2.05	1.22	0.809	0.643
S 4 × 9.5	2.79	4.00	0.326	2.796	0.293	0.75	6.79	3.39	1.56	0.903	0.646	0.569
× 7.7	2.26	4.00	0.193	2.663	0.293	0.75	6.08	3.04	1.64	0.764	0.574	0.581
S 3 × 7.5	2.21	3.00	0.349	2.509	0.260	0.69	2.93	1.95	1.15	0.586	0.468	0.516
× 5.7	1.67	3.00	0.170	2.330	0.260	0.69	2.52	1.68	1.23	0.455	0.390	0.522

Source: Adapted from data in the *Manual of Steel Construction*, 8th ed., with permission of the publishers, American Institute of Steel Construction.

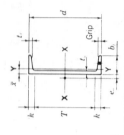

TABLE 4-3. Properties of American Standard Channel (C) Shapes

Designation	Area A	Depth d	Web Thickness t_w	Flange Width b_f	Flange Thickness t_f	k	Elastic Properties Axis X-X I	S	r	Axis Y-Y I	S	r	$\bar{x}$
	in²	in.	in.	in.	in.	in.	in⁴	in³	in.	in⁴	in³	in.	in.
C 15 × 50	14.7	15.00	0.716	3.716	0.650	1.44	404	53.8	5.24	11.0	3.78	0.867	0.798
× 40	11.8	15.00	0.520	3.520	0.650	1.44	349	46.5	5.44	9.23	3.37	0.886	0.777
× 33.9	9.96	15.00	0.400	3.400	0.650	1.44	315	42.0	5.62	8.13	3.11	0.904	0.787
C 12 × 30	8.82	12.00	0.510	3.170	0.501	1.13	162	27.0	4.29	5.14	2.06	0.763	0.674
× 25	7.35	12.00	0.387	3.047	0.501	1.13	144	24.1	4.43	4.47	1.88	0.780	0.674
× 20.7	6.09	12.00	0.282	2.942	0.501	1.13	129	21.5	4.61	3.88	1.73	0.799	0.698
C 10 × 30	8.82	10.00	0.673	3.033	0.436	1.00	103	20.7	3.42	3.94	1.65	0.669	0.649
× 25	7.35	10.00	0.526	2.886	0.436	1.00	91.2	18.2	3.52	3.36	1.48	0.676	0.617

Designation													
× 20	5.88	10.00	0.379	2.739	0.436	1.00	78.9	15.8	3.66	2.81	1.32	0.692	0.606
× 15.3	4.49	10.00	0.240	2.600	0.436	1.00	67.4	13.5	3.87	2.28	1.16	0.713	0.634
C 9 × 20	5.88	9.00	0.448	2.648	0.413	0.94	60.9	13.5	3.22	2.42	1.17	0.642	0.583
× 15	4.41	9.00	0.285	2.485	0.413	0.94	51.0	11.3	3.40	1.93	1.01	0.661	0.586
× 13.4	3.94	9.00	0.233	2.433	0.413	0.94	47.9	10.6	3.48	1.76	0.962	0.669	0.601
C 8 × 18.75	5.51	8.00	0.487	2.527	0.390	0.94	44.0	11.0	2.82	1.98	1.01	0.599	0.565
× 13.75	4.04	8.00	0.303	2.343	0.390	0.94	36.1	9.03	2.99	1.53	0.854	0.615	0.553
× 11.5	3.38	8.00	0.220	2.260	0.390	0.94	32.6	8.14	3.11	1.32	0.781	0.625	0.571
C 7 × 14.75	4.33	7.00	0.419	2.299	0.366	0.88	27.2	7.78	2.51	1.38	0.779	0.564	0.532
× 12.25	3.60	7.00	0.314	2.194	0.366	0.88	24.2	6.93	2.60	1.17	0.703	0.571	0.525
× 9.8	2.87	7.00	0.210	2.090	0.366	0.88	21.3	6.08	2.72	0.968	0.625	0.581	0.540
C 6 × 13	3.83	6.00	0.437	2.157	0.343	0.81	17.4	5.80	2.13	1.05	0.642	0.525	0.514
× 10.5	3.09	6.00	0.314	2.034	0.343	0.81	15.2	5.06	2.22	0.866	0.564	0.529	0.499
× 8.2	2.40	6.00	0.200	1.920	0.343	0.81	13.1	4.38	2.34	0.693	0.492	0.537	0.511
C 5 × 9	2.64	5.00	0.325	1.885	0.320	0.75	8.90	3.56	1.83	0.632	0.450	0.489	0.478
× 6.7	1.97	5.00	0.190	1.750	0.320	0.75	7.49	3.00	1.95	0.479	0.378	0.493	0.484
C 4 × 7.25	2.13	4.00	0.321	1.721	0.296	0.69	4.59	2.29	1.47	0.433	0.343	0.450	0.459
× 5.4	1.59	4.00	0.184	1.584	0.296	0.69	3.85	1.93	1.56	0.319	0.283	0.449	0.457
C 3 × 6	1.76	3.00	0.356	1.596	0.273	0.69	2.07	1.38	1.08	0.305	0.268	0.416	0.455
× 5	1.47	3.00	0.258	1.498	0.273	0.69	1.85	1.24	1.12	0.247	0.233	0.410	0.438
× 4.1	1.21	3.00	0.170	1.410	0.273	0.69	1.66	1.10	1.17	0.197	0.202	0.404	0.436

Source: Adapted from data in the *Manual of Steel Construction,* 8th ed., with permission of the publishers, American Institute of Steel Construction.

TABLE 4-4. Properties of Angles

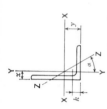

Size and Thickness in.	k in.	Weight lb	Area in²	Axis X-X				Axis Y-Y				Axis Z-Z	
				I in⁴	S in³	r in.	y in.	I in⁴	S in³	r in.	x in.	r in.	Tan α
8 × 8 × 1 1/8	1.75	56.9	16.7	98.0	17.5	2.42	2.41	98.0	17.5	2.42	2.41	1.56	1.000
1	1.625	51.0	15.0	89.0	15.8	2.44	2.37	89.0	15.8	2.44	2.37	1.56	1.000
7/8	1.50	45.0	13.2	79.6	14.0	2.45	2.32	79.6	14.0	2.45	2.32	1.57	1.000
3/4	1.375	38.9	11.4	69.7	12.2	2.47	2.28	69.7	12.2	2.47	2.28	1.58	1.000
5/8	1.25	32.7	9.61	59.4	10.3	2.49	2.23	59.4	10.3	2.49	2.23	1.58	1.000
1/2	1.125	26.4	7.75	48.6	8.36	2.50	2.19	48.6	8.36	2.50	2.19	1.59	1.000
8 × 6 × 1	1.50	44.2	13.0	80.8	15.1	2.49	2.65	38.8	8.92	1.73	1.65	1.28	0.543
3/4	1.25	33.8	9.94	63.4	11.7	2.53	2.56	30.7	6.92	1.76	1.56	1.29	0.551
1/2	1.0	23.0	6.75	44.3	8.02	2.56	2.47	21.7	4.79	1.79	1.47	1.30	0.558
8 × 4 × 1	1.50	37.4	11.0	69.6	14.1	2.52	3.05	11.6	3.94	1.03	1.05	0.846	0.247
3/4	1.25	28.7	8.44	54.9	10.9	2.55	2.95	9.36	3.07	1.05	0.953	0.852	0.258
1/2	1.0	19.6	5.75	38.5	7.49	2.59	2.86	6.74	2.15	1.08	0.859	0.865	0.267
7 × 4 × 3/4	1.25	26.2	7.69	37.8	8.42	2.22	2.51	9.05	3.03	1.09	1.01	0.860	0.324
1/2	1.0	17.9	5.25	26.7	5.81	2.25	2.42	6.53	2.12	1.11	0.917	0.872	0.335
3/8	0.875	13.6	3.98	20.6	4.44	2.27	2.37	5.10	1.63	1.13	0.870	0.880	0.340

Size													
6 × 6 × 1	1.50	37.4	11.0	35.5	8.57	1.80	1.86	35.5	8.57	1.80	1.86	1.17	1.000
7/8	1.375	33.1	9.73	31.9	7.63	1.81	1.82	31.9	7.63	1.81	1.82	1.17	1.000
3/4	1.25	28.7	8.44	28.2	6.66	1.83	1.78	28.2	6.66	1.83	1.78	1.17	1.000
5/8	1.125	24.2	7.11	24.2	5.66	1.84	1.73	24.2	5.66	1.84	1.73	1.18	1.000
1/2	1.0	19.6	5.75	19.9	4.61	1.86	1.68	19.9	4.61	1.86	1.68	1.18	1.000
3/8	0.875	14.9	4.36	15.4	3.53	1.88	1.64	15.4	3.53	1.88	1.64	1.19	1.000
6 × 4 × 3/4	1.25	23.6	6.94	24.5	6.25	1.88	2.08	8.68	2.97	1.12	1.08	0.860	0.428
5/8	1.125	20.0	5.86	21.1	5.31	1.90	2.03	7.52	2.54	1.13	1.03	0.864	0.435
1/2	1.00	16.2	4.75	17.4	4.33	1.91	1.99	6.27	2.08	1.15	0.987	0.870	0.440
3/8	0.875	12.3	3.61	13.5	3.32	1.93	1.94	4.90	1.60	1.17	0.941	0.877	0.446
6 × 3 1/2 × 3/8	0.875	11.7	3.42	12.9	3.24	1.94	2.04	3.34	1.23	0.988	0.787	0.767	0.350
5/16	0.8125	9.8	2.87	10.9	2.73	1.95	2.01	2.85	1.04	0.996	0.763	0.772	0.352
5 × 5 × 7/8	1.375	27.2	7.98	17.8	5.17	1.49	1.57	17.8	5.17	1.49	1.57	0.973	1.000
3/4	1.25	23.6	6.94	15.7	4.53	1.51	1.52	15.7	4.53	1.51	1.52	0.975	1.000
1/2	1.0	16.2	4.75	11.3	3.16	1.54	1.43	11.3	3.16	1.54	1.43	0.983	1.000
5 × 5 × 3/8	0.875	12.3	3.61	8.74	2.42	1.56	1.39	8.74	2.42	1.56	1.39	0.990	1.000
5/16	0.8125	10.3	3.03	7.42	2.04	1.57	1.37	7.42	2.04	1.57	1.37	0.994	1.000
5 × 3 1/2 × 3/4	1.25	19.8	5.81	13.9	4.28	1.55	1.75	5.55	2.22	0.977	0.996	0.748	0.464
1/2	1.0	13.6	4.00	9.99	2.99	1.58	1.66	4.05	1.56	1.01	0.906	0.755	0.479
3/8	0.875	10.4	3.05	7.78	2.29	1.60	1.61	3.18	1.21	1.02	0.861	0.762	0.486
5/16	0.8125	8.7	2.56	6.60	1.94	1.61	1.59	2.72	1.02	1.03	0.838	0.766	0.489
5 × 3 × 1/2	1.0	12.8	3.75	9.45	2.91	1.59	1.75	2.58	1.15	0.829	0.750	0.648	0.357
3/8	0.875	9.8	2.86	7.37	2.24	1.61	1.70	2.04	0.888	0.845	0.704	0.654	0.364
5/16	0.8125	8.2	2.40	6.26	1.89	1.61	1.68	1.75	0.753	0.853	0.681	0.658	0.368
1/4	0.75	6.6	1.94	5.11	1.53	1.62	1.66	1.44	0.614	0.861	0.657	0.663	0.371

TABLE 4-4. (Continued)

Size and Thickness in.	k in.	Weight lb	Area in²	Axis X-X				Axis Y-Y				Axis Z-Z	
				I in⁴	S in³	r in.	y in.	I in⁴	S in³	r in.	x in.	r in.	Tan α
4 × 4 × 3/4	1.125	18.5	5.44	7.67	2.81	1.19	1.27	7.67	2.81	1.19	1.27	0.778	1.000
5/8	1.0	15.7	4.61	6.66	2.40	1.20	1.23	6.66	2.40	1.20	1.23	0.779	1.000
1/2	0.875	12.8	3.75	5.56	1.97	1.22	1.18	5.56	1.97	1.22	1.18	0.782	1.000
3/8	0.75	9.8	2.86	4.36	1.52	1.23	1.14	4.36	1.52	1.23	1.14	0.788	1.000
5/16	0.6875	8.2	2.40	3.71	1.29	1.24	1.12	3.71	1.29	1.24	1.12	0.791	1.000
1/4	0.625	6.6	1.94	3.04	1.05	1.25	1.09	3.04	1.05	1.25	1.09	0.795	1.000
4 × 3 1/2 × 1/2	0.9375	11.9	3.50	5.32	1.94	1.23	1.25	3.79	1.52	1.04	1.00	0.722	0.750
3/8	0.8125	9.1	2.67	4.18	1.49	1.25	1.21	2.95	1.17	1.06	0.955	0.727	0.755
5/16	0.75	7.7	2.25	3.56	1.26	1.26	1.18	2.55	0.994	1.07	0.932	0.730	0.757
1/4	0.6875	6.2	1.81	2.91	1.03	1.27	1.16	2.09	0.808	1.07	0.909	0.734	0.759
4 × 3 × 1/2	0.9375	11.1	3.25	5.05	1.89	1.25	1.33	2.42	1.12	0.864	0.827	0.639	0.543
3/8	0.8125	8.5	2.48	3.96	1.46	1.26	1.28	1.92	0.866	0.879	0.782	0.644	0.551
5/16	0.75	7.2	2.09	3.38	1.23	1.27	1.26	1.65	0.734	0.887	0.759	0.647	0.554
1/4	0.6875	5.8	1.69	2.77	1.00	1.28	1.24	1.36	0.599	0.896	0.736	0.651	0.558
3 1/2 × 3 1/2 × 3/8	0.75	8.5	2.48	2.87	1.15	1.07	1.01	2.87	1.15	1.07	1.01	0.687	1.000
5/16	0.6875	7.2	2.09	2.45	0.976	1.08	0.990	2.45	0.976	1.08	0.990	0.690	1.000
1/4	0.625	5.8	1.69	2.01	0.794	1.09	0.968	2.01	0.794	1.09	0.968	0.694	1.000
3 1/2 × 3 × 3/8	0.8125	7.9	2.30	2.72	1.13	1.09	1.08	1.85	0.851	0.897	0.830	0.625	0.721
5/16	0.75	6.6	1.93	2.33	0.954	1.10	1.06	1.58	0.722	0.905	0.808	0.627	0.724
1/4	0.6875	5.4	1.56	1.91	0.776	1.11	1.04	1.30	0.589	0.914	0.785	0.631	0.727
3 1/2 × 2 1/2 × 3/8	0.8125	7.2	2.11	2.56	1.09	1.10	1.16	1.09	0.592	0.719	0.660	0.537	0.496
5/16	0.75	6.1	1.78	2.19	0.927	1.11	1.14	0.939	0.504	0.727	0.637	0.540	0.501

Section	1/4											
1/4	0.506	0.544	0.614	0.735	0.412	0.777	1.11	0.755	1.80	1.44	4.9	0.6875
3 × 3 × 1/2	1.000	0.584	0.932	0.898	1.07	2.22	0.932	1.07	2.22	2.75	9.4	0.8125
3/8	1.000	0.587	0.888	0.913	0.833	1.76	0.888	0.833	1.76	2.11	7.2	0.6875
5/16	1.000	0.589	0.865	0.922	0.707	1.51	0.865	0.707	1.51	1.78	6.1	0.625
1/4	1.000	0.592	0.842	0.930	0.577	1.24	0.842	0.577	1.24	1.44	4.9	0.5625
3/16	1.000	0.596	0.820	0.939	0.441	0.962	0.820	0.441	0.962	1.09	3.71	0.5000
3 × 2 1/2 × 3/8	0.676	0.522	0.706	0.736	0.581	1.04	0.956	0.810	1.66	1.92	6.6	0.75
1/4	0.684	0.528	0.661	0.753	0.404	0.743	0.911	0.561	1.17	1.31	4.5	0.625
3/16	0.688	0.533	0.638	0.761	0.310	0.577	0.888	0.430	0.907	0.996	3.39	0.5625
3 × 2 × 3/8	0.428	0.430	0.539	0.559	0.371	0.543	1.04	0.781	1.53	1.73	5.9	0.6875
5/16	0.435	0.432	0.516	0.567	0.317	0.470	1.02	0.664	1.32	1.46	5.0	0.625
1/4	0.440	0.435	0.493	0.574	0.260	0.392	0.993	0.542	1.09	1.19	4.1	0.5625
3/16	0.446	0.439	0.470	0.583	0.200	0.307	0.970	0.415	0.842	0.902	3.07	0.5000
2 1/2 × 2 1/2 × 3/8	1.000	0.487	0.762	0.753	0.566	0.984	0.762	0.566	0.984	1.73	5.9	0.6875
5/16	1.000	0.489	0.740	0.761	0.482	0.849	0.740	0.482	0.849	1.46	5.0	0.625
1/4	1.000	0.491	0.717	0.769	0.394	0.703	0.717	0.394	0.703	1.19	4.1	0.5625
3/16	1.000	0.495	0.694	0.778	0.303	0.547	0.694	0.303	0.547	0.902	3.07	0.5000
2 1/2 × 2 × 3/8	0.614	0.420	0.581	0.577	0.363	0.514	0.831	0.547	0.912	1.55	5.3	0.6875
5/16	0.620	0.422	0.559	0.584	0.310	0.446	0.809	0.466	0.788	1.31	4.5	0.625
1/4	0.626	0.424	0.537	0.592	0.254	0.372	0.787	0.381	0.654	1.06	3.62	0.5625
3/16	0.631	0.427	0.514	0.600	0.196	0.291	0.764	0.293	0.509	0.809	2.75	0.5000
2 × 2 × 3/8	1.000	0.389	0.636	0.594	0.351	0.479	0.636	0.351	0.479	1.36	4.7	0.6875
5/16	1.000	0.390	0.614	0.601	0.300	0.416	0.614	0.300	0.416	1.15	3.92	0.625
1/4	1.000	0.391	0.592	0.609	0.247	0.348	0.592	0.247	0.348	0.938	3.19	0.5625
3/16	1.000	0.394	0.569	0.617	0.190	0.272	0.569	0.190	0.272	0.715	2.44	0.5000

Source: Abstracted from the Manual of Steel Construction, 8th ed., with permission of the publishers, American Institute of Steel Construction.

TABLE 4-5. Properties of Round Steel Pipe

Dimensions				Weight per Foot	Properties			
Nominal Diameter In.	Outside Diameter In.	Inside Diameter In.	Wall Thickness In.	Lbs. Plain Ends	A In.^{2}	I In.^{4}	S In.^{3}	r In.
Standard Weight								
½	.840	.622	.109	.85	.250	.017	.041	.261
¾	1.050	.824	.113	1.13	.333	.037	.071	.334
1	1.315	1.049	.133	1.68	.494	.087	.133	.421
1¼	1.660	1.380	.140	2.27	.669	.195	.235	.540
1½	1.900	1.610	.145	2.72	.799	.310	.326	.623
2	2.375	2.067	.154	3.65	1.07	.666	.561	.787
2½	2.875	2.469	.203	5.79	1.70	1.53	1.06	.947
3	3.500	3.068	.216	7.58	2.23	3.02	1.72	1.16
3½	4.000	3.548	.226	9.11	2.68	4.79	2.39	1.34
4	4.500	4.026	.237	10.79	3.17	7.23	3.21	1.51
5	5.563	5.047	.258	14.62	4.30	15.2	5.45	1.88
6	6.625	6.065	.280	18.97	5.58	28.1	8.50	2.25
8	8.625	7.981	.322	28.55	8.40	72.5	16.8	2.94
10	10.750	10.020	.365	40.48	11.9	161	29.9	3.67
12	12.750	12.000	.375	49.56	14.6	279	43.8	4.38
Extra Strong								
½	.840	.546	.147	1.09	.320	.020	.048	.250
¾	1.050	.742	.154	1.47	.433	.045	.085	.321
1	1.315	.957	.179	2.17	.639	.106	.161	.407
1¼	1.660	1.278	.191	3.00	.881	.242	.291	.524
1½	1.900	1.500	.200	3.63	1.07	.391	.412	.605
2	2.375	1.939	.218	5.02	1.48	.868	.731	.766
2½	2.875	2.323	.276	7.66	2.25	1.92	1.34	.924
3	3.500	2.900	.300	10.25	3.02	3.89	2.23	1.14
3½	4.000	3.364	.318	12.50	3.68	6.28	3.14	1.31
4	4.500	3.826	.337	14.98	4.41	9.61	4.27	1.48
5	5.563	4.813	.375	20.78	6.11	20.7	7.43	1.84
6	6.625	5.761	.432	28.57	8.40	40.5	12.2	2.19
8	8.625	7.625	.500	43.39	12.8	106	24.5	2.88
10	10.750	9.750	.500	54.74	16.1	212	39.4	3.63
12	12.750	11.750	.500	65.42	19.2	362	56.7	4.33
Double-Extra Strong								
2	2.375	1.503	.436	9.03	2.66	1.31	1.10	.703
2½	2.875	1.771	.552	13.69	4.03	2.87	2.00	.844
3	3.500	2.300	.600	18.58	5.47	5.99	3.42	1.05
4	4.500	3.152	.674	27.54	8.10	15.3	6.79	1.37
5	5.563	4.063	.750	38.55	11.3	33.6	12.1	1.72
6	6.625	4.897	.864	53.16	15.6	66.3	20.0	2.06
8	8.625	6.875	.875	72.42	21.3	162	37.6	2.76

The listed sections are available in conformance with ASTM Specification A53 Grade B or A501. Other sections are made to these specifications. Consult with pipe manufacturers or distributors for availability.

Source: Reprinted from the *Manual of Steel Construction*, 8th ed., with permission of the publishers, American Institute of Steel Construction.

TABLE 4-6. Properties of Structural Steel Tubing

Square

DIMENSIONS				PROPERTIES**			
Nominal* Size	Wall Thickness		Weight per Foot	Area	I	S	r
In.	In.		Lb.	In.^{2}	In.^{4}	In.^{3}	In.
16 x 16	.5000	1/2	103.30	30.4	1200	150	6.29
	.3750	3/8	78.52	23.1	931	116	6.35
	.3125	5/16	65.87	19.4	789	98.6	6.38
14 x 14	.5000	1/2	89.68	26.4	791	113	5.48
	.3750	3/8	68.31	20.1	615	87.9	5.54
	.3125	5/16	57.36	16.9	522	74.6	5.57
12 x 12	.5000	1/2	76.07	22.4	485	80.9	4.66
	.3750	3/8	58.10	17.1	380	63.4	4.72
	.3125	5/16	48.86	14.4	324	54.0	4.75
	.2500	1/4	39.43	11.6	265	44.1	4.78
10 x 10	.6250	5/8	76.33	22.4	321	64.2	3.78
	.5000	1/2	62.46	18.4	271	54.2	3.84
	.3750	3/8	47.90	14.1	214	42.9	3.90
	.3125	5/16	40.35	11.9	183	36.7	3.93
	.2500	1/4	32.63	9.59	151	30.1	3.96
8 x 8	.6250	5/8	59.32	17.4	153	38.3	2.96
	.5000	1/2	48.85	14.4	131	32.9	3.03
	.3750	3/8	37.69	11.1	106	26.4	3.09
	.3125	5/16	31.84	9.36	90.9	22.7	3.12
	.2500	1/4	25.82	7.59	75.1	18.8	3.15
	.1875	3/16	19.63	5.77	58.2	14.6	3.18
7 x 7	.5000	1/2	42.05	12.4	84.6	24.2	2.62
	.3750	3/8	32.58	9.58	68.7	19.6	2.68
	.3125	5/16	27.59	8.11	59.5	17.0	2.71
	.2500	1/4	22.42	6.59	49.4	14.1	2.74
	.1875	3/16	17.08	5.02	38.5	11.0	2.77
6 x 6	.5000	1/2	35.24	10.4	50.5	16.8	2.21
	.3750	3/8	27.48	8.08	41.6	13.9	2.27
	.3125	5/16	23.34	6.86	36.3	12.1	2.30
	.2500	1/4	19.02	5.59	30.3	10.1	2.33
	.1875	3/16	14.53	4.27	23.8	7.93	2.36
5 x 5	.5000	1/2	28.43	8.36	27.0	10.8	1.80
	.3750	3/8	22.37	6.58	22.8	9.11	1.86
	.3125	5/16	19.08	5.61	20.1	8.02	1.89
	.2500	1/4	15.62	4.59	16.9	6.78	1.92
	.1875	3/16	11.97	3.52	13.4	5.36	1.95

* Outside dimensions across flat sides.
** Properties are based upon a nominal outside corner radius equal to two times the wall thickness.

TABLE 4-6. (Continued)

Square

Nominal* Size	Wall Thickness		Weight per Foot	Area	I	S	r
In.	In.		Lb	In.^{2}	In.^{4}	In.^{3}	In.
4 x 4	.5000	$^1/_2$	21.63	6.36	12.3	6.13	1.39
	.3750	$^3/_8$	17.27	5.08	10.7	5.35	1.45
	.3125	$^5/_{16}$	14.83	4.36	9.58	4.79	1.48
	.2500	$^1/_4$	12.21	3.59	8.22	4.11	1.51
	.1875	$^3/_{16}$	9.42	2.77	6.59	3.30	1.54
3.5 x 3.5	.3125	$^5/_{16}$	12.70	3.73	6.09	3.48	1.28
	.2500	$^1/_4$	10.51	3.09	5.29	3.02	1.31
	.1875	$^3/_{16}$	8.15	2.39	4.29	2.45	1.34
3 x 3	.3125	$^5/_{16}$	10.58	3.11	3.58	2.39	1.07
	.2500	$^1/_4$	8.81	2.59	3.16	2.10	1.10
	.1875	$^3/_{16}$	6.87	2.02	2.60	1.73	1.13
2.5 x 2.5	.2500	$^1/_4$	7.11	2.09	1.69	1.35	.899
	.1875	$^3/_{16}$	5.59	1.64	1.42	1.14	.930
2 x 2	.2500	$^1/_4$	5.41	1.59	.766	.766	.694
	.1875	$^3/_{16}$	4.32	1.27	.668	.668	.726

Outside dimensions across flat sides
Properties are based upon a nominal outside corner radius equal to two times the wall thickness

Rectangular

Nominal* Size	Wall Thickness		Weight per Foot	Area	X X AXIS				Y-Y AXIS		
					I_x	S_x	r_x		I_y	S_y	r_y
In.	In.		Lb	In.^{2}	In.^{4}	In.^{3}	In.		In.^{4}	In.^{3}	In.
20 x 12	.5000	$^1/_2$	103.30	30.4	1650	165	7.37		750	125.0	4.97
	.3750	$^3/_8$	78.52	23.1	1280	128	7.44		583	97.2	5.03
	.3125	$^5/_{16}$	65.87	19.4	1080	108	7.47		495	82.5	5.06
20 x 8	.5000	$^1/_2$	89.68	26.4	1270	127	6.94		300	75.1	3.38
	.3750	$^3/_8$	68.31	20.1	988	98.8	7.02		236	59.1	3.43
	.3125	$^5/_{16}$	57.36	16.9	838	83.8	7.05		202	50.4	3.46
20 x 4	.5000	$^1/_2$	76.07	22.4	889	88.9	6.31		61.6	30.8	1.66
	.3750	$^3/_8$	58.10	17.1	699	69.9	6.40		50.3	25.1	1.72
	.3125	$^5/_{16}$	48.86	14.4	596	59.6	6.44		43.7	21.8	1.74

* Outside dimensions across flat sides.
** Properties are based upon a nominal outside corner radius equal to two times the wall thickness.

TABLE 4-6. (Continued)

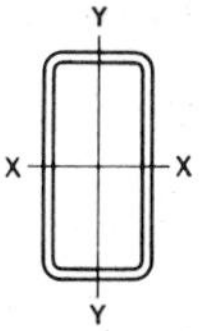

Rectangular

DIMENSIONS			PROPERTIES**						
Nominal* Size	Wall Thickness	Weight per Foot	Area	X-X AXIS			Y-Y AXIS		
				I_x	S_x	r_x	I_y	S_y	r_y
In.	In.	In.	In.^{2}	In.^{4}	In.^{3}	In.	In.^{4}	In.^{3}	In.
18 x 6	.5000 ½	76.07	22.4	818	90.9	6.05	141	47.2	2.52
	.3750 ⅜	58.10	17.1	641	71.3	6.13	113	37.6	2.57
	.3125 ⁵⁄₁₆	48.86	14.4	546	60.7	6.17	97.0	32.3	2.60
16 x 12	.5000 ½	89.68	26.4	962	120	6.04	618	103.0	4.84
	.3750 ⅜	68.31	20.1	748	93.5	6.11	482	80.3	4.90
	.3125 ⁵⁄₁₆	57.36	16.9	635	79.4	6.14	409	68.2	4.93
16 x 8	.5000 ½	76.07	22.4	722	90.2	5.68	244	61.0	3.30
	.3750 ⅜	58.10	17.1	565	70.6	5.75	193	48.2	3.36
	.3125 ⁵⁄₁₆	48.86	14.4	481	60.1	5.79	165	41.2	3.39
16 x 4	.5000 ½	62.46	18.4	481	60.2	5.12	49.3	24.6	1.64
	.3750 ⅜	47.90	14.1	382	47.8	5.21	40.4	20.2	1.69
	.3125 ⁵⁄₁₆	40.35	11.9	327	40.9	5.25	35.1	17.6	1.72
14 x 10	.5000 ½	76.07	22.4	608	86.9	5.22	361	72.3	4.02
	.3750 ⅜	58.10	17.1	476	68.0	5.28	284	56.8	4.08
	.3125 ⁵⁄₁₆	48.86	14.4	405	57.9	5.31	242	48.4	4.11
14 x 6	.5000 ½	62.46	18.4	426	60.8	4.82	111	37.1	2.46
	.3750 ⅜	47.90	14.1	337	48.1	4.89	89.1	29.7	2.52
	.3125 ⁵⁄₁₆	40.35	11.9	288	41.2	4.93	76.7	25.6	2.54
	.2500 ¼	32.63	9.59	237	33.8	4.97	63.4	21.1	2.57
14 x 4	.5000 ½	55.66	16.4	335	47.8	4.52	43.1	21.5	1.62
	.3750 ⅜	42.79	12.6	267	38.2	4.61	35.4	17.7	1.68
	.3125 ⁵⁄₁₆	36.10	10.6	230	32.8	4.65	30.9	15.4	1.71
	.2500 ¼	29.23	8.59	189	27.0	4.69	25.8	12.9	1.73
12 x 8	.6250 ⅝	76.33	22.4	418	69.7	4.32	221	55.3	3.14
	.5000 ½	62.46	18.4	353	58.9	4.39	188	46.9	3.20
	.3750 ⅜	47.90	14.1	279	46.5	4.45	149	37.3	3.26
	.3125 ⁵⁄₁₆	40.35	11.9	239	39.8	4.49	128	32.0	3.28
	.2500 ¼	32.63	9.59	196	32.6	4.52	105	26.3	3.31
12 x 6	.5000 ½	55.66	16.4	287	47.8	4.19	96.0	32.0	2.42
	.3750 ⅜	42.79	12.6	228	38.1	4.26	77.2	25.7	2.48
	.3125 ⁵⁄₁₆	36.10	10.6	196	32.6	4.30	66.6	22.2	2.51
	.2500 ¼	29.23	8.59	161	26.9	4.33	55.2	18.4	2.53
	.1875 ³⁄₁₆	22.18	6.52	124	20.7	4.37	42.8	14.3	2.56
12 x 4	.5000 ½	48.85	14.4	221	36.8	3.92	36.9	18.5	1.60
	.3750 ⅜	37.69	11.1	178	29.6	4.01	30.5	15.2	1.66
	.3125 ⁵⁄₁₆	31.84	9.36	153	25.5	4.05	26.6	13.3	1.69
	.2500 ¼	25.82	7.59	127	21.1	4.09	22.3	11.1	1.71
	.1875 ³⁄₁₆	19.63	5.77	98.2	16.4	4.13	17.5	8.75	1.74
12 x 2	.2500 ¼	22.42	6.59	92.2	15.4	3.74	4.62	4.62	.837
	.1875 ³⁄₁₆	17.08	5.02	72.0	12.0	3.79	3.76	3.76	.865

* Outside dimensions across flat sides.
** Properties are based upon a nominal outside corner radius equal to two times the wall thickness.

TABLE 4-6. (*Continued*)

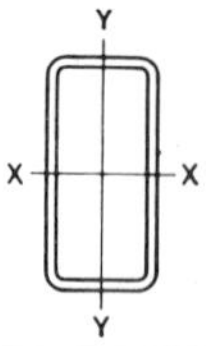

Rectangular

DIMENSIONS			PROPERTIES**							
Nominal* Size	Wall Thickness		Weight per Foot	Area	X-X AXIS			Y-Y AXIS		
					I_x	S_x	r_x	I_y	S_y	r_y
In.	In.		Lb.	In.²	In.⁴	In.³	In.	In.⁴	In.³	In.
10 x 6	.5000	½	48.85	14.4	181	36.2	3.55	80.8	26.9	2.37
	.3750	⅜	37.69	11.1	145	29.0	3.62	65.4	21.8	2.43
	.3125	⁵⁄₁₆	31.84	9.36	125	25.0	3.65	56.5	18.8	2.46
	.2500	¼	25.82	7.59	103	20.6	3.69	46.9	15.6	2.49
	.1875	³⁄₁₆	19.63	5.77	79.8	16.0	3.72	36.5	12.2	2.51
10 x 4	.5000	½	42.05	12.4	136	27.1	3.31	30.8	15.4	1.58
	.3750	⅜	32.58	9.58	110	22.0	3.39	25.5	12.8	1.63
	.3125	⁵⁄₁₆	27.59	8.11	95.0	19.1	3.43	22.4	11.2	1.66
	.2500	¼	22.42	6.59	79.3	15.9	3.47	18.8	9.39	1.69
	.1875	³⁄₁₆	17.08	5.02	61.7	12.3	3.51	14.8	7.39	1.72
10 x 2	.3750	⅜	27.48	8.08	75.4	15.1	3.06	4.85	4.85	.775
	.3125	⁵⁄₁₆	23.34	6.86	66.1	13.2	3.10	4.42	4.42	.802
	.2500	¼	19.02	5.59	55.5	11.1	3.15	3.85	3.85	.830
	.1875	³⁄₁₆	14.53	4.27	43.7	8.74	3.20	3.14	3.14	.858
8 x 6	.5000	½	42.05	12.4	103	25.8	2.89	65.7	21.9	2.31
	.3750	⅜	32.58	9.58	83.7	20.9	2.96	53.5	17.8	2.36
	.3125	⁵⁄₁₆	27.59	8.11	72.4	18.1	2.99	46.4	15.5	2.39
	.2500	¼	22.42	6.59	60.1	15.0	3.02	38.6	12.9	2.42
	.1875	³⁄₁₆	17.08	5.02	46.8	11.7	3.05	30.1	10.0	2.45
8 x 4	.5000	½	35.24	10.4	75.1	18.8	2.69	24.6	12.3	1.54
	.3750	⅜	27.48	8.08	61.9	15.5	2.77	20.6	10.3	1.60
	.3125	⁵⁄₁₆	23.34	6.86	53.9	13.5	2.80	18.1	9.05	1.62
	.2500	¼	19.02	5.59	45.1	11.3	2.84	15.3	7.63	1.65
	.1875	³⁄₁₆	14.53	4.27	35.3	8.83	2.88	12.0	6.02	1.68
8 x 3	.3750	⅜	24.93	7.33	51.0	12.7	2.64	10.4	6.92	1.19
	.3125	⁵⁄₁₆	21.21	6.23	44.7	11.2	2.68	9.25	6.16	1.22
	.2500	¼	17.32	5.09	37.6	9.40	2.72	7.90	5.26	1.25
	.1875	³⁄₁₆	13.25	3.89	29.6	7.40	2.76	6.31	4.21	1.27
8 x 2	.3750	⅜	22.37	6.58	40.1	10.0	2.47	3.85	3.85	.765
	.3125	⁵⁄₁₆	19.08	5.61	35.5	8.87	2.51	3.52	3.52	.792
	.2500	¼	15.62	4.59	30.1	7.52	2.56	3.08	3.08	.819
	.1875	³⁄₁₆	11.97	3.52	23.9	5.97	2.60	2.52	2.52	.847
7 x 5	.5000	½	35.24	10.4	63.5	18.1	2.48	37.2	14.9	1.90
	.3750	⅜	27.48	8.08	52.2	14.9	2.54	30.8	12.3	1.95
	.3125	⁵⁄₁₆	23.34	6.86	45.5	13.0	2.58	26.9	10.8	1.98
	.2500	¼	19.02	5.59	38.0	10.9	2.61	22.6	9.04	2.01
	.1875	³⁄₁₆	14 53	4.27	29.8	8.50	2.64	17.7	7.10	2.04
7 x 4	.3750	⅜	24.93	7.33	44.0	12.6	2.45	18.1	9.06	1.57
	.3125	⁵⁄₁₆	21.21	6.23	38.5	11.0	2.49	16.0	7.98	1.60
	.2500	¼	17.32	5.09	32.3	9.23	2.52	13.5	6.75	1.63
	.1875	³⁄₁₆	13.25	3.89	25.4	7.26	2.55	10.7	5.34	1.66

* Outside dimensions across flat sides.
** Properties are based upon a nominal outside corner radius equal to two times the wall thickness.

TABLE 4-6. (Continued)

Rectangular

Nominal* Size	Wall Thickness		Weight per Foot	Area	X X AXIS I_x	S_x	r_x	Y Y AXIS I_y	S_y	r_y
In	In		In	In.^{2}	In.^{4}	In.^{3}	In	In.^{4}	In.^{3}	In
7 x 3	.3750	$^3/_8$	22.37	6.58	35.7	10.2	2.33	9.08	6.05	1.18
	.3125	$^5/_{16}$	19.08	5.61	31.5	9.00	2.37	8.11	5.41	1.20
	.2500	$^1/_4$	15.62	4.59	26.6	7.61	2.41	6.95	4.63	1.23
	.1875	$^3/_{16}$	11.97	3.52	21.1	6.02	2.45	5.57	3.71	1.26
6 x 4	.5000	$^1/_2$	28.43	8.36	35.3	11.8	2.06	18.4	9.21	1.48
	.3750	$^3/_8$	22.37	6.58	29.7	9.90	2.13	15.6	7.82	1.54
	.3125	$^5/_{16}$	19.08	5.61	26.2	8.72	2.16	13.8	6.92	1.57
	.2500	$^1/_4$	15.62	4.59	22.1	7.36	2.19	11.7	5.87	1.60
	.1875	$^3/_{16}$	11.97	3.52	17.4	5.81	2.23	9.32	4.66	1.63
6 x 3	.3750	$^3/_8$	19.82	5.83	23.8	7.92	2.02	7.78	5.19	1.16
	.3125	$^5/_{16}$	16.96	4.98	21.1	7.03	2.06	6.98	4.65	1.18
	.2500	$^1/_4$	13.91	4.09	17.9	5.98	2.09	6.00	4.00	1.21
	.1875	$^3/_{16}$	10.70	3.14	14.3	4.76	2.13	4.83	3.22	1.24
6 x 2	.3750	$^3/_8$	17.27	5.08	17.8	5.94	1.87	2.84	2.84	.748
	.3125	$^5/_{16}$	14.83	4.36	16.0	5.34	1.92	2.62	2.62	.775
	.2500	$^1/_4$	12.21	3.59	13.8	4.60	1.96	2.31	2.31	.802
	.1875	$^3/_{16}$	9.42	2.77	11.1	3.70	2.00	1.90	1.90	.829
5 x 4	.3750	$^3/_8$	19.82	5.83	18.7	7.50	1.79	13.2	6.58	1.50
	.3125	$^5/_{16}$	16.96	4.98	16.6	6.65	1.83	11.7	5.85	1.53
	.2500	$^1/_4$	13.91	4.09	14.1	5.65	1.86	9.98	4.99	1.56
	.1875	$^3/_{16}$	10.70	3.14	11.2	4.49	1.89	7.96	3.98	1.59
5 x 3	.5000	$^1/_2$	21.63	6.36	16.9	6.75	1.63	7.33	4.88	1.07
	.3750	$^3/_8$	17.27	5.08	14.7	5.89	1.70	6.48	4.32	1.13
	.3125	$^5/_{16}$	14.83	4.36	13.2	5.27	1.74	5.85	3.90	1.16
	.2500	$^1/_4$	12.21	3.59	11.3	4.52	1.77	5.05	3.37	1.19
	.1875	$^3/_{16}$	9.42	2.77	9.06	3.62	1.81	4.08	2.72	1.21
5 x 2	.3125	$^5/_{16}$	12.70	3.73	9.74	3.90	1.62	2.16	2.16	.76
	.2500	$^1/_4$	10.51	3.09	8.48	3.39	1.66	1.92	1.92	.78
	.1875	$^3/_{16}$	8.15	2.39	6.89	2.75	1.70	1.60	1.60	.81
4 x 3	.3125	$^5/_{16}$	12.70	3.73	7.45	3.72	1.41	4.71	3.14	1.12
	.2500	$^1/_4$	10.51	3.09	6.45	3.23	1.45	4.10	2.74	1.15
	.1875	$^3/_{16}$	8.15	2.39	5.23	2.62	1.48	3.34	2.23	1.18
4 x 2	.3125	$^5/_{16}$	10.58	3.11	5.32	2.66	1.31	1.71	1.71	.743
	.2500	$^1/_4$	8.81	2.59	4.69	2.35	1.35	1.54	1.54	.770
	.1875	$^3/_{16}$	6.87	2.02	3.87	1.93	1.38	1.29	1.29	.798
3 x 2	.2500	$^1/_4$	7.11	2.09	2.21	1.47	1.03	1.15	1.15	.742
	.1875	$^3/_{16}$	5.59	1.64	1.86	1.24	1.06	.977	.977	.771

* Outside dimensions across flat sides.
** Properties are based upon a nominal outside corner radius equal to two times the wall thickness.

Source: Reprinted from the *Manual of Steel Construction*, 8th ed., with permission of the publishers, American Institute of Steel Construction.

TABLE 4-7. Properties of Structural Lumber

Dimensions (in.)				Area A in^2	Section Modulus S in^3	Moment of Inertia I in^4	Weight[a] lb/ft
Nominal b	h	Actual b	h				
2 × 3		1.5 × 2.5		3.75	1.563	1.953	0.9
2 × 4		1.5 × 3.5		5.25	3.063	5.359	1.3
2 × 6		1.5 × 5.5		8.25	7.563	20.797	2.0
2 × 8		1.5 × 7.25		10.875	13.141	47.635	2.6
2 × 10		1.5 × 9.25		13.875	21.391	98.932	3.4
2 × 12		1.5 × 11.25		16.875	31.641	177.979	4.1
2 × 14		1.5 × 13.25		19.875	43.891	290.775	4.8
3 × 2		2.5 × 1.5		3.75	0.938	0.703	0.9
3 × 4		2.5 × 3.5		8.75	5.104	8.932	2.1
3 × 6		2.5 × 5.5		13.75	12.604	34.661	3.3
3 × 8		2.5 × 7.25		18.125	21.901	79.391	4.4
3 × 10		2.5 × 9.25		23.125	35.651	164.886	5.6
3 × 12		2.5 × 11.25		28.125	52.734	296.631	6.8
3 × 14		2.5 × 13.25		33.125	73.151	484.625	8.1
3 × 16		2.5 × 15.25		38.125	96.901	738.870	9.3
4 × 2		3.5 × 1.5		5.25	1.313	0.984	1.3
4 × 3		3.5 × 2.5		8.75	3.646	4.557	2.1
4 × 4		3.5 × 3.5		12.25	7.146	12.505	3.0
4 × 6		3.5 × 5.5		19.25	17.646	48.526	4.7
4 × 8		3.5 × 7.25		23.375	30.661	111.148	6.2
4 × 10		3.5 × 9.25		32.375	49.911	230.840	7.9
4 × 12		3.5 × 11.25		39.375	73.828	415.283	9.6
4 × 14		3.5 × 13.25		46.375	102.411	678.475	11.3
4 × 16		3.5 × 15.25		53.375	135.661	1034.418	13.0
6 × 2		5.5 × 1.5		8.25	2.063	1.547	2.0
6 × 3		5.5 × 2.5		13.75	5.729	7.161	3.3
6 × 4		5.5 × 3.5		19.25	11.229	19.651	4.7
6 × 6		5.5 × 5.5		30.25	27.729	76.255	7.4

TABLE 4-7. (*Continued*)

Dimensions (in.)			Area A in^2	Section Modulus S in^3	Moment of Inertia I in^4	Weight[a] lb/ft
Nominal b h		Actual b h				
6 × 8		5.5 × 7.5	41.25	51.563	193.359	10.0
6 × 10		5.5 × 9.5	52.25	82.729	392.963	12.7
6 × 12		5.5 × 11.5	63.25	121.229	697.068	15.4
6 × 14		5.5 × 13.5	74.25	167.063	1127.672	18.0
6 × 16		5.5 × 15.5	85.25	220.229	1706.776	20.7
8 × 2		7.25 × 1.5	10.875	2.719	2.039	2.6
8 × 3		7.25 × 2.5	18.125	7.552	9.440	4.4
8 × 4		7.25 × 3.5	25.375	14.802	25.904	6.2
8 × 6		7.5 × 5.5	41.25	37.813	103.984	10.0
8 × 8		7.5 × 7.5	56.25	70.313	263.672	13.7
8 × 10		7.5 × 9.5	71.25	112.813	535.859	17.3
8 × 12		7.5 × 11.5	86.25	165.313	950.547	21.0
8 × 14		7.5 × 13.5	101.25	227.813	1537.734	24.6
8 × 16		7.5 × 15.5	116.25	300.313	2327.422	28.3
8 × 18		7.5 × 17.5	131.25	382.813	3349.609	31.9
8 × 20		7.5 × 19.5	146.25	475.313	4634.297	35.5
10 × 10		9.5 × 9.5	90.25	142.896	678.755	21.9
10 × 12		9.5 × 11.5	109.25	209.396	1204.026	26.6
10 × 14		9.5 × 13.5	128.25	288.563	1947.797	31.2
10 × 16		9.5 × 15.5	147.25	380.396	2948.068	35.8
10 × 18		9.5 × 17.5	166.25	484.896	4242.836	40.4
10 × 20		9.5 × 19.5	185.25	602.063	5870.109	45.0
12 × 12		11.5 × 11.5	132.25	253.479	1457.505	32.1
12 × 14		11.5 × 13.5	155.25	349.313	2357.859	37.7
12 × 16		11.5 × 15.5	178.25	460.479	3568.713	43.3
12 × 18		11.5 × 17.5	201.25	586.979	5136.066	48.9
12 × 20		11.5 × 19.5	224.25	728.813	7105.922	54.5
12 × 22		11.5 × 21.5	247.25	885.979	9524.273	60.1

TABLE 4-7.   (*Continued*)

Dimensions (in.)			Area A in^2	Section Modulus S in^3	Moment of Inertia I in^4	Weight[a] lb/ft
Nominal b h		Actual b h				
12 × 24		11.5 × 23.5	270.25	1058.479	12437.129	65.7
14 × 14		13.5 × 13.5	182.25	410.063	2767.922	44.3
16 × 16		15.5 × 15.5	240.25	620.646	4810.004	58.4

Source: Compiled from data in the *National Design Specification for Wood Construction,* 1982 ed., with permission of the publishers, National Forest Products Association.

[a] Based on an assumed average weight of 35 lb/ft^3.

lated for the commonly used structural steel shapes in Tables 4-1 through 4-6, and the properties of structural lumber are given in Table 4-7.

Reference to the tables will show that structural steel shapes have two major axes, designated *X-X* and *Y-Y*. The position in which the member is placed determines the axis to be considered. For example, the wide flange sections and I-beams (Tables 4-1 and 4-2, respectively) are nearly always used with the web vertical; hence the *X-X*, or horizontal, axis determines the applicable properties. The rectangular cross sections characteristic of structural lumber also have a vertical as well as a horizontal axis, but Table 4-7 records properties about the *X-X* axis only, since beams and joists are always used with the longer side vertical and planks with the shorter side vertical.

Several of the properties listed in the tables are discussed below, and others are considered under the design of structural members in Parts 2 and 3.

4-4   Centroids

The centroid of a plane surface is a point that corresponds to the center of gravity of a very thin homogeneous plate of the same area and shape. It can be shown that the neutral axis of a beam cross section passes through its centroid; consequently it is necessary to know its exact position. For symmetrical sections such as rectangles and I-shape sections, it can be seen by inspection that the centroid lies at a point midway between the upper and lower surfaces of the section, at the intersection of the *X-X* and *Y-Y* axes referred to at the heads of Tables 4-1 and 4-2. The distance *c* from the extreme (most remote) fiber to the neutral axis is, therefore, half the depth of symmetrical sections. For unsymmetrical sections the position of the centroid must be computed. This is accomplished by using the principle of *statical moments*.

The statical moment of a plane area with respect to an axis is the area multiplied by the perpendicular distance from the centroid to the axis. *If an area is divided into a number of parts, the statical moment of the entire area is equal to the sum of the statical moments of the parts*. This is our key for locating the centroid. The following example shows how the distance *c* is computed.

Example. Figure 4-3 is a beam cross section, unsymmetrical with respect to the horizontal axis. Find the value of *c*, the distance of the neutral axis from the most remote fiber.

Solution: (1) It is not always possible to tell by observation whether the centroid is nearer the top or bottom edge of the area.

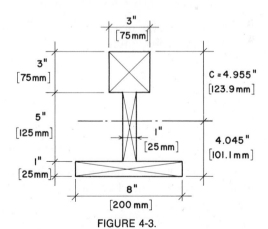

FIGURE 4-3.

In this instance let us write an equation of moments about an axis through the uppermost edge. First divide the total area into some number of simple shapes, in this case the three rectangles shown by the diagonals. The area of the upper part is 9 in^2 [5625 mm^2], and its centroid is 1.5 in. [37.5 mm] from the reference axis. The center part has an area of 5 in^2 [3125 mm^2], and its centroid is 5.5 in. [137.5 mm] from the axis. The bottom part contains 8 in^2 [5000 mm^2], and its centroid is 8.5 in. [212.5 mm] from the axis. Considering these individual areas as forces, their combined moment about the reference axis is

$$M = (9 \times 1.5) + (5 \times 5.5) + (8 \times 8.5)$$

$$= 13.5 + 27.5 + 68 = 109 \text{ in}^3$$

Since this is the moment of the entire area about the axis, it may be equated to the product of the entire area (22 in^2) times the single centroid distance c, as shown in the figure. Thus

$$M = 109 = 22 \times c$$

$$c = 109 \div 22 = 4.955 \text{ in. } [123.9 \text{ mm}]$$

which is the distance of the centroid from the reference axis.

(2) The depth of the section is 9 in. [225 mm]; hence the distance of the centroid from the bottom edge is

$$9 - 4.955 = 4.045 \text{ in. } [101.1 \text{ mm}]$$

Call this distance c_1. The value of c_1 may be checked by writing an equation of moments about a reference axis through the bottom edge; thus

$$(9 \times 7.5) + (5 \times 3.5) + (8 \times 0.5) = (22 \times c_1)$$

from which

$$c_1 = \frac{89}{22} = 4.045 \text{ in.}$$

Remember that the centroid of the section is the point through which the neutral axis (zero bending stress point) passes. The position of the centroid for structural steel angles may be found directly from Table 4-4. The locating dimensions from the backs of the angle legs are given as y and x distances.

Problems 4-4-A*-B-C-D*-E-F. Find the location of the centroid for the cross-sectional areas shown in Fig. 4-4. Use the references and indicate the distances as c_x and c_y, as shown in Fig. 4-4b.

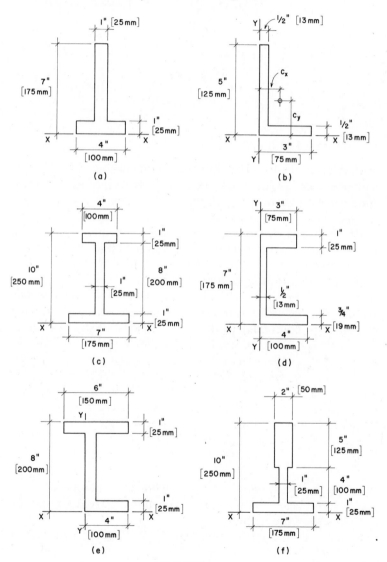

FIGURE 4-4.

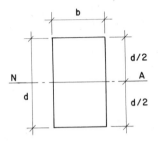

FIGURE 4-5.

4-5 Moment of Inertia

In developing the flexure formula, we found the quantity $\Sigma a \times z^2$. This is the *moment of inertia*, which may be defined as *the sum of the products obtained by multiplying all the infinitely small areas by the square of their distances to the neutral axis*. It is represented by I. The elementary areas, though extremely small, are in units of square inches, and square inches multiplied by a distance squared gives inches to the fourth power. For instance, 24 in^4 is read "24 inches to the fourth power."

In Fig. 4-5 we have a rectangular cross section of breadth b and depth d. The neutral axis passes through the centroid of the cross section; it is represented by the line NA and is a distance $d/2$ from the upper and lower edges. It can be shown that *the moment of inertia of a rectangular cross section about an axis passing through its centroid, parallel to the base, is $bd^3/12$*. By the use of this formula, $I = bd^3/12$, the value of the moment of inertia for rectangular cross sections is quickly found. The values of I for standard timber sizes are given in Table 4-7; they are computed for actual dressed sizes (Section 10-2).

Example 1. Find the value of the moment of inertia for a 6 × 12-in. wood beam about an axis through its centroid and parallel to the base of the section.

Solution: Referring to Table 4-7, we find that a nominal 6 × 12-in. section has standard dressed dimensions of 5.5 × 11.5 in.

[139.7 × 292.1 mm]. Then

$$I = \frac{bd^3}{12} = \frac{(5.5)(11.5)^3}{12} = 697.07 \text{ in}^4 \ [290.1 \times 10^6 \text{ mm}^4]$$

which is in agreement with the value of I listed in the table.

The moment of inertia of a rectangular area with respect to an axis of reference passing through its base may be found as $bd^3/3$. This is sometimes useful in finding the moment of inertia of complex sections, as is demonstrated in the following example.

Example 2. Find the moment of inertia of a 6 × 4 × $\frac{1}{2}$-in. angle [152.4 × 101.6 × 12.7 mm] (Fig. 4-6) about its neutral axis, which has been determined to be 2 in. above its base (the back of the 4-in. leg).
Solution: (1) We first divide the section into three parts. The first part is the portion of the 6-in. leg above the neutral axis. The second part is assumed to be a solid rectangle of 2 in. by 4 in. as shown in Fig. 4-6b. The third part is a negative area, which is what must be subtracted from the second part; this area is shown in Fig. 4-6c. The sum of these parts will produce the net form of the angle.
(2) We now proceed to find the total moment of inertia for the angle by finding the individual moments of inertia of these three parts about their edges. For the first part: (See Table 4-8.)

$$I = \frac{bd^3}{3} = \frac{(0.5)(4)^3}{3} = 10.667 \text{ in}^4 \ [4.440 \times 10^6 \text{ mm}^4]$$

For the second part:

$$I = \frac{(4)(2)^3}{3} = 10.667 \text{ in}^4 \ [4.440 \times 10^6 \text{ mm}^4]$$

For the third part:

$$I = -\frac{(3.5)(1.5)^3}{3} = 3.938 \text{ in}^4 \ [1.639 \times 10^6 \text{ mm}^4]$$

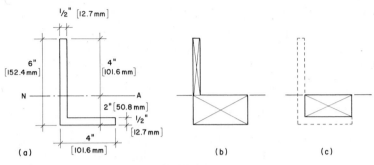

FIGURE 4-6.

And for the entire angle

$$I = 10.667 + 10.667 - 3.938 = 17.396 \text{ in}^4 \ [7.240 \times 10^6 \text{ mm}^4]$$

which may be compared to the value listed for I_x in Table 4-4.

The preceding example illustrates one method of finding the moment of inertia of an unsymmetrical section, but it is not necessary to compute I for the commonly used structural steel angles, since these may be found directly from Table 4-4 and from the more extensive tables in the AISC Manual.

Problems 4-5-A-B*-C-D-E-F*. Compute the moment of inertia about the neutral axes of the beam sections shown in Fig. 4-7.

4-6 Section Modulus

As noted in Section 4-2, the term I/c in the flexure formula is called the *section modulus*. It is defined as the moment of inertia divided by the distance of the most remote fiber from the neutral axis and is denoted by the symbol S. Since I and c always have the same values for any given cross section, values of S may be computed and tabulated for structural shapes. With I expressed in inches to the fourth power and c a linear dimension in inches, S is in units of inches to the third power, written *inches*^{3}. Section moduli are among the properties tabulated for structural steel shapes in Tables 4-1 through 4-6 and for structural lumber cross sections in Table 4-7.

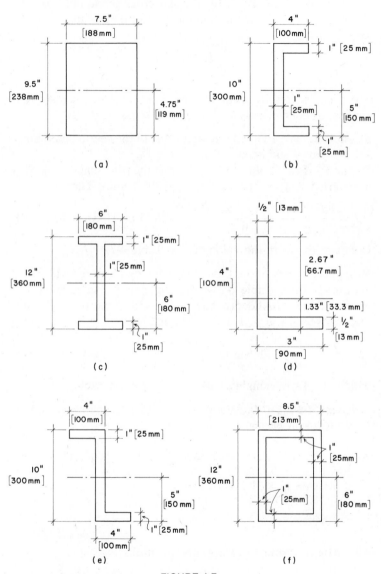

FIGURE 4-7.

109

Referring to Fig. 4-5, a rectangular cross section of breadth b and depth d, we know that the moment of inertia about the X-X axis is $bd^3/12$ and that $c = d/2$. Therefore

$$\frac{I}{c} \text{ or } S = \frac{bd^3/12}{d/2} = \frac{bd^3}{12} \times \frac{2}{d}$$

or $S = bd^2/6$. It is often convenient to use this formula directly. It applies, of course, only to rectangular cross sections.

Example 1.   Verify the tabulated value of the section modulus of a 6×12-in. wood beam.
Solution: From Table 4-7 we find that the true dimensions of the beam are 5.5 in. $\times$ 11.5 in. [139.7 $\times$ 292.1 mm]. Then

$$S = \frac{bd^2}{6} = \frac{5.5 \times 11.5 \times 11.5}{6} = 121.23 \text{ in}^3 \text{ [1987} \times 10^3 \text{ mm}^3]$$

Compare this with the value of S in Table 4-7.

Example 2.   Verify the tabulated value of S_x for a W 18×46.
Solution: From Table 4-1 we find that $I_x = 712$ in^4 [296.33 $\times$ 10^6 mm^4] and the actual depth is 18.06 in. [458.72 mm]. For the symmetrical section, $c = d \div 2 = 9.03$ in. Then

$$S = \frac{I}{c} = \frac{712}{9.03} = 78.848 \text{ in}^3 \left[\frac{296.33 \times 10^6}{229.36} = 1{,}292 \times 10^3 \text{ mm}^3 \right]$$

which checks reasonably with the value in the table.

Problems 4-6-A-B-C-D.   Verify the tabulated section modulus values for the following elements.

A   A 6×8 in. timber beam. Actually 5.5×7.5 in. [139.7 $\times$ 190.5 mm.]
B S_x for an S 12×31.8 rolled shape.
C S_x for an L $5 \times 3\frac{1}{2} \times \frac{1}{2}$ rolled steel angle.
D S_y for an L $4 \times 4 \times \frac{1}{2}$ steel angle.

4-7 Application of the Flexure Formula

Now that we have discussed the properties of structural sections in some detail, let us return to the flexure formula and consider its application. The expression $M = fS$ may be stated in three different forms depending upon the information desired. These are given below using a nomenclature which makes a distinction with respect to f as the *allowable* bending stress (F_b) and f as the *computed* bending stress (f_b).

$$(1) \ M = F_b S \qquad (2) \ f_b = \frac{M}{S} \qquad (3) \ S = \frac{M}{F_b}$$

Form (1) gives the maximum potential resisting moment when the section modulus of the beam and the maximum allowable bending stress are known. Form (2) gives the computed bending stress when the maximum bending moment due to the loading is known, together with the section modulus of the beam. These are the two forms used when investigating the adequacy of given beams.

Form (3) is the one used in design. It gives the *required* section modulus when the maximum bending moment and the allowable bending stress are known. When the required section modulus has been determined, a beam having an S equal to or greater than the computed value is selected from tables giving properties of the various structural shapes.

When using the beam formula, exercise care with respect to the units in which the terms are expressed. Bending stress values F_b and f_b may be written in pounds per square inch (psi) or kips per square inch (ksi); S is stated in inches^3 and I in inches^4. Therefore, M must be written in inch-pounds or kip-inches. As customarily computed from the loads and reactions, M is given in foot-pounds or kip-feet and must be converted to inch-pounds or kip-inches; this is accomplished by multiplying its value by 12 before it is used in the formula.

Example. Select a wide flange steel beam to support a uniformly distributed load of 48 kips [213.5 kN], including its own weight, on a span of 16 ft [4.88 m]. The allowable extreme fiber stress in bending is 24 ksi [165 MPa].

Solution: (1) Computing the bending moment (Case 2, Table 3-1), we find

$$M = \frac{WL}{8} = \frac{48 \times 16}{8} = 96 \text{ kip-ft} \left[\frac{213.5 \times 4.88}{8} = 130.2 \text{ kN-m} \right]$$

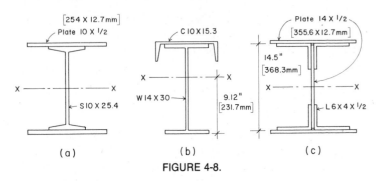

FIGURE 4-8.

(2) Substituting in the flexure formula, we find

$$S = \frac{M}{F_b} = \frac{96 \times 12}{24} = 48 \text{ in}^3 \left[\frac{130.2 \times 10^6}{165} = 789 \times 10^3 \text{ mm}^3 \right]$$

(3) Referring to Table 4-1, we find a W 14 × 34 has an S of 48.6 in³; therefore this beam may be used. A W 12 × 40 ($S = 51.9$ in³) is also acceptable and is not quite so deep. Although any steel beam with S greater than 48 in³ will safely support the load, the lightest weight beam is usually the most economical, since it represents a savings of material.

4-8 Transferring Moments of Inertia

When rolled steel shapes are combined to form built-up structural sections similar to those illustrated in Fig. 4-8, it is necessary to determine the moment of inertia of the built-up section about its neutral axis. This requires transferring the moments of inertia of some of the individual parts from one axis to another and is accomplished by means of the transfer-of-axis equation, which may be stated as follows:

The moment of inertia of a cross section about any axis parallel to an axis through its own centroid is equal to the moment of inertia of the cross section about its own gravity axis, plus its area times the square of the distance between the two axes. Expressed mathematically,

$$I = I_0 + Az^2$$

In this formula,

> I = moment of inertia of the cross section about the required axis,
>
> I_0 = moment of inertia of the cross section about its own gravity axis parallel to the required axis,
>
> A = area of the cross section,
>
> z = distance between the two parallel axes.

These relationships are indicated in Fig. 4-9, where X-X is the gravity axis of the angle (passing through its centroid) and Y-Y is the axis about which the moment of inertia is to be found.

To illustrate the use of the equation, we may prove the proposition stated in Section 4-5 that the value of I for a rectangle about an axis through its base is $bd^3/3$. Since I for a rectangle about its gravity axis is known to be $bd^3/12$, and z in this instance is $d/2$, we may write

$$I = I_0 + Az^2$$

$$I = \frac{bd^3}{12} + \left[bd \times \left(\frac{d}{2}\right)^2 \right]$$

$$I = \frac{bd^3}{12} + \frac{bd^3}{4} = \frac{bd^3}{3}$$

The application of the transfer formula to the steel built-up section shown in Fig. 4-10 is illustrated in the following example.

Example.   Compute the moment of inertia about the X-X axis of a built-up section composed of two C 12 × 30 channels and two 16 × ½-in. [406.4 × 12.7-mm] plates. (See Fig. 4-10.)

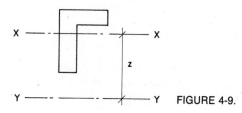

FIGURE 4-9.

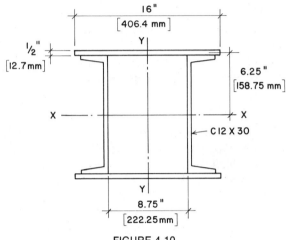

FIGURE 4-10.

Solution: (1) From Table 4-3 we find that I_x for the channel is 162 in^4 [67.42 × 10^6 mm^4], so the value for the two channels is twice this, or 324 in^4 [134.84 × 10^6 mm^4].

(2) For one plate the moment of inertia about an axis through its own centroid is

$$I_0 = \frac{bd^3}{12} = \frac{16 \times 0.5^3}{12} = 0.1667 \text{ in}^4 \ [0.0694 \times 10^6 \text{ mm}^4]$$

The distance between the centroid of the plate and the X-X axis is 6.25 in. [158.75 mm], and the area of one plate is 8 in^2 [5161.3 mm^2]. Therefore the I of one plate about the X-X axis of the combined section is

$$I = I_0 + Az^2 = 0.1667 + (8)(6.25)^2 = 312.7 \text{ in}^4$$

$$[0.0694 \times 10^6 + (5161.3)(158.75)^2 = 130.14 \times 10^6 \text{ mm}^4]$$

and the value for two plates is twice this, or 625 in^4 [260.3 × 10^6 mm^4].

(3) Adding the moments of inertia of the channels and plates, we obtain the I for the entire cross section as 324 + 625 = 949 in^4 [(134.84 + 260.3) × 10^6 = 395.14 × 10^6 mm^4].

Problems 4-8-A*-B*-C. Compute I with respect to the X-X axes for the built-up sections shown in Fig. 4-8. Make use of any appropriate data given in Tables 4-1 through 4-7.

4-9 Radius of Gyration

This property of a cross section is related to the design of compression members rather than beams and will be discussed in more detail under the design of columns in subsequent sections of the book. Radius of gyration will be considered briefly here, however, because it is a property listed in Tables 4-1 through 4-5.

Just as the section modulus is a measure of the resistance of a beam section to bending, the radius of gyration (which is also related to the size and shape of the cross section) is an index of the stiffness of a structural section when used as a column or other compression member. The radius of gyration is found from the formula

$$r = \sqrt{\frac{I}{A}}$$

and is expressed in inches, since the moment of inertia is in inches^4 and the cross-sectional area is in square inches.

If a section is symmetrical about both major axes, the moment of inertia, and consequently the radius of gyration, is the same for each axis. But most column sections, particularly steel columns, are not symmetrical about the two major axes, and *in the design of columns the least moment of inertia, and therefore the least radius of gyration, is the one used in computations.* By *least* we mean the smallest in magnitude. Note in Table 4-4 that the least radius of gyration of angle sections occurs about the Z-Z axes.

Example. Verify the tabulated values of radii of gyration for a W 12 × 58, as given in Table 4-1.
Solution: (1) The table shows the area of this section to be 17.0 in^2 [10968 mm^2] and I with respect to the X-X axis to be 475 in^4 [197.7 × 10^6 mm^4]. Then

$$r_x = \sqrt{\frac{I}{A}} = \sqrt{\frac{475}{17}} = \sqrt{27.94} = 5.286 \text{ in.}$$

$$\left[\sqrt{\frac{197.7 \times 10^6}{10968}} = \sqrt{18025} = 134.26 \text{ mm} \right]$$

(2) The table value for I with respect to the Y-Y axis is 107 in^4 [44.53×10^6 mm^4]. Therefore

$$r_y = \sqrt{\frac{I}{A}} = \sqrt{\frac{107}{17}} = \sqrt{6.294} = 2.51 \text{ in.}$$

$$\left[\sqrt{\frac{44.53 \times 10^6}{10968}} = \sqrt{4060} = 63.72 \text{ mm} \right]$$

(3) Compare these values with those listed in Table 4-1.

Properties of several common geometric shapes are given in Table 4-8.

TABLE 4-8.　Properties of Common Geometric Shapes

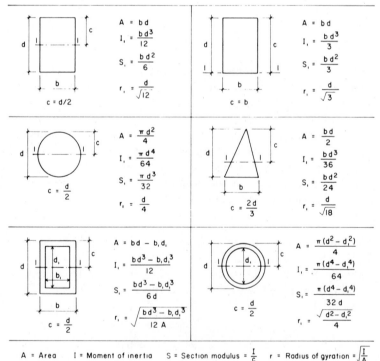

A = Area　　I = Moment of inertia　　S = Section modulus = $\frac{I}{c}$　　r = Radius of gyration = $\sqrt{\frac{I}{A}}$

Problem 4-9-A.   Verify the value of r_y for an S 12 × 40.8.

Problem 4-9-B.   Verify the value of r_x for an L 4 × 3 × $\frac{5}{8}$.

Problem 4-9-C*.   Compute the radius of gyration with respect to the X-X axis for the built-up section shown in Fig. 4-10. (See example calculation in Section 4-8.)

II

STEEL CONSTRUCTION

5

Steel Beams

||||||||||||||||||||||||||||||||||||||||||||||||||||||||||||||||||||

5-1 Sources of Design Information

The most widely used source for information for design of steel
structures is the *Manual of Steel Construction* published by the
American Institute of Steel Construction. It is commonly referred
to simply as the AISC Manual, and we will refer to it as such
here. The current edition is the eighth edition, copyrighted in
1980, and it contains the 1978 Specification for the Design, Fabrication, and Erection of Structural Steel for Buildings. This specification, in its various editions, has been generally adopted for
reference by code-enforcing agencies in the United States. When
referring to this document, we will call it simply the AISC Specification.

Although the AISC is the principal service organization in the
area of steel construction, there are many other industry and
professional organizations that provide materials for the designer.
The American Society for Testing and Materials (ASTM) establishes widely used standard specifications for types of steels, for
welding and connector materials, and for various production and
fabrication processes. Standard grades of steel and other materials are quite commonly referred to by short versions of their
ASTM designation or identification codes. Some other organizations of note are the Steel Deck Institute (SDI), the Steel Joist
Institute (SJI), and the American Iron and Steel Institute (AISI).

5-2 Structural Steel

Steel meeting the requirements of ASTM Specification A36 is the grade commonly used for rolled steel elements utilized in building construction. This steel is required to have a minimum yield point of 36 ksi and a minimum ultimate tensile strength of 58–80 ksi. It may be used for bolted, riveted, or welded fabrication. Other steels are available when special properties are desired, such as higher strength or increased corrosion resistance.

Table 5-1 gives the critical stress properties of the steels currently used for rolled products and indicates the various elements that are produced from them. Table 5-2 gives the groupings of rolled shapes referred to in Table 5-1. The structural property of primary concern is the yield point, or yield stress, which is designated F_y. Most allowable design stresses are based on this value. The other limiting stress is the ultimate tensile stress, designated F_u, on which a few design stresses are based. For some grades the ultimate stress is given as a range rather than as a single value, in which case it is advisable to use the lower value for design unless a higher value can be verified by a specific supplier for a particular rolled product.

5-3 Structural Shapes

The rolled structural steel shapes that are most commonly used in building construction are the wide flanges (W shapes), American Standard I-beams (S shapes), channels (C shapes), angles (L shapes), and plates. Tables 4-1 through 4-4 give properties for some of these products. Complete tables for all available rolled shapes are given in the AISC Manual.

A study of the tables will reveal that wide flange shapes have greater flange widths and thinner webs than Standard I-beams and are further characterized by parallel flange surfaces, as contrasted with the tapered inside flange surfaces of the S shapes. In W, M, and S shapes most of the material is concentrated in the flanges, which makes them especially efficient in resisting bending. W shapes are specified by a nominal dimension of depth (out-to-out of the flanges) which usually varies slightly from the actual

TABLE 5-1. ASTM Structural Steel Grades for Rolled Products

Steel Type	ASTM Designation	F_y Minimum Yield Stress (ksi)	F_u Tensile Stress[b] (ksi)	Shapes[a] Group per ASTM A6 1	2	3	4	5	To 1/2" Incl.	Over 1/2" to 3/4" Incl.	Over 3/4" to 1 1/4" Incl.	Over 1 1/4" to 1 1/2" Incl.	Over 1 1/2" to 2" Incl.	Over 2" to 2 1/2" Incl.	Over 2 1/2" to 4" Incl.	Over 4" to 5" Incl.	Over 5" to 6" Incl.	Over 6" to 8" Incl.	Over 8"
Carbon	A36	32	58-80																
	A36	36	58-80[d]																
	A529	42	60-85																
High-Strength Low-Alloy	A441	40	60																
	A441	42	63																
	A441	46	67																
	A441	50	70																
	A572—Grade	42	42	60															
	A572—Grade	50	50	65															
	A572—Grade	60	60	75															
	A572—Grade	65	65	80															
Corrosion-Resistant High-Strength Low-Alloy	A242	42	63																
	A242	46	67																
	A242	50	70																
	A588	42	63																
	A588	46	67																
	A588	50	70																
Quenched & Tempered Alloy	A514[e]	90	100-130																
	A514[e]	100	110-130																

Source: Reprinted from the *Manual of Steel Construction,* 8th ed., with permission of the publishers, American Institute of Steel Construction.

Note: Shaded portion in table indicates availability of products for each grade.

[a] See Table 5-2.
[b] Minimum unless a range is shown.
[c] Includes bar-size shapes.
[d] For shapes over 426 lb/ft minimum of 58 ksi only applies.
[e] Plates only.

TABLE 5-2. Structural Shape Size Groupings for Tensile
Property Classification

Structural Shape	Group 1	Group 2	Group 3	Group 4	Group 5
W Shapes	W 24x55, 62	W 36x135 to 210 incl	W 36x230 to 300 incl	W 14x233 to 550 incl	W 14x605 to 730 incl
	W 21x44 to 57 incl	W 33x118 to 152 incl	W 33x201 to 241 incl	W 12x210 to 336 incl	
	W 18x35 to 71 incl	W 30x99 to 211 incl	W 14x145 to 211 incl		
	W 16x26 to 57 incl	W 27x84 to 178 incl	W 12x120 to 190 incl		
	W 14x22 to 53 incl	W 24x68 to 162 incl			
	W 12x14 to 58 incl	W 21x62 to 147 incl			
	W 10x12 to 45 incl	W 18x76 to 119 incl			
	W 8x10 to 48 incl	W 16x67 to 100 incl			
	W 6x9 to 25 incl	W 14x61 to 132 incl			
	W 5x16, 19	W 12x65 to 106 incl			
	W 4x13	W 10x49 to 112 incl			
		W 8x58, 67			
M Shapes	to 20 lb/ft incl				
S Shapes	to 35 lb/ft incl	over 35 lb/ft			
HP Shapes		to 102 lb/ft incl	over 102 lb/ft		
American Standard Channels (C)	to 20.7 lb/ft incl	over 20.7 lb/ft			
Miscellaneous Channels (MC)	to 28.5 lb/ft incl	over 28.5 lb/ft			
Angles (L), Structural & Bar-Size	to ½ in. incl	over ½ to ¾ in. incl	over ¾ in.		

Notes: Structural tees from W, M and S shapes fall in the same group as the structural shape from which they are cut.

Group 4 and Group 5 shapes are generally contemplated for application as compression members. When used in other applications or when subject to welding or thermal cutting, the material specification should be reviewed to determine if it adequately covers the properties and quality appropriate for the particular application. Where warranted, the use of killed steel or special metallurgical requirements should be considered.

Source: Reprinted from the *Manual of Steel Construction,* 8th ed., with permission of the publishers, American Institute of Steel Construction.

depth. S and C shapes have their nominal and actual depths the same.

5-4 Nomenclature

The standard symbols—called the general nomenclature—that are used in steel design work are those established in the AISC Specification. Although these are in part special to the area of steel, the attempt is being made increasingly to standardize the nomenclature among all fields. The following is a list of the symbols used in this section, which is an abridged version of a more extensive list in the AISC Manual.

A Cross-sectional area (sq in.)
Gross area of an axially loaded compression member (sq in.)

A_e Effective net area of an axially loaded tension member (sq in.)

A_n Net area of an axially loaded tension member (sq in.)

C_c Column slenderness ratio separating elastic and inelastic buckling

E Modulus of elasticity of steel (29,000 ksi)

F_a Axial compressive stress permitted in a prismatic member in the absence of a bending moment (ksi)

F_b Bending stress permitted in a prismatic member in the absence of axial force (ksi)

F_e' Euler stress for a prismatic member divided by the factor of safety (ksi)

F_p Allowable bearing stress (ksi)

F_t Allowable axial tensile stress (ksi)

F_u Specified minimum tensile strength of the type of steel or fastener being used (ksi)

F_v Allowable shear stress (ksi)

F_y Specified minimum yield stress of the type of steel being used (ksi); as used in this book, *yield stress* denotes either the specified minimum yield point (for those steels that have a yield point) or specified minimum yield strength (for those steels that have no yield point)

I Moment of inertia of a section (in⁴)

I_x Moment of inertia of a section about the X-X axis (in⁴)

I_y Moment of inertia of a section about the Y-Y axis (in⁴)

J Torsional constant of a cross section (in⁴)

K Effective length factor for a prismatic member

L Span length (ft)
       Length of connection angles (in.)

L_c Maximum unbraced length of the compression flange at which the allowable bending stress may be taken at $0.66F_y$ or as determined by AISC Specification Formula (1.5-5a) or Formula (1.5-5b), when applicable (ft)
       Unsupported length of a column section (ft)

L_u Maximum unbraced length of the compression flange at which the allowable bending stress may be taken at $0.6F_y$ (ft)

M Moment (k-ft)

M_D Moment produced by a dead load

M_L Moment produced by a live load

M_p Plastic moment (k-ft)

M_R Beam resisting moment (k-ft)

N Length of base plate (in.)
       Length of bearing of applied load (in.)

P Applied load (kips)
       Force transmitted by a fastener (k)

S Elastic section modules (in³)

S_x Elastic section modulus about the X-X (major) axis (in³)

S_y Elastic section modulus about the Y-Y (minor) axis (in³)

V Maximum permissible web shear (k)
       Statical shear on a beam (k)

Z Plastic section modulus (in³)

Z_x Plastic section modulus with respect to the major (X-X) axis (in³)

Z_y Plastic section modulus with respect to the minor (Y-Y) axis (in³)

b_f Flange width of a rolled beam or plate girder (in.)

d Depth of a column, beam, or girder (in.)
Nominal diameter of a fastener (in.)

e_o Distance from the outside face of a web to the shear center of a channel section (in.)

f_a Computed axial stress (ksi)

f_b Computed bending stress (ksi)

f_c' Specified compression strength of concrete at 28 days (ksi)

f_p Actual bearing pressure on a support (ksi)

f_t Computed tensile stress (ksi)

f_v Computed shear stress (ksi)

g Transverse spacing locating fastener gage lines (in.)

k Distance from the outer face of a flange to the web toe of a fillet of rolled shape or equivalent distance on a welded section (in.)

l For beams, the distance between cross sections braced against twist or lateral displacement of the compression flange (in.)
For columns, the actual unbraced length of a member (in.)
Length of a weld (in.)

m Cantilever dimension of a base plate (in.)

n Number of fasteners in one vertical row
Cantilever dimension of a base plate (in.)

r Governing radius of gyration (in.)

r_x Radius of gyration with respect to the X-X axis (in.)

r_y Radius of gyration with respect to the Y-Y axis (in.)

s Longitudinal center-to-center spacing (pitch) of any two consecutive holes (in.)

t Girder, beam, or column web thickness (in.)
Thickness of a connected part (in.)
Wall thickness of a tubular member (in.)
Angle thickness (in.)

t_f Flange thickness (in.)

t_w Web thickness (in.)

x Subscript relating a symbol to strong axis bending

y Subscript relating a symbol to weak axis bending

D   Beam deflection (in.)

5-5 Allowable Stresses for Structural Steel

For structural steel the AISC Specification expresses the allowable unit stresses in terms of some percent of the yield stress F_y or the ultimate tensile strength F_u. Selected allowable unit stresses used in design are shown in Table 5-3, with specific values given for ASTM A36 steel with F_y of 36 ksi and F_u of 58 ksi. This is not a complete list, but it includes the stresses used in the examples in this section. Reference is made to the more complete descriptions in the AISC Specification which is contained in the AISC Manual.

5-6 Materials and Stresses for Connectors

In addition to the steels used for structural shapes, the AISC Specification designates the type of material permitted in bolted, riveted, and welded connections. Material specifications and allowable stresses for bolts are discussed in Chapter 7. Filler materials for welding and allowable stresses in welds are discussed in Chapter 8.

5-7 Factors in Beam Design

The complete design of a beam includes considerations of bending strength, shear resistance, deflection, lateral support, web crippling, and support details. Design for bending is usually the primary concern, but all other factors must be considered.

The current AISC Specification includes special factors that influence design for bending. The availability of higher strength steels and attendant higher allowable stresses has led to the classification of structural shapes for beams as *compact* or *noncompact* sections. Consequently, the shape employed, the laterally unsupported length of the span, and the grade of steel must be known in order to establish the allowable stress for bending. Unless otherwise stated, the examples and problems presented in this book are based on the assumption that A36 steel is used. The design procedures and specification formulas, however, are applicable to any grade of steel by selecting the appropriate values for the yield stress F_y or the ultimate strength F_u.

5-8   Compact and Noncompact Sections

In order to qualify for use of the maximum allowable bending stress of 0.66 F_y, a beam consisting of a rolled section must satisfy several qualifications. The principal ones are the following:

1. The beam section must be symmetrical about its minor (Y-Y) axis, and the plane of the loading must coincide with the plane of this axis; otherwise a torsional twist will be developed along with the bending.
2. The web and flanges of the section must have width–thickness ratios that qualify the section as *compact*.
3. The compression flange of the beam must be adequately braced against lateral buckling.

The criteria for establishing whether or not a section is compact include as a variable the F_y of the steel. It is therefore not possible to identify sections for this condition strictly on the basis of their geometric properties. The yield stress limits for the qualification of sections as compact for bending and those for combined actions of compression and bending are given, together with other properties, in the tables in the *Manual of Steel Construction* published by the American Institute of Steel Construction. For beams of A36 steel, all S and M sections and all wide flange sections, except the W 6 × 15, qualify as compact when used for bending alone. When sections do not qualify as compact, the allowable bending stress must be reduced with the use of formulas given in the AISC Specification.

5-9   Lateral Support of Beams

A beam may fail by sideways (lateral) buckling of the compression flange when lateral deflection is not prevented. The tendency to buckle increases as the compressive bending stress in the

TABLE 5-3. Allowable Unit Stresses for Structural Steel: ASTM A36

Type of Stress and Conditions	See Discussion in This Book in	Stress Designation	AISC Specification	Allowable Stress (ksi)	(MPa)
Tension					
1. On the gross (unreduced) area	Section 7-4	F_t	$0.60 F_y$	22	150
2. On the effective net area, except at pinholes			$0.50 F_u$	29	125
3. Threaded rods on net area at thread			$0.33 F_u$	19	80
Compression	Chapter 6	F_a	See discussion		
Shear					
1. Except at reduced sections	Section 5-11	F_v	$0.40 F_y$	14.5	100
2. At reduced sections	Section 7-5		$0.30 F_u$	17.4	120
Bending					
1. Tension and compression on extreme fibers of compact members braced laterally,	Chapter 5	F_b	$0.66 F_y$	24	165

	symmetrical about and loaded in the plane of their minor axis			
2.	Tension and compression on extreme fibers of other rolled shapes braced laterally	$0.60\,F_y$	22	150
3.	Tension and compression on extreme fibers of solid round and square bars, on solid rectangular sections bent on their weak axis, on qualified doubly symmetrical I & H shapes bent about their minor axis	$0.75\,F_y$	27	188
Bearing				
1.	On contact area of milled surfaces	F_p $\quad$ $0.90\,F_y$	32.4	225
2.	On projected area of bolts and rivets in shear connections	Chapter 7 $\quad$ $1.50\,F_u$	87	600

Note: $F_y = 36$ ksi; assume that $F_u = 58$ ksi; some table values are rounded off as permitted in the AISC Manual. For SI units $F_y = 250$ MPa, $F_u = 400$ MPa.

flange increases and also as the unbraced length of the span increases. The full value of the allowable bending stress, $F_b = 0.66$ F_y, can be used only when the compression flange is adequately braced. As for the compact section, the value of this required bracing length includes the variable of the F_y of the beam steel.

When the compression flanges of compact beams are supported laterally at intervals not greater than L_c, the full allowable stress of 0.66 F_y may be used. For laterally unsupported lengths greater than L_c but not greater than L_u, the allowable bending stress is reduced to 0.60 F_y. When the laterally unsupported length exceeds L_u, the allowable stress is reduced by a formula that includes the specific value of the unsupported length. Design of beams by use of these requirements is not a simple matter, and the AISC Manual gives supplementary charts to aid the designer. A reproduction of one of these charts and an illustration of its use are given in Section 5-15.

5-10 Flexure or Bending

As demonstrated in the example of Section 4-7, the design of a beam for bending consists of applying form (3) of the flexure formula, $S = M/F_b$. Before starting computations, it is necessary to decide on the grade of steel to be used and to determine the laterally unsupported length of the compression flange. Because the allowable bending stress given in the example is 24 ksi, we may assume that A36 steel was specified and that adequate lateral support for the beam was provided. With these items established, the design procedure followed involved the steps listed below:

1. Determine the maximum bending moment.
2. Compute the required section modulus.
3. Refer to the tables of properties of steel sections, and select a beam with a section modulus equal to or greater than that which is required. In general, the lightest weight section is the most economical. Table 5-4 will be found very useful in this operation because the beams are listed in the order of decreasing section moduli. This arrangement, and the listing of corresponding values of L_c and L_u, is pre-

sented in the AISC Manual in a much more extensive table called Allowable Stress Design Selection Table for Shapes Used as Beams. The reader should verify the beam sections selected in Step (3) of the Section 4-7 example by checking them out in Table 5-4.

The beam selected in the Section 4-7 example (W 14 × 34) was designed for bending only. A complete design would require investigation for shear and deflection. In the following example the procedure outlined above is applied to a simple beam with a more complicated loading.

Example. A girder has a span of 18 ft with floor beams framing into it from both sides at 6-ft intervals, similar to the arrangement shown in Fig. 5-4a. The reaction of each floor beam is 4000 lb, so the girder receives two concentrated loads of 8000 lb each at the third points of the span. The framing of the beams provides lateral support for the girder at 6-ft intervals. In addition, the girder has a uniformly distributed load of 400 lb/ft, including its own weight, extending over the full span. Design the girder for bending, assuming that A36 steel is used.

Solution: (1) Because of the symmetry of loading, we know that the maximum bending moment will occur at the center of the span and that each reaction will equal half the total load. Then

$$R = \tfrac{1}{2}[8000 + 8000 + (400 \times 18)] = 11,600 \text{ lb}$$

$$M_{(x=9)} = (11,600 \times 9) - [(8000 \times 3) + (400 \times 9 \times 4.5)]$$
$$= 64,200 \text{ ft-lb}$$

(2) Assuming a compact shape will be used, and therefore $F_b = 24$ ksi (Table 5-3), the required section modulus is

$$S = \frac{M}{F_b} = \frac{64,200 \times 12}{24,000} = 32.1 \text{ in}^3$$

(3) From Table 4-1 we determine some possible choices to be

W 10 × 30, $S = 32.4 \text{ in}^3$

W 12 × 26, $S = 33.4 \text{ in}^3$

W 14 × 26, $S = 35.3 \text{ in}^3$

TABLE 5-4. Section Modulus and Moment of Resistance for Selected Rolled Structural Shapes

S_x	Shape	$F_y = 36$ ksi			S_x	Shape	$F_y = 36$ ksi		
		L_c	L_u	M_R			L_c	L_u	M_R
In.³		Ft.	Ft.	Kip-ft.	In.³		Ft.	Ft.	Kip-ft.
1110	W 36x300	17.6	35.3	2220	269	W 30x 99	10.9	11.4	538
					267	W 27x102	10.6	14.2	534
1030	W 36x280	17.5	33.1	2060	258	W 24x104	13.5	18.4	516
					249	W 21x111	13.0	23.3	498
953	W 36x260	17.5	30.5	1910					
					243	W 27x 94	10.5	12.8	486
895	W 36x245	17.4	28.6	1790	231	W 18x119	11.9	29.1	462
					227	W 21x101	13.0	21.3	454
837	W 36x230	17.4	26.8	1670					
829	W 33x241	16.7	30.1	1660	222	W 24x 94	9.6	15.1	444
757	W 33x221	16.7	27.6	1510	213	W 27x 84	10.5	11.0	426
					204	W 18x106	11.8	26.0	408
719	W 36x210	12.9	20.9	1440					
					196	W 24x 84	9.5	13.3	392
684	W 33x201	16.6	24.9	1370	192	W 21x 93	8.9	16.8	384
					190	W 14x132	15.5	44.1	380
664	W 36x194	12.8	19.4	1330	188	W 18x 97	11.8	24.1	376
663	W 30x211	15.9	29.7	1330					
					176	W 24x 76	9.5	11.8	352
623	W 36x182	12.7	18.2	1250	175	W 16x100	11.0	28.1	350
598	W 30x191	15.9	26.9	1200	173	W 14x109	15.4	40.6	346
					171	W 21x 83	8.8	15.1	342
580	W 36x170	12.7	17.0	1160	166	W 18x 86	11.7	21.5	332
					157	W14x 99	15.4	37.0	314
542	W 36x160	12.7	15.7	1080	155	W 16x 89	10.9	25.0	310
539	W 30x173	15.8	24.2	1080					
					154	W 24x 68	9.5	10.2	308
504	W 36x150	12.6	14.6	1010	151	W 21x 73	8.8	13.4	302
502	W 27x178	14.9	27.9	1000	146	W 18x 76	11.6	19.1	292
487	W 33x152	12.2	16.9	974	143	W 14x 90	15.3	34.0	286
455	W 27x161	14.8	25.4	910					
					140	W 21x 68	8.7	12.4	280
448	W 33x141	12.2	15.4	896	134	W 16x 77	10.9	21.9	268
439	W 36x135	12.3	13.0	878	131	W 24x 62	7.4	8.1	262
414	W 24x162	13.7	29.3	828					
411	W 27x146	14.7	23.0	822	127	W 21x 62	8.7	11.2	254
					127	W 18x 71	8.1	15.5	.254
406	W 33x130	12.1	13.8	812	123	W 14x 82	10.7	28.1	246
380	W 30x132	11.1	16.1	760	118	W 12x 87	12.8	36.2	236
371	W 24x146	13.6	26.3	742	117	W 18x 65	8.0	14.4	234
					117	W 16x 67	10.8	19.3	234
359	W 33x118	12.0	12.6	718					
355	W 30x124	11.1	15.0	710	114	W 24x 55	7.0	7.5	228
					112	W 14x 74	10.6	25.9	224
329	W 30x116	11.1	13.8	658	111	W 21x 57	6.9	9.4	222
329	W 24x131	13.6	23.4	658	108	W 18x 60	8.0	13.3	216
329	W 21x147	13.2	30.3	658	107	W 12x 79	12.8	33.3	214
					103	W 14x 68	10.6	23.9	206
299	W 30x108	11.1	12.3	598					
299	W 27x114	10.6	15.9	598	98.3	W 18x 55	7.9	12.1	197
295	W 21x132	13.1	27.2	590	97.4	W 12x 72	12.7	30.5	195
291	W 24x117	13.5	20.8	582					
273	W 21x122	13.1	25.4	546					

TABLE 5-4. (Continued)

S_x	Shape	F_y = 36 ksi				S_x	Shape	F_y = 36 ksi		
		L_c	L_u	M_R				L_c	L_u	M_R
In.³		Ft.	Ft.	Kip·ft.		In.³		Ft.	Ft.	Kip·ft.
94.5	**W 21x50**	**6.9**	**7.8**	**189**		**29.0**	**W 14x22**	**5.3**	**5.6**	**58**
92.2	W 16x57	7.5	14.3	184		27.9	W 10x26	6.1	11.4	56
92.2	W 14x61	10.6	21.5	184		27.5	W 8x31	8.4	20.1	55
88.9	**W 18x50**	**7.9**	**11.0**	**178**		**25.4**	**W 12x22**	**4.3**	**6.4**	**51**
87.9	W 12x65	12.7	27.7	176		24.3	W 8x28	6.9	17.5	49
81.6	**W 21x44**	**6.6**	**7.0**	**163**		**23.2**	**W 10x22**	**6.1**	**9.4**	**46**
81.0	W 16x50	7.5	12.7	162						
78.8	W 18x46	6.4	9.4	158		**21.3**	**W 12x19**	**4.2**	**5.3**	**43**
78.0	W 12x58	10.6	24.4	156						
77.8	W 14x53	8.5	17.7	156		**21.1**	**M 14x18**	**3.6**	**4.0**	**42**
72.7	W 16x45	7.4	11.4	145		20.9	W 8x24	6.9	15.2	42
70.6	W 12x53	10.6	22.0	141		18.8	W 10x19	4.2	7.2	38
70.3	W 14x48	8.5	16.0	141		18.2	W 8x21	5.6	11.8	36
68.4	**W 18x40**	**6.3**	**8.2**	**137**		**17.1**	**W 12x16**	**4.1**	**4.3**	**34**
66.7	W 10x60	10.6	31.1	133		16.7	W 6x25	6.4	20.0	33
						16.2	W 10x17	4.2	6.1	32
64.7	**W 16x40**	**7.4**	**10.2**	**129**		15.2	W 8x18	5.5	9.9	30
64.7	W 12x50	8.5	19.6	129						
62.7	W 14x43	8.4	14.4	125		**14.9**	**W 12x14**	**3.5**	**4.2**	**30**
60.0	W 10x54	10.6	28.2	120		13.8	W 10x15	4.2	5.0	28
58.1	W 12x45	8.5	17.7	116		13.4	W 6x20	6.4	16.4	27
						13.0	M 6x20	6.3	17.4	26
57.6	**W 18x35**	**6.3**	**6.7**	**115**						
56.5	W 16x36	7.4	8.8	113		**12.0**	**M 12x11.8**	**2.7**	**3.0**	**24**
54.6	W 14x38	7.1	11.5	109		11.8	W 8x15	4.2	7.2	24
54.6	W 10x49	10.6	26.0	109		10.9	W 10x12	3.9	4.3	22
51.9	W 12x40	8.4	16.0	104		10.2	W 6x16	4.3	12.0	20
49.1	W 10x45	8.5	22.8	98		10.2	W 5x19	5.3	19.5	20
						9.91	W 8x13	4.2	5.9	20
48.6	**W 14x34**	**7.1**	**10.2**	**97**		9.72	W 6x15	6.3	12.0	19
						9.63	M 5x18.9	5.3	19.3	19
47.2	**W 16x31**	**5.8**	**7.1**	**94**		8.51	W 5x16	5.3	16.7	17
45.6	W 12x35	6.9	12.6	91						
42.1	W 10x39	8.4	19.8	84		**7.81**	**W 8x10**	**4.2**	**4.7**	**16**
42.0	**W 14x30**	**7.1**	**8.7**	**84**		7.76	M 10x 9	2.6	2.7	16
						7.31	W 6x12	4.2	8.6	15
38.6	**W 12x30**	**6.9**	**10.8**	**77**						
						5.56	W 6x 9	4.2	6.7	11
38.4	**W 16x26**	**5.6**	**6.0**	**77**		5.46	W 4x13	4.3	15.6	11
						5.24	M 4x13	4.2	16.9	10
35.3	**W 14x26**	**5.3**	**7.0**	**71**						
35.0	W 10x33	8.4	16.5	70		**4.62**	**M 8x 6.5**	**2.4**	**2.5**	**9**
33.4	**W 12x26**	**6.9**	**9.4**	**67**						
32.4	W 10x30	6.1	13.1	65		**2.40**	**M 6x 4.4**	**1.9**	**2.4**	**5**
31.2	W 8x35	8.5	22.6	62						

Selection on the basis of required section modulus is made easier by the use of Table 5-4, in which the shapes most used for beams are listed in order of their S_x values. Moment values given in the table represent the total resisting moments of the shapes for A36 steel, based on a maximum bending stress of 24 ksi [165 MPa] for compact sections and 22 ksi [150 MPa] for noncompact shapes.

To assist in the identification of situations in which buckling is a concern, the table gives the limiting values L_c and L_u—discussed in Section 5-9—for each shape. The maximum bending stress of 0.66 F_y is permitted for unsupported lengths only up to L_c. For lengths between L_c and L_u, the allowable stress is reduced to 0.60 F_y. For laterally unsupported lengths in excess of L_u, design is more practically achieved by use of the charts from the AISC Manual, of which Fig. 5-2 is an example.

In the preceding example the three possible choices could have been more quickly determined from Table 5-4 than from scanning Table 4-1.

In the following problems assume that A36 steel is used and that the beams are supported laterally throughout their length, permitting the use of the full allowable bending stress of 24 ksi.

Problem 5-10-A.   Design for flexure a simple beam 14 ft [4.3 m] in length and having a total uniformly distributed load of 19.8 kips [88 kN].

Problem 5-10-B*.   Design for flexure a beam having a span of 16 ft [4.9 m] with a concentrated load of 12.4 kips [55 kN] at the center of the span.

Problem 5-10-C.   A beam 15 ft [4.6 m] in length has three concentrated loads of 4 kips, 5 kips, and 6 kips at 4 ft, 10 ft, and 12 ft [17.8 kN, 22.2 kN, and 26.7 kN at 1.2 m, 3 m, and 3.6 m], respectively, from the left-hand support. Design the beam for flexure.

Problem 5-10-D*.   A beam 30 ft [9 m] long has concentrated loads of 9 kips [40 kN] each at the third points and also a total uniformly distributed load of 30 kips [133 kN]. Design the beam for flexure.

Problem 5-10-E.   Design for flexure a beam 12 ft [3.6 m] in length, having a uniformly distributed load of 2 kips/ft [29 kN/m] and a concentrated load of 8.4 kips [37.4 kN] a distance of 5 ft [1.5 m] from one support.

Problem 5-10-F. A beam of 19 ft [5.8 m] in length has concentrated loads of 6 kips [26.7 kN] and 9 kips [40 kN] at 5 ft [1.5 m] and 13 ft [4 m], respectively, from the left-hand support. In addition, there is a uniformly distributed load of 1.2 k/ft [17.5 kN/m] beginning 5 ft [1.5 m] from the left support and continuing to the right support. Design the beam for flexure.

Problem 5-10-G*. A steel beam 16 ft [4.9 m] long has two uniformly distributed loads, one of 200 lb/ft [2.92 kN/m] extending 10 ft [3 m] from the left support and the other of 100 lb/ft [1.46 kN/m] extending over the remainder of the beam. In addition, there is a concentrated load of 8 kips [35.6 kN] at 10 ft [3 m] from the left support. Design the beam for flexure.

Problem 5-10-H. Design for flexure a simple beam 12 ft [3.7 m] in length, having two concentrated loads of 12 kips [53.4 kN] each, one 4 ft [1.2 m] from the left end and the other 4 ft [1.2 m] from the right end.

Problem 5-10-I. A cantilever beam 8 ft [2.4 m] long has a uniformly distributed load of 1600 lb/ft [23.3 kN/m]. Design the beam for flexure.

Problem 5-10-J*. A cantilever beam 6 ft [1.8 m] long has a concentrated load of 12.3 kips [54.7 kN] at its unsupported end. Design the beam for flexure.

5-11 Shear

Shear stress in a steel beam is seldom a factor in determining its size. If, however, the beam has a relatively short span with a large load or has any large load placed near one support, the bending moment may be low and the shear relatively high. For W, S, and C shape beams, the shear stress is assumed to be developed by the beam web only, with the web area being determined as the product of the beam depth times the web thickness. For this situation the AISC Specification allows a maximum stress of 0.40 F_y rounded off to 14.5 ksi [100 MPa] for A36 steel. Shear stress is thus investigated with the following formula:

$$f_v = \frac{V}{A_w} = \frac{V}{dt_w}$$

in which f_v = actual unit shear stress,
  V = maximum vertical shear,
  A_w = gross area of the beam web,
  d = overall depth of the beam,
  t_w = thickness of the beam web.

Example. A simple beam of A36 steel is 6 ft [1.83 m] long and has a concentrated load of 36 kips [160 kN] applied 1 ft [0.3 m] from the left end. It is found that a W 8 × 24 is large enough to sustain the resulting bending moment. Investigate the beam for shear.

Solution: We find the two reactions for this loading to be 30 kips [133 kN] and 6 kips [27 kN]. The maximum vertical shear is thus equal to the value of the larger reaction—the usual case for a simple span beam.

From Table 4-1 we find that $d = 7.93$ in. and $t_w = 0.245$ in. for the W 8 × 24. Then

$$A_w = d \times t_w = 7.93 \times 0.245 = 1.94 \text{ in}^2$$

and

$$f_v = \frac{V}{A_w} = \frac{30}{1.94} = 15.5 \text{ ksi}$$

Since this exceeds the allowable value of 14.5 ksi, the W 8 × 24 is not acceptable.

It may be noted that S shapes in general have rather thicker webs than W shapes of corresponding depth. We may thus consider the use of an S shape with the necessary value for S_x, for which an inspection of Table 4-2 will yield an S 8 × 23. From Table 4-2 we find the depth of this shape to be 8 in. and the web thickness 0.441 in. Then

$$f_v = \frac{V}{A_w} = \frac{30}{8 \times 0.441} = 8.50 \text{ ksi}$$

which is less than the allowable stress.

Problems 5-11-A*-B-C-D. Compute the maximum permissible web shears for the following beams of A36 steel:

A*   S   12 × 40.8
B   W 12 × 40
C   W 10 × 22
D   C 10 × 20

5-12   Deflection

In addition to resisting bending and shear, beams must not deflect excessively. Floor and ceiling cracks may result if the beams are not stiff enough, and beams should be investigated to see that the deflection does not exceed $\frac{1}{360}$ of the span, the generally accepted limit with plastered ceilings. The current AISC Specification requirement is that steel beams and girders supporting plastered ceilings be of such dimensions that the maximum *live load* deflection will not exceed $\frac{1}{360}$ of the span. It frequently happens that a beam may be of adequate dimensions to resist bending and shear but, on investigation, may be found to deflect more than the maximum permitted by building codes.

For typical beams and loads the actual deflection may be computed from the formulas given in Fig. 3-19, but in using these formulas note carefully that *l*, in the term l^3, is in inches, not feet. For a simple beam with a uniformly distributed load, the deflection is found by the formula

$$D = \frac{5}{384} \times \frac{Wl^3}{EI} \qquad \text{(Case 2, Fig. 3-19)}$$

in which D = maximum deflection in inches,
$\quad\quad W$ = total uniformly distributed load in pounds or kips,
$\quad\quad l$ = length of the span *in inches,*
$\quad\quad E$ = modulus of elasticity of the beam in psi or ksi (for structural steel $E = 29{,}000{,}000$ psi),
$\quad\quad I$ = moment of inertia of the cross section of the beam in inches to the fourth power.

For a beam on which the loading is not typical, we may find W, the uniformly distributed load that would produce the same bending moment, and then apply the foregoing formula to find the approximate deflection. When the maximum deflection occurs at the center of the span, it is sometimes convenient to compute the deflections due to individual loads on the beam; their sum will be the total deflection.

For a uniformly loaded simple span beam (Case 2, Fig. 3-19) with a fixed value for the maximum bending stress, a formula may be derived that expresses the maximum deflection in terms of the span and the beam depth. For a stress value of 24 ksi, the formula has the form

$$D = \frac{0.02483 \times L^2}{d}$$

in which D = maximum deflection in inches,

　　　　L = span length in feet,

　　　　d = beam depth in inches.

Note: For a derivation of this formula, see *Simplified Design of Structural Steel,* 5th ed., by Harry Parker (New York, Wiley, 1983).

Figure 5-1 presents graphs of this equation for various values

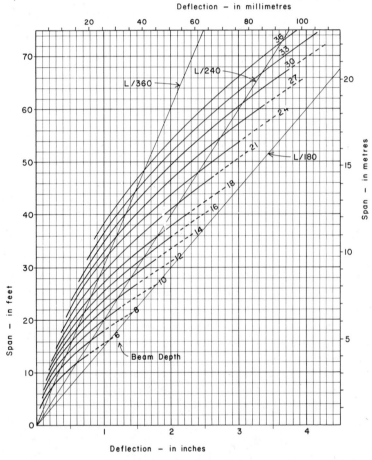

FIGURE 5-1. Deflection of steel beams with bending stress of 24 ksi [165 MPa].

of beam depth from 6 in. to 36 in. Deflections may be obtained directly from these graphs for the particular case of a beam of A36 steel with maximum bending stress of 24 ksi. Since both the bending stress and the deflection are directly proportional to the load, deflections for stress values of other magnitudes may be obtained by simple proportion, as indicated in the following example.

Example. A W 12 × 26 is used for a simple span beam to carry a total uniformly distributed load of 10 kips, including the beam weight. If the span is 16 ft, find the maximum deflection.
Solution: We first determine the actual maximum bending stress as follows:

$$M = \frac{W \times L}{8} = \frac{10 \times 16}{8} = 20 \text{ k-ft}$$

From Table 4.1 S_x = 33.4 in^3 for the shape. Then

$$f_b = \frac{M}{S} = \frac{20 \times 12}{33.4} = 7.186 \text{ ksi}$$

We next use Fig. 5-1 to obtain a deflection value of approximately 0.53 in. for the 12-in.-deep beam. The true deflection is then determined as

$$D = \frac{7.186}{24} \times 0.53 = 0.16 \text{ in.}$$

Not much accuracy can be expected with this process, but it is usually sufficient, since many conditions in real design situations make deflection calculations subject to speculation.

When SI units are used, the formula used for the graphs in Fig. 5-1 takes the form

$$D = \frac{0.1719 \times L^2}{d}$$

in which L = the span length in meters,
d = the beam depth in millimeters,

and the deflection is determined in millimeters.

Problems 5-12-A*-B-C. Find the maximum deflection for the following uniformly loaded beams of A36 steel. Find deflection values in inches, using Fig. 5-1, and verify the values by use of the equation for Case 2, Fig. 3-19.

A   W 10 × 30, span = 18 ft, total load = 30 kips [5.5 m, 133 kN]
B   W 16 × 36, span = 20 ft, total load = 50 kips [6 m, 222 kN]
C   W 18 × 40, span = 24 ft, total load = 55 kips [7.3 m, 245 kN]

5-13 Safe-Load Tables

The simple span beam loaded entirely with uniformly distributed load occurs so frequently in steel structural systems that it is useful to have a rapid design method for quick selection of shapes for a given load and span condition. Use of Table 5-5 allows for a simple design procedure where design conditions permit its use. When beams of A36 steel are loaded in the plane of their minor axis (Y-Y) and have lateral bracing spaced not farther than L_c, they may be selected from Table 5-5 after determination of only the total load and the beam span.

For a check, the values of L_c are given for each shape in the table. If the actual distance between points of lateral support for the compression (top) flange exceeds L_c, the table values must be reduced, as discussed in Section 5-15.

Deflections may be determined by using the factors given in the table or the graphs in Fig. 5-1. Both of these are based on a maximum bending stress of 24 ksi [165 MPa], and deflections for other stresses may be proportioned from these values if necessary.

The loads in the tables will not result in excessive shear stress on the beam webs if the full beam depth is available for stress development. Where end framing details result in some reduction of the web area, investigation of the stress on the reduced section may be required.

The following examples illustrate the use of Table 5-5 for some common design situations.

Example 1.   A simple span beam of A36 steel is required to carry a total uniformly distributed load of 40 kips [178 kN] on a span of 30 ft [9.14 m]. Find (1) the lightest shape permitted and (2) the shallowest (least deep) shape permitted.

TABLE 5-5. Load-Span Values for Beams

Shape	L_c^{**} (ft)	8	10	12	14	16	18	20	22	24	26	28	30
Span (ft) → *Deflection Factor* →*		1.59	2.48	3.58	4.87	6.36	8.05	9.93	12.0	14.3	16.8	19.5	22.3
M 8 × 6.5	2.4	9.24	7.39	6.16	5.28	4.62	4.11						
M 10 × 9	2.6	15.5	12.4	10.3	8.87	7.76	6.90	6.21	5.64				
W 8 × 10	4.2	15.6	12.5	10.4	8.92	7.81	6.94						
W 8 × 13	4.2	19.8	15.9	13.2	11.3	9.91	8.81						
W 10 × 12	3.9	21.8	17.4	14.5	12.5	10.9	9.69	8.72	7.93				
W 8 × 15	4.2	23.6	18.9	15.7	13.5	11.8	10.5						
M 12 × 11.8	2.7	24.0	19.2	16.0	13.7	12.0	10.7	9.60	8.73	8.00	7.38	6.86	
W 10 × 15	4.2	27.6	22.1	18.4	15.8	13.8	12.3	11.0	10.0				
W 12 × 14	3.5	29.8	23.8	19.9	17.0	14.9	13.2	11.9	10.8	9.93	9.17	8.51	
W 8 × 18	5.5	30.4	24.3	20.3	17.4	15.2	13.5						
W 10 × 17	4.2	32.4	25.9	21.6	18.5	16.2	14.4	13.0	11.8				
W 12 × 16	4.1	34.2	27.4	22.8	19.5	17.1	15.2	13.7	12.4	11.4	10.5	9.77	
W 8 × 21	5.6	36.4	29.1	24.3	20.8	18.2	16.2						
W 10 × 19	4.2	37.6	30.1	25.1	21.5	18.8	16.7	15.0	13.7				
W 8 × 24	6.9	41.8	33.4	27.9	23.9	20.9	18.6						
M 14 × 18	3.6	42.2	33.8	28.1	24.1	21.1	18.7	16.9	15.3	14.1	13.0	12.0	11.2
W 12 × 19	4.2	42.6	34.1	28.4	24.3	21.3	18.9	17.0	15.5	14.2	13.1	12.2	
W 10 × 22	6.1	46.4	37.1	30.9	26.5	23.2	20.6	18.5	16.9				
W 8 × 28	6.9	48.6	38.9	32.4	27.8	24.3	21.6						

TABLE 5-5. (Continued)

Shape	L_c** (ft)	Span (ft) 12	14	16	18	20	22	24	26	28	30	32	34
Deflection Factor*		3.58	4.87	6.36	8.05	9.93	12.0	14.3	16.8	19.5	22.3	25.4	28.7
W 12 × 22	4.3	33.9	29.0	25.4	22.6	20.3	18.5	16.9	15.6	14.5			
W 10 × 26	6.1	37.2	31.9	27.9	24.8	22.3	20.3						
W 14 × 22	5.3	38.7	33.1	29.0	25.8	23.2	21.1	19.3	17.8	16.6	15.5	14.5	
W 10 × 30	6.1	43.2	37.0	32.4	28.8	25.9	23.6						
W 12 × 26	6.9	44.5	38.2	33.4	29.7	26.7	24.3	22.3	20.5	19.1			
W 10 × 33	8.4	46.7	40.0	35.0	31.0	28.0	25.4						
W 14 × 26	5.3	47.1	40.3	35.3	31.4	28.2	25.7	23.5	21.7	20.2	18.8	17.6	
W 16 × 26	5.6	51.2	43.9	38.4	34.1	30.7	27.9	25.6	23.6	21.9	20.5	19.2	18.1
W 12 × 30	6.9	51.5	44.1	38.6	34.3	30.9	28.1	25.7	23.8	22.0			
W 14 × 30	7.1	56.0	48.0	42.0	37.3	33.6	30.5	28.0	25.8	24.0	22.4	21.0	
W 10 × 39	8.4	56.1	48.1	42.1	37.4	33.7	30.6						
W 12 × 35	6.9	60.8	52.1	45.6	40.5	36.5	33.2	30.4	28.1	26.0	25.2	23.6	22.2
W 16 × 31	5.8	62.9	53.9	47.2	41.9	37.8	34.3	31.5	29.0	27.0	25.9	24.3	
W 14 × 34	7.1	64.8	55.5	48.6	43.2	38.9	35.3	32.4	29.9	27.8			
W 10 × 45	8.5	65.5	56.1	49.1	43.6	39.3	35.7						

Shape	Span (ft)	16	18	20	22	24	26	28	30	32	34	36	38
	Deflection Factor* L_c** (ft)	6.36	8.05	9.93	12.0	14.3	16.8	19.5	22.3	25.4	28.7	32.2	35.9
W 12 × 40	8.4	51.9	46.1	41.5	37.7	34.6	31.9	29.6					
W 14 × 38	7.1	54.6	48.5	43.7	39.7	36.4	33.6	31.2	29.1	27.3			
W 16 × 36	7.4	56.5	50.2	45.2	41.1	37.7	34.8	32.3	30.1	28.2	26.6	25.1	
W 18 × 35	6.3	57.8	51.4	46.2	42.0	38.5	35.6	33.0	30.8	28.9	27.2	25.7	24.3
W 12 × 45	8.5	58.1	51.6	46.5	42.2	38.7	35.7	33.2					
W 14 × 43	8.4	62.7	55.7	50.1	45.6	41.8	38.6	35.8	33.4	31.3			
W 12 × 50	8.5	64.7	57.5	51.7	47.0	43.1	39.8	37.0					
W 16 × 40	7.4	64.7	57.5	51.7	47.0	43.1	39.8	37.0	34.5	32.3	30.4	28.7	
W 18 × 40	6.3	68.4	60.8	54.7	49.7	45.6	42.1	39.1	36.5	34.2	32.2	30.4	28.8
W 14 × 48	8.5	70.3	62.5	56.2	51.1	46.9	43.3	40.2	37.5	35.1			
W 12 × 53	10.6	70.6	62.7	56.5	51.3	47.1	43.4	40.3					
W 16 × 45	7.4	72.7	64.6	58.2	52.9	48.5	44.7	41.5	38.8	36.3	34.2	32.3	
W 14 × 53	8.5	77.8	69.1	62.2	56.6	51.9	47.9	44.4	41.5	38.9			
W 18 × 46	6.4	78.8	70.0	63.0	57.3	52.5	48.5	45.0	42.0	39.4	37.1	35.0	33.2
W 16 × 50	7.5	81.0	72.0	64.8	58.9	54.0	49.8	46.3	43.2	40.5	38.1	36.0	33.2

TABLE 5-5. (Continued)

Span (ft) →	45	42	39	36	33	30	27	24	22	20	18	16	Deflection Factor* — L_c** (ft)	Shape
Deflection Factor*	50.3	43.8	37.8	32.2	27.0	22.3	18.1	14.3	12.0	9.93	8.05	6.36		
	29.0	31.1	33.5	36.3	39.6	43.5	48.3	54.4	59.3	65.3	72.5	81.6	6.6	W 21 × 44
			36.5	39.5	43.1	47.4	52.7	59.3	64.6	71.1	79.0	88.9	7.9	W 18 × 50
				41.0	44.7	49.2	54.6	61.5	67.0	73.8	81.9	92.2	10.6	W 14 × 61
	33.6			41.0	44.7	49.2	54.6	61.5	67.0	73.8	81.9	92.2	7.5	W 16 × 57
	33.6	36.0	38.8	42.0	45.8	50.4	56.0	63.0	68.7	75.6	84.0	94.5	6.9	W 21 × 50
			40.3	43.7	47.7	52.4	58.2	65.5	71.5	78.6	87.4	98.3	7.9	W 18 × 55
			44.3	48.0	52.4	57.6	64.0	72.0	78.5	86.4	96.0	108	8.0	W 18 × 60
	39.5	42.3	45.5	49.3	53.8	59.2	65.8	74.0	80.7	88.6	98.7	111	6.9	W 21 × 57
	40.5	43.4	46.8	50.7	55.3	60.8	67.5	76.0	82.9	91.2	101	114	7.0	W 24 × 55
				52.0	56.7	62.4	69.3	78.0	85.1	93.6	104	117	10.8	W 16 × 67
			48.0	52.0	56.7	62.4	69.3	78.0	85.1	93.6	104	117	8.0	W 18 × 65
			52.1	56.4	61.5	67.7	72.2	84.7	92.4	102	113	127	8.1	W 18 × 71
	45.1	48.4	52.1	56.4	61.5	67.7	72.2	84.7	92.4	102	113	127	8.7	W 21 × 62
	46.6	49.9	53.7	58.2	63.5	69.9	77.6	87.3	95.3	105	116	131	7.4	W 24 × 62
				59.5	65.0	71.5	79.4	89.3	97.4	107	119	134	10.9	W 16 × 77
	49.8	53.3	57.4	62.2	67.9	74.7	83.0	93.3	102	112	124	140	8.7	W 21 × 68
			59.9	64.9	70.8	77.9	86.5	97.3	106	117	130	146	11.6	W 18 × 76
	53.7	57.5	61.9	67.1	73.2	80.5	89.5	101	110	121	134	151	8.8	W 21 × 73
	54.7	58.7	63.2	68.4	74.7	82.1	91.2	103	112	123	137	154	9.5	W 24 × 68
			68.1	73.8	80.5	88.5	98.4	111	121	133	147	166	11.7	W 18 × 86
	60.8	65.1	70.1	76.0	82.9	91.2	101	114	124	137	152	171	8.8	W 21 × 83

Shape	L_c** (ft)	over Span (ft) / Deflection Factor* →											
		Span (ft) 24	27	30	33	36	39	42	45	48	52	56	60
Deflection Factor*		14.3	18.1	22.3	27.0	32.2	37.8	43.8	50.3	57.2	67.1	77.9	89.4
W 24 × 76	9.5	117	104	93.9	85.3	78.2	72.2	67.0	62.6	58.7			
W 21 × 93	8.9	128	114	102	93.1	85.3	78.8	73.1	68.3				
W 24 × 84	9.5	131	116	104	95.0	87.1	80.4	74.7	69.7	65.3			
W 27 × 84	10.5	142	126	114	103	94.7	87.4	81.1	75.7	71.0	65.5		
W 24 × 94	9.6	148	131	118	108	98.7	91.1	84.6	78.9	74.0		60.8	
W 21 × 101	13.0	151	134	121	110	101	93.1	86.5	80.7				
W 27 × 94	10.5	162	144	130	118	108	99.7	92.6	86.4	81.0	74.8	69.4	
W 24 × 104	13.5	172	153	138	125	115	106	98.3	91.7	86.0			
W 27 × 102	10.6	178	158	142	129	119	109	102	94.9	89.0	82.1	76.3	
W 30 × 99	10.9	179	159	143	130	120	110	102	95.6	89.7	82.8	76.9	71.7
W 24 × 117	13.5	194	172	155	141	129	119	111	103	97.0			
W 27 × 114	10.6	199	177	159	145	133	123	114	106	99.7	92.0	85.4	79.7
W 30 × 108	11.1	199	177	159	145	133	123	114	106	99.7	92.0	85.4	87.7
W 30 × 116	11.1	219	195	175	159	146	135	125	117	110	101	94.0	94.7
W 30 × 124	11.1	237	210	189	172	158	146	135	126	118	109	101	

147

TABLE 5-5. (Continued)

Shape	L_c** (ft)	Span (ft) 70	65	60	56	52	48	45	42	39	36	33	30
Deflection Factor*		122	105	89.4	77.9	67.1	57.2	50.3	43.8	37.8	32.2	27.0	22.3
W 33 × 118	12.0		88.4	95.7	103	110	120	128	137	147	159	174	191
W 30 × 132	11.1			101	109	117	127	135	145	156	169	184	203
W 33 × 130	12.1		99.9	108	116	125	135	144	155	166	180	197	216
W 27 × 146	14.7				117	126	137	146	156	169	183	199	219
W 36 × 135	12.3	100	108	117	125	135	146	156	167	180	195	213	234
W 33 × 141	12.2		110	119	128	138	149	159	171	184	199	217	239
W 33 × 152	12.2		120	130	139	150	162	173	185	200	216	236	260
W 36 × 150	12.6	115	124	134	144	155	168	179	192	207	224	244	269
W 30 × 173	15.8			144	154	166	180	192	205	221	239	261	287
W 36 × 160	12.7	124	133	144	155	167	181	193	206	222	241	263	289
W 36 × 170	12.7	132	143	155	166	178	193	206	221	238	258	281	309
W 30 × 191	15.9			159	171	184	199	213	228	245	268	290	319
W 36 × 182	12.7	142	153	166	178	192	208	221	237	256	277	302	332
W 36 × 194	12.8	152	163	177	190	204	221	236	253	272	295	322	354
W 33 × 201	16.6		168	182	195	210	228	243	260	281	304	332	365
W 36 × 210	12.9	164	177	192	205	221	240	256	274	295	319	349	383
W 33 × 221	16.7		186	202	216	233	252	269	288	310	336	367	404
W 33 × 241	16.7		204	221	237	255	276	295	316	340	368	402	442
W 36 × 230	17.4	191	206	223	239	257	279	298	319	343	372	406	446
W 36 × 245	17.4	204	220	239	256	275	298	318	341	367	398	434	477
W 36 × 260	17.5	218	234	254	272	293	318	339	363	391	423	462	508
W 36 × 280	17.5	235	253	275	294	317	343	366	392	422	458	499	549
W 36 × 300	17.6	254	273	296	317	341	370	395	423	455	493	538	592

Note: Total allowable uniformly distributed load is in kips for simple span beams of A36 steel with yield stress of 36 ksi [250 MPa]. This is based on a maximum bending stress of 24 ksi [165 MPa].

* Maximum deflection in inches at the center of the span may be obtained by dividing this factor by the depth of the beam in inches. (See Fig. 5-2 for example.)

** Maximum permitted distance between points of lateral support. If distance exceeds this, use the charts in the AISC Manual.

Solution: From Table 5-5 we find the following:

Shape	Allowable Load
W21 × 44	43.5 kips
W18 × 46	42.0
W16 × 50	43.2
W14 × 53	41.5

Thus the lightest shape is the W 21 × 44, and the shallowest is the W 14 × 53.

Example 2.   A simple span beam of A36 steel is required to carry a total uniformly distributed load of 25 kips [111 kN] on a span of 24 ft [7.32 m] while sustaining a maximum deflection of not more than $\frac{1}{360}$ of the span. Find the lightest shape permitted.
Solution: From Table 5-5 we find the lightest shape that will carry this load to be the W 16 × 26. For this beam the deflection will be

$$D = \frac{25}{25.6} \times \frac{14.3}{16} = \frac{\text{(actual load)}}{\text{(table load)}} \times \frac{\text{(deflection factor)}}{\text{(beam depth)}}$$

$$= 0.873 \text{ in.}$$

which exceeds the allowable of $\dfrac{24 \times 12}{360} = 0.80$ in.

The next heaviest shape from Table 5-5 is a W 16 × 31, for which the deflection will be

$$D = \frac{25}{31.5} \times \frac{14.3}{16} = 0.709 \text{ in.}$$

which is less than the limit, so this shape is the lightest choice.

Problems 5-13-A*-B-C-D-E*-F.   For each of the following conditions find (1) the lightest permitted shape and (2) the shallowest permitted shape of A36 steel.

	Span in ft	Total uniformly distributed load in kips	Deflection limited to $\frac{1}{360}$ of the span
A	16	10	no
B	20	30	no
C	36	40	no
D	18	16	yes
E	32	20	yes
F	42	50	yes

5-14 Equivalent Tabular Loads

The safe loads shown in Table 5-5 are uniformly distributed loads on simple span beams. By use of coefficients it is possible to convert other types of loading to equivalent uniform loads and thereby greatly extend the usefulness of Table 5-5.

For a simple beam with a uniformly distributed load, the maximum moment is $WL/8$. For a simple beam with two equal loads at the third points of the span, the maximum moment is $PL/3$. These values are shown in Fig. 3-19, Case 2 and Case 3, respectively. Equating these moment values, we can write

$$\frac{WL}{8} = \frac{PL}{3} \quad \text{and} \quad W = 2.67 \times P$$

which demonstrates that a total uniformly distributed load 2.67 times the value of one of the concentrated loads will produce the same maximum moment on the beam as that caused by the concentrated loads. Thus in order to use Table 5-5 for design of a beam with this loading, we merely multiply the concentrated load by the ETL (equivalent tabular load) factor and look for the new load value in the table.

Coefficients for other loadings are also given in Fig. 3-19. It is important to remember that an ETL loading does not include the weight of the beam, for which an estimated amount should be added. Also the shear values and reactions for the beam must be determined from the actual loading and not from the ETL.

5-15 Design of Laterally Unsupported Beams

As discussed in Section 5-9, the allowable bending stress for beams of A36 steel is 24 ksi [165 MPa] (0.66 F_y) when the compression flanges of compact shapes are braced laterally at intervals not greater than L_c. For beams supported laterally at intervals greater than L_c but not greater than L_u, the allowable bending stress is reduced to 22 ksi [152 MPa] (0.60 F_y, rounded off). Both L_c and L_u values are given in Table 5-4, so when the lateral support interval falls between these values, the various table values based on a stress of 24 ksi may simply be reduced by the proportion 22/24 = 0.917.

When the distance between points of lateral support exceeds L_u, the AISC specification provides an equation for the determination of the allowable stress that includes the value of the unsupported length. This is quite cumbersome to use for design, so a series of charts is provided in the AISC Manual that contain plots of the relationship of the moment capacity to the laterally unsupported length for commonly used beam shapes. Figure 5-2 is a reproduction of one of these charts. In order to use the chart, you must determine the maximum bending moment in the beam and the distance between points of lateral support. These two values are used to find a point on the chart; any beam whose graph lies above or to the right of this point is adequate, the nearest solid-line graph representing the shape of least weight. The following example illustrates the use of the chart in Fig. 5-2.

Example. A simple beam carries a total uniform load of 22 kips [98 kN], including its own weight, over a span of 24 ft [7.32 m]. Other elements of the framing provide lateral support at intervals of 8 ft [2.44 m]. Select a beam of A36 steel for these conditions. *Solution:* For this loading the maximum moment is

$$M = \frac{WL}{8} = \frac{22 \times 24}{8} = 66 \text{ k-ft} \left[\frac{98 \times 7.32}{8} = 89.7 \text{ kN-m} \right]$$

Using this moment and the unsupported length, we locate the point on the chart in Fig. 5-2. The nearest line above and to the right of this point is that for a W 10 × 33, which is an acceptable beam. However, the line for this beam is dashed, indicating that there is a lighter possible choice. Proceeding on the chart, upward and to the right, the first solid line we encounter is that for a W 12 × 30, which is also acceptable and represents the lightest possible shape.

ALLOWABLE MOMENTS IN BEAMS

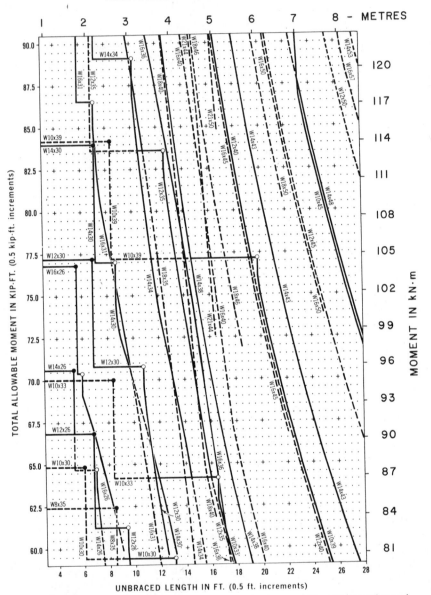

FIGURE 5-2. Allowable bending moment for beams with various unbraced lengths. F_y = 36 ksi [250 MPa].

The chart in Fig. 5-2 does not incorporate consideration of shear or deflection, which should also be investigated in real design situations.

Problem 5-15-A*.   A W 14 × 34 is used as a simple beam to carry a total uniformly distributed load, including the beam weight, of 40 kips [178 kN] on a span of 16 ft [4.88 m]. Lateral support is provided only at the ends and at the midspan. Is the shape an adequate choice for these conditions?

Problem 5-15-B*.   A simple beam is required for a span of 24 ft [7.32 m]. The load, including the beam weight, is uniformly distributed and consists of a total of 28 kips [125 kN]. If lateral support is provided only at the ends and at the midspan, and the deflection under full load is limited to 0.6 in. [15.2 mm], find an adequate rolled shape for the beam.

5-16   Fireproofing for Beams

In fire resistive construction, some insulating material must be placed around the structural steel. Materials commonly used for this purpose are concrete, masonry, or lath and plaster. There are also fibrous and cementitious coatings that can be sprayed directly on the surfaces of steel members or on the undersides of steel roofs and floor decks. When a poured concrete deck is used with a steel framing system, the steel beams are sometimes fireproofed by encasement in poured concrete, as shown in Fig. 5-3. This results in considerable dead weight for the beam, which should be given consideration in the design of the beam.

Building codes differ in the thickness of fireproofing material required. A common specification for beams and girders calls for 2 in. of concrete on the flat surfaces and 1.5 in. on the edges of flanges, as shown in Fig. 5-3. With these thicknesses the cross-sectional area of the concrete (drawn hatched in the figure) is $d \times (b + 3)$ in.^{2}. The number of cubic feet of concrete per linear foot of beam is the area in square inches divided by 144. Taking an average value for the weight of the concrete as 144 lb/ft^3, the weight of the fireproofing per linear foot of beam becomes

$$W_{FP} = \frac{d \times (b + 3)}{144} \times 144 = d \times (b + 3) \text{ lb/ft}^3$$

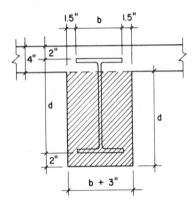

FIGURE 5-3. Fireproofing of a steel beam with concrete.

It should be noted that this expression depends on the thickness of the structural slab, the covering thicknesses of the fireproofing, and the distance from the top of the steel beam to the top of the slab. It also ignores the reduction in concrete volume due to the displacement of the encased beam. It is reasonably accurate, however, for the purpose of making a preliminary allowance for the weight when designing the beam.

Table 5-6, based on the preceding formula, gives the weight of concrete fireproofing for some typical beam sections. When the beam size has finally been determined, the true weight of the beam and its fireproofing may be checked to see that an adequate allowance was made in the design.

Fireproofing is presently more often accomplished by use of sprayed materials or encasement in plaster or dry-wall materials, resulting in weights considerably less than those obtained with concrete encasement.

5-17 Roof and Floor Framing Systems

Flat-spanning roof and floor structures are commonly formed by arrangements of beam and girder systems, ordinarily supported by steel columns. The area of the building, the floor plan, and the occupancy requirements determine the location of columns or

TABLE 5-6. Approximate Weight of Concrete Fireproofing for Wide Flange Beams

Sections	Weight	Sections	Weight
W 8 × 10 to 15	56	W 18 × 35 to 46	162
W 8 × 18, 21	66	W 18 × 50 to 71	189
W 8 × 24, 28	76	W 18 × 76 to 119	252
W 10 × 12 to 19	70	W 21 × 44 to 57	200
W 10 × 22 to 30	88	W 21 × 62 to 93	236
W 10 × 33 to 45	110	W 21 × 101 to 147	323
W 12 × 14 to 22	84	W 24 × 55, 62	240
W 12 × 26 to 35	114	W 24 × 68 to 94	288
W 12 × 40 to 50	132	W 24 × 104 to 162	381
W 12 × 53, 58	156	W 27 × 84 to 114	351
W 14 × 22, 26	112	W 27 × 146 to 178	459
W 14 × 30 to 38	137	W 30 × 99 to 132	405
W 14 × 43 to 53	154	W 30 × 173 to 211	540
W 14 × 61 to 82	182	W 33 × 118 to 152	479
W 16 × 26, 31	136	W 33 × 201 to 241	619
W 16 × 36 to 57	160	W 36 × 135 to 210	540
W 16 × 67 to 100	216	W 36 × 230 to 300	702

Note: Weight in lb/ft of beam length; does not include weight of steel beam.

other means of support. The layout of beams and girders depends mostly on the column spacing and the type of deck employed. Figure 5-4 illustrates two common arrangements of framing when one of the shorter-span deck systems is used, such as the solid concrete slab or one of the corrugated or ribbed steel deck systems.

The area of floor supported by any one beam is equal to the beam span length multiplied by the sum of half the distances to adjacent beams. The span lengths used in design are generally taken from center to center of supporting members. The total uniformly distributed load carried by a beam is found by multiplying the supported area by the sum of the dead and live loads per unit of the supported area.

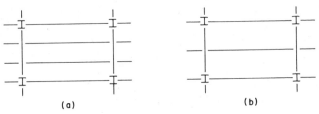

FIGURE 5-4.   Framing layouts.

5-18   Dead Load

Dead load consists of the weight of the materials of which the building is constructed: walls, partitions, columns, framing, floors, roofs, and ceilings. In the design of a beam, the dead load must include an allowance for the weight of the beam itself. Table 5-7, which lists the weights of many construction materials, may be used in the computation of dead loads. Dead loads are due to gravity, and they result in downward vertical forces.

TABLE 5-7.   Weights of Building Construction

	lb/ft^2	kN/m^2
Roofs		
3-ply ready roofing (roll, composition)	1	0.05
3-ply felt and gravel	5.5	0.26
5-ply felt and gravel	6.5	0.31
Shingles		
wood	2	0.10
asphalt	2–3	0.10–0.15
clay tile	9–12	0.43–0.58
concrete tile	8–12	0.38–0.58
slate, 1/4 in.	10	0.48
fiber glass	2–3	0.10–0.15
aluminum	1	0.05
steel	2	0.10
Insulation		
fiber glass batts	0.5	0.025

TABLE 5-7. (*Continued*)

	lb/ft^2	kN/m^2
rigid foam plastic	1.5	0.075
foamed concrete, mineral aggregate	2.5/in.	0.0047/mm
Wood rafters		
2 × 6 at 24 in.	1.0	0.05
2 × 8 at 24 in.	1.4	0.07
2 × 10 at 24 in.	1.7	0.08
2 × 12 at 24 in.	2.1	0.10
Steel deck, painted		
22 ga	1.6	0.08
20 ga	2.0	0.10
18 ga	2.6	0.13
Skylight		
glass with steel frame	6–10	0.29–0.48
plastic with aluminum frame	3–6	0.15–0.29
Plywood or softwood board sheathing	3.0/in.	0.0057/mm
Ceilings		
Suspended steel channels	1	0.05
Lath		
steel mesh	0.5	0.025
gypsum board, 1/2 in.	2	0.10
Fiber tile	1	0.05
Dry wall, gypsum board, 1/2 in.	2.5	0.12
Plaster		
gypsum, acoustic	5	0.24
cement	8.5	0.41
Suspended lighting and air distribution systems, average	3	0.15
Floors		
Hardwood, 1/2 in.	2.5	0.12
Vinyl tile, 1/8 in.	1.5	0.07
Asphalt mastic	12/in.	0.023/mm
Ceramic tile		
3/4 in.	10	0.48
thin set	5	0.24

TABLE 5-7. (Continued)

	lb/ft^2	kN/m^2
Fiberboard underlay, 5/8 in.	3	0.15
Carpet and pad, average	3	0.15
Timber deck	2.5/in.	0.0047/mm
Steel deck, stone concrete fill, average	35–40	1.68–1.92
Concrete deck, stone aggregate	12.5/in.	0.024/mm
Wood joists		
2 × 8 at 16 in.	2.1	0.10
2 × 10 at 16 in.	2.6	0.13
2 × 12 at 16 in.	3.2	0.16
Lightweight concrete fill	8.0/in.	0.015/mm
Walls		
2 × 4 studs at 16 in., average	2	0.10
Steel studs at 16 in., average	4	0.20
Lath, plaster; see Ceilings		
Gypsum dry wall, 5/8 in. single	2.5	0.12
Stucco, 7/8 in., on wire and paper or felt	10	0.48
Windows, average, glazing + frame		
small pane, single glazing, wood or metal frame	5	0.24
large pane, single glazing, wood or metal frame	8	0.38
increase for double glazing	2–3	0.10–0.15
curtain walls, manufactured units	10–15	0.48–0.72
Brick veneer		
4 in., mortar joints	40	1.92
1/2 in., mastic	10	0.48
Concrete block		
lightweight, unreinforced—4 in.	20	0.96
6 in.	25	1.20
8 in.	30	1.44
heavy, reinforced, grouted—6 in.	45	2.15
8 in.	60	2.87
12 in.	85	4.07

5-19   Roof Loads

In addition to the dead loads they support, roofs are designed for a uniformly distributed live load that includes snow accumulation and the general loadings that occur during construction and maintenance of the roof. Snow loads are based on local snowfalls and are specified by local building codes.

Table 5-8 gives the minimum roof live load requirements specified by the 1982 edition of the *Uniform Building Code*. Note the adjustments for roof slope and for the total area of roof surface supported by a structural element. The latter accounts for the increase in probability of the lack of total surface loading as the size of the surface area increases.

Roof surfaces must also be designed for wind pressure, the magnitude and manner of application of which is specified by local building codes, based on local wind histories. For very light roof construction, a critical problem is sometimes that of the upward (suction) effect of the wind, which may exceed the dead load and result in a net upward lifting force.

Although the term *flat roof* is often used, there is generally no such thing; all roofs must be designed for some water drainage. The minimum required pitch is usually $\frac{1}{4}$ in/ft, or a slope of approximately 1 : 50. With roof surfaces that are this close to flat, a potential problem is that of *ponding*, a phenomenon in which the weight of water on the surface causes deflection of the supporting structure, which in turn allows for more water accumulation (in a pond) causing more deflection, and so on, resulting in an accelerated collapse condition. Investigation of this condition is required for nearly flat roofs, and a procedure for the investigation is specified in the AISC Manual.

5-20   Floor Live Loads

The live load on a floor represents the probable effects created by the occupancy. It includes the weights of human occupants, furniture, equipment, stored materials, and so on. All building codes provide minimum live loads to be used in the design of buildings for various occupancies. Since there is a lack of uniformity among different codes in specifying live loads, the local code

TABLE 5-8. Minimum Roof Live Loads

Roof Slope Conditions	Minimum Uniformly Distributed Load (lb/ft²) Tributary Loaded Area for Structural Member (ft²)			Minimum Uniformly Distributed Load (kN/m²) (m²)		
	0–200	201–600	Over 600	0–18.6	18.7–55.7	Over 55.7
1. Flat or rise less than 4 in./ft (1:3). Arch or dome with rise less than 1/8 span.	20	16	12	0.96	0.77	0.575
2. Rise 4 in./ft (1:3) to less than 12 in./ft (1:1). Arch or dome with rise 1/8 of span to less than 3/8 of span.	16	14	12	0.77	0.67	0.575
3. Rise 12 in./ft (1:1) or greater. Arch or dome with rise 3/8 of span or greater.	12	12	12	0.575	0.575	0.575
4. Awnings, except cloth covered.	5	5	5	0.24	0.24	0.24
5. Greenhouses, lath houses, and agricultural buildings.	10	10	10	0.48	0.48	0.48

Source: Adapted from the *Uniform Building Code*, 1982 ed., with permission of the publishers, International Conference of Building Officials.

should always be used. Table 5-9 contains values for floor live loads as given by the 1982 edition of the *Uniform Building Code*.

Although expressed as uniform loads, code-required values are usually established large enough to account for ordinary concentrations that occur. For offices, parking garages, and some other occupancies, codes often require the consideration of a specified concentrated load as well as the distributed loading. Where buildings are to contain heavy machinery, stored materials, or other contents of unusual weight, these must be provided for individually in the design of the structure.

When structural framing members support large areas, most codes allow some reduction in the total live load to be used for design. These reductions, in the case of roof loads, are incorporated into the data in Table 5-8. The following is the method given in the 1982 edition of the *Uniform Building Code* for determining the reduction permitted for beams, trusses, or columns that support large floor areas.

Except for floors in places of assembly (theaters, etc.), and except for live loads greater than 100 psf [4.79 kN/m^2], the design live load on a member may be reduced in accordance with the formula

$$R = 0.08 \, (A - 150)$$

$$[R = 0.86 \, (A - 14)]$$

The reduction shall not exceed 40% for horizontal members or for vertical members receiving load from one level only, 60% for other vertical members, nor R as determined by the formula

$$R = 23.1 \, (1 + D/L)$$

In these formulas

R = reduction in percent,
A = area of floor supported by a member,
D = unit dead load/sq ft of supported area,
L = unit live load/sq ft of supported area.

In office buildings and certain other building types, partitions may not be permanently fixed in location but may be erected or moved from one position to another in accordance with the re-

TABLE 5-9. Minimum Floor Live Loads

Use or Occupancy		Uniform Load		Concentrated Load	
Description	Description	(psf)	(kN/m²)	(lb)	(kN)
Armories		150	7.2		
Assembly areas and auditoriums and balconies therewith	Fixed seating areas	50	2.4		
	Movable seating and other areas	100	4.8		
	Stages and enclosed platforms	125	6.0		
Cornices, marquees, and residential balconies		60	2.9		
Exit facilities		100	4.8		
Garages	General storage, repair	100	4.8	*	
	Private pleasure car	50	2.4	*	
Hospitals	Wards and rooms	40	1.9	1000	4.5
Libraries	Reading rooms	60	2.9	1000	4.5
	Stack rooms	125	6.0	1500	6.7
Manufacturing	Light	75	3.6	2000	9.0
	Heavy	125	6.0	3000	13.3
Offices		50	2.4	2000	9.0
Printing plants	Press rooms	150	7.2	2500	11.1
	Composing rooms	100	4.8	2000	9.0
Residential		40	1.9		
Rest rooms		**			
Reviewing stands, grandstands, and bleachers		100	4.8		
Roof decks (occupied)	Same as area served				
Schools	Classrooms	40	1.9	1000	4.5
Sidewalks and driveways	Public access	250	12.0	*	
Storage	Light	125	6.0		
	Heavy	250	12.0		
Stores	Retail	75	3.6	2000	9.0
	Wholesale	100	4.8	3000	13.3

Source: Adapted from the *Uniform Building Code,* 1982 ed., with permission of the publishers, International Conference of Building Officials.

* Wheel loads related to size of vehicles that have access to the area.
** Same as the area served or minimum of 50 psf.

quirements of the occupants. In order to provide for this flexibility, it is customary to require an allowance of 15 to 20 psf [0.72 to 0.96 kN/m^2], which is usually added to other dead loads.

5-21   Decks

Figure 5-5 shows four possibilities for a floor deck, all of which may be used with a steel framing system. When a wood deck is used, it is usually nailed to a series of wood joists or trusses, which are supported by the steel beams. However, in some cases wood nailers may be bolted to the tops of steel beams, and the deck can then be directly attached to the beams.

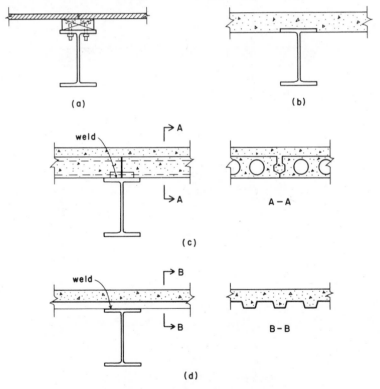

FIGURE 5-5.   Typical floor decks.

When a concrete slab is poured at the site, it may be combined with poured concrete fireproofing, as shown in Fig. 5-3. When the beams are fireproofed by other means, it is common to use the detail shown in Fig. 5-5*b*. Concrete may also be used in the form of precast deck units that are welded to the steel beams using steel devices cast into the units. A concrete fill is normally used on top of precast units in order to provide a smooth top surface.

A floor deck system used widely with steel framing is that of corrugated or fluted steel units with concrete fill, as shown in Fig. 5-5*d*. For low-slope roofs the decking problem is essentially similar to that for floors, and many of the same types of deck may be used. However, there are some different issues involved in roof construction, and therefore some products are used only for roofs. (See Fig. 5-6.)

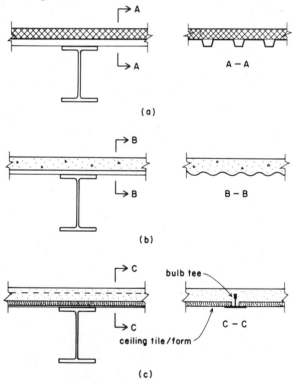

FIGURE 5-6.   Typical roof decks.

Steel deck units are available from a large number of manufacturers. Specific information regarding the type of deck available,
possible range of sizes, rated load-carrying capacities, and so on,
should be obtained directly from those who supply these products
to the region in which a proposed building is to be built. The
information in Table 5-10 is provided by the national organization
that provides standards to the deck manufacturers and indicates
one widely used type of deck.

5-22   Design of Typical Framing

The following example illustrates the design of beams and girders
for a typical bay of floor framing.

Example.   The floor construction of a building is as shown in
Fig. 5-7. The live load is 60 psf [2.87 kN/m²], and an allowance of
15 psf [0.72 kN/m²] is to be included for movable partitions. Design the framing using A36 steel.

Solution for the beam: Using data from Table 5-7, we determine
the floor load as follows:

Dead load
4-in. concrete slab	=	50 psf
2-in. lightweight concrete fill	=	16
carpet underlay (fiberboard)	=	3
carpet and pad	=	3
suspended ceiling	=	10
movable partitions	=	15
Total dead load	=	97 psf
Live load	=	60 psf
Total floor load	=	157 psf [7.52 kN/m²]

The total superimposed load on one beam is the panel area
times this unit load, or

$$W = 7 \times 20 \times 0.157 = 22.0 \text{ kips}$$

$$[W = 2.13 \times 6.1 \times 7.52 = 97.7 \text{ kN}]$$

The maximum bending moment due to this total uniformly

TABLE 5-10. Load Capacity of Steel Roof Deck

Deck[a] Type	Span Condition	Weight[b] (psf)	Total (Dead & Live) Safe Load[c] for Spans Indicated in ft-in.												
			4-0	4-6	5-0	5-6	6-0	6-6	7-0	7-6	8-0	8-6	9-0	9-6	10-0
NR22	Simple	1.6	73	58	47										
NR20		2.0	91	72	58	48	40								
NR18		2.7	121	95	77	64	54	46							
NR22	Two	1.6	80	63	51	42									
NR20		2.0	96	76	61	51	43								
NR18		2.7	124	98	79	66	55	47	41						
NR22	Three or More	1.6	100	79	64	53	44								
NR20		2.0	120	95	77	63	53	45							
NR18		2.7	155	123	99	82	69	59	51	44					
IR22	Simple	1.6	86	68	55	45									
IR20		2.0	106	84	68	56	47	40							
IR18		2.7	142	112	91	75	63	54	46	40					
IR22	Two	1.6	93	74	60	49	41								
IR20		2.0	112	88	71	59	50	42							
IR18		2.7	145	115	93	77	64	55	47	41					
IR22	Three or More	1.6	117	92	75	62	52	44							
IR20		2.0	140	110	89	74	62	53	46	40					
IR18		2.7	181	143	116	96	81	69	59	52	45	40			
WR22	Simple	1.6			(89)	(70)	(56)	(46)							
WR20		2.0			(112)	(87)	(69)	(57)	(47)	(40)					
WR18		2.7			(154)	(119)	(94)	(76)	(63)	(53)	(45)				
WR22	Two	1.6			98	81	68	58	50	43					
WR20		2.0			125	103	87	74	64	55	49	43			
WR18		2.7			165	137	115	98	84	73	65	57	51	46	41
WR22	Three or More	1.6			122	101	85	72	62	54	(46)	(40)			
WR20		2.0			156	129	108	92	80	(67)	(57)	(49)	(43)		
WR18		2.7			207	171	144	122	105	(91)	(76)	(65)	(57)	(50)	(44)

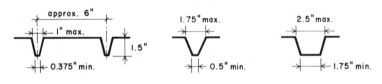

Narrow Rib Deck — NR	Intermediate Rib Deck - IR	Wide Rib Deck - WR

Source: Adapted from the *Steel Deck Institute Design Manual for Composite Decks, Form Decks, and Roof Decks,* 1981–82 issue, with permission of the publishers, the Steel Deck Institute. May not be reproduced without express permission of the Steel Deck Institute.

[a] Letters refer to rib type (see key). Numbers indicate gage (thickness) of steel.
[b] Approximate weight with paint finish; other finishes also available.
[c] Total safe allowable load in lb/sq ft. Loads in parentheses are governed by live load deflection not in excess of 1/240 of the span, assuming a dead load of 10 psf.

166

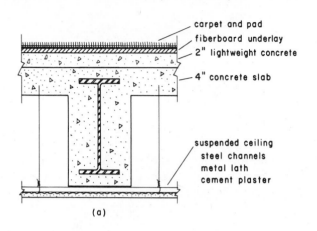

(a)

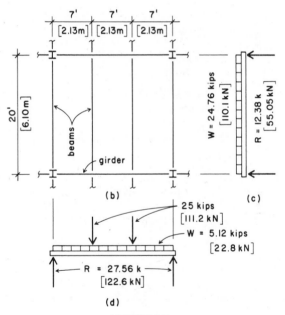

FIGURE 5-7.

distributed load is

$$M = \frac{WL}{8} = \frac{22.0 \times 20}{8} = 55.0 \text{ kip-ft}$$

$$\left[M = \frac{97.7 \times 6.10}{8} = 74.5 \text{ kN-m} \right]$$

Because the top flange of the beam is adequately supported laterally by the encasing concrete, the allowable bending stress is 24 ksi [165 MPa], and the required section modulus is

$$S = \frac{M}{F_b} = \frac{55.0 \times 12}{24} = 27.5 \text{ in}^3$$

$$\left[S = \frac{74.5 \times 10^6}{165} = 452 \times 10^3 \text{ mm}^3 \right]$$

Bearing in mind that so far we have not taken into account the weight of the beam and its fireproofing, we refer to Table 5-4 and look for a beam with a slightly greater section modulus than that computed. A possible consideration is a W 14 × 26, with $S_x = 35.3 \text{ in}^3$. From Table 5-6 we find that the fireproofing for this section weighs 112 lb/ft, making a total weight of 26 + 112 = 138 lb/ft for the beam and fireproofing. This will add a total uniformly distributed load of

$$W = 20 \times 0.138 = 2.76 \text{ kips [12.28 kN]}$$

which is added to the superimposed load for a revised load of

$$W = 2.76 + 22.0 = 24.76 \text{ kips [110.1 kN]}$$

and results in a new moment and section modulus of

$$M = \frac{WL}{8} = \frac{24.76 \times 20}{8} = 61.9 \text{ kip-ft}$$

$$[83.9 \text{ kN-m}]$$

$$S = \frac{M}{F_b} = \frac{61.9 \times 12}{24} = 31.0 \text{ in}^3$$

$$[508 \times 10^3 \text{ mm}^3]$$

Returning to Table 5-4, we find that no section that will fulfill this requirement, is lighter than the W 14 × 26, although there are alternatives if a shallower beam is desired. If we stay with our choice, it is necessary to consider the problems of shear and deflection. It is unlikely that shear will be critical in the typical uniformly loaded floor and roof beam under normal loading. If any doubt exists, however, an investigation should be made by using the procedures developed in Section 5-11.

Although the requirements of the actual building code of jurisdiction must be used for any real design work, the following are typical deflection criteria:

Loading condition	Maximum deflection
Live load only	$L/360$
Total load, roofs	$L/180$
Total load, floors	$L/240$

Because the dead load is high in our case, we use the total load limit; thus the maximum deflection to be allowed is

$$D = \frac{L}{240} = \frac{20 \times 12}{240} = 1.0 \text{ in.}$$

The actual deflection may be determined in several ways—for example, by the formula for Case 2 in Fig. 3-19 or the deflection factor from Table 5-5. Using the latter procedure, we find from the table that the load capacity of the W 14 × 26 is 28 kips and the deflection factor for the 20-ft span is 9.93. Then the actual deflection under the total load is

$$D = \frac{9.93}{14} \times \frac{24.76}{28} = 0.63 \text{ in.}$$

which is less than that allowed.

If the end of the beam is connected to the girder in the usual way (Fig. 7-11e), it may be necessary to investigate the condition of shear across the net area of the web. This type of investigation is discussed in Section 7-5.

An alternative to the procedure of using the superimposed load to find a required section modulus would have been to compare the load directly with Table 5-5. Scanning the table for a beam to carry slightly more than 22 kips on a 20-ft span produces the same results as the computations that were performed. This is the simpler procedure because the beam carries only uniform loads.

Solution for the girder: The girder loading takes the form shown in Fig. 5-7d. The superimposed loading consists of the concentrated loads due to the end reactions of the beams. With two beams being supported at each point, each load is equal to the total load on one beam, or approximately 25 kips. The maximum bending moment due to this loading (Case 3, Fig. 3-19) is

$$M = \frac{PL}{3} = \frac{25 \times 21}{3} = 175 \text{ kip-ft}$$

[237 kN-m]

and the required section modulus, assuming that the concrete encasement provides full lateral support and using $F_b = 24$ ksi, is

$$S = \frac{M}{F_b} = \frac{175 \times 12}{24} = 87.5 \text{ in}^3$$

[1434 × 10³ mm³]

We now consider the section listings in Table 5-4, bearing in mind that the beam weight and fireproofing have not been taken into account. One possible choice is the W 18 × 55 with S_x of 98.3 in³. From Table 5-6 the fireproofing will add 189 lb/ft to the beam weight, making a total of 244 lb/ft and producing an additional moment of

$$M = \frac{WL}{8} = \frac{(0.244 \times 21 \times 21)}{8} = \frac{5.12 \times 21}{8} = 13.4 \text{ kip-ft}$$

which, when added to the moment due to the superimposed load, produces a total moment on the beam of

$$M = 175 + 13.4 = 188.4 \text{ kip-ft [255 kN-m]}$$

and requires a section modulus of

$$S = \frac{M}{F_b} = \frac{188.4 \times 12}{24} = 94.2 \text{ in}^3 \text{ [1544 × 10}^3 \text{ mm}^3\text{]}$$

From Table 5-4 we find that the W 21 × 50 just satisfies this requirement with an S of 94.5. The deeper section, however, will have slightly more fireproofing concrete; therefore the margin of excess is probably not as indicated. If the deeper section is acceptable, the W 21 × 50 would be the lightest selection. Let us consider the case for deflection with the W 18 × 55. As for the beam, we consider the limit to be that under total load; thus

$$D = \frac{L}{240} = \frac{21 \times 12}{240} = 1.05 \text{ in.}$$

For the actual deflection we consider the two loadings separately, compute two deflections, and add the results. From Fig. 3-19, Case 2, the deflection due to the uniformly distributed load is

$$D = \frac{5WL^3}{384EI} = \frac{5 \times 5.12 \times (21 \times 12)^3}{384 \times 29,000 \times 890} = 0.041 \text{ in.}$$

and from Fig. 3-19, Case 3, the deflection due to the beam loads is

$$D = \frac{23PL^3}{648EI} = \frac{23 \times 25 \times (21 \times 12)^3}{648 \times 29,000 \times 890} = 0.550 \text{ in.}$$

The total deflection is thus $0.041 + 0.550 = 0.591$ in., which is considerably less than the limit of 1.05 in.

Shear stress on the gross web (depth times web thickness) will be low on the girder. Other possible concerns for shear on a net web section or for web crippling may exist if the end framing of the girder is known. In this example we finish with a consideration of bending and deflection.

It is common with framing systems of this type to provide for cambering of the beams to compensate for the dead-load deflection. This consists of cold-bending the beams to produce a residual upward deflection (crowning) so that the beam top is flat under the dead load. In this example, which concerns relatively short spans, this is not really critical, but when spans are greater it is well advised.

Problem 5-22-A.   The floor framing for a typical interior bay of a building is shown in Fig. 5-8; the layout is indicated in Fig. 5-8b and the floor construction in Fig. 5-8a. Design the beam and girder for a live load of

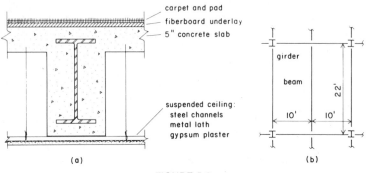

FIGURE 5-8.

100 psf using A-36 steel. Assume that the steel deck provides full lateral support for the beams but that the girders are braced only at the column and beam intersections.

5-23   Crippling of Beam Webs

An excessive end reaction on a beam or an excessive concentrated load at some point along the interior of the span may cause crippling, or localized yielding, of the beam web. The AISC Specification requires that end reactions or concentrated loads for beams without stiffeners or other web reinforcement shall not exceed the following (Fig. 5-9):

$$\text{maximum end reaction} = 0.75F_y t(N + k)$$

$$\text{maximum interior load} = 0.75F_y t(N + 2k)$$

where t = thickness of the beam *web,* in inches,

N = length of the bearing or length of the concentrated load (not less than k for end reactions), in inches,

k = distance from the outer face of the flange to the web toe of the fillet, in inches,

$0.75F_y$ = 27 ksi for A36 steel [186 MPa].

When these values are exceeded, the webs of the beams should be reinforced with stiffeners, the length of bearing increased, or a beam with a thicker web selected.

Example 1.   A W 21 × 57 beam of A36 steel has an end reaction that is developed in bearing over a length of N = 10 in. [254 mm].

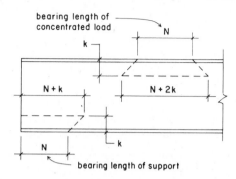

bearing length of
concentrated load

N

k

N + k

N + 2k

N

k

bearing length of support

FIGURE 5-9. Determination
of effective length for computation of web crippling.

Check the beam for web cripplng if the reaction is 44 kips [196 kN].

Solution: In Table 4-1 we find that $k = 1.385$ in. [35 mm] and the web thickness is 0.405 in. [10 mm]. To check for web crippling, we find the maximum end reaction permitted and compare it with the actual value for the reaction. Thus

$$R = F_p \times t \times (N + k) = 27 \times 0.405 \times (10 + 1.385)$$

$$= 124 \text{ kips (the allowable reaction)}$$

$$\left[R = \frac{186 \times 10 \times (254 + 35)}{10^3} = 538 \text{ kN} \right]$$

Because this is greater than the actual reaction, the beam is not critical with regard to web crippling.

Example 2. A W 12 × 26 of A36 steel supports a column load of 70 kips [311 kN] at the center of the span. The bearing length of the column on the beam is 10 in. [254 mm]. Investigate the beam for web crippling under this concentrated load.

Solution: In Table 4-1 we find that $k = 0.885$ in. [22 mm] and the web thickness is 0.230 in. [5.84 mm]. The allowable load that can be supported on the given bearing length is

$$P = F_p \times t \times (N + 2k) = 27 \times 0.230 \times \{10 + (2 \times 0.885)\}$$

$$= 73 \text{ kips}$$

$$\left[P = \frac{186 \times 5.84}{10^3} \times \{254 + (2 \times 22)\} = 324 \text{ kN} \right]$$

which exceeds the required load. Because the column load is less than this value, the beam web is safe from web crippling.

Problem 5-23-A. Compute the maximum allowable reaction with respect to web crippling for a W 14 × 30 of A36 steel with an 8-in. [203-mm] bearing-plate length.

Problem 5-23-B. A column load of 81 kips [360 kN] with a bearing-plate length of 11 in. [279 mm] is placed on top of the beam in the preceding problem. Are web stiffeners required to prevent the web crippling?

5-24   Beam Bearing Plates

Beams that are supported on walls or piers of masonry or concrete usually rest on steel bearing plates. The purpose of the plate is to provide an ample bearing area. The plate also helps to seat the beam at its proper elevation. Bearing plates provide a level surface for a support and, when properly placed, afford a uniform distribution of the beam reaction over the area of contact with the supporting material.

By reference to Fig. 5-10, the area of the bearing plate is $B \times N$. It is found by dividing the beam reaction by F_p, the allowable bearing value of the supporting material. Then

$$A = \frac{R}{F_p}$$

where $A = B \times N$, the area of the plate in sq in.,

$\quad R$ = reaction of the beam in pounds or kips,

$\quad F_p$ = allowable bearing pressure on the supporting material in psi or ksi (see Table 5-11).

The thickness of the wall generally determines N, the dimension of the plate parallel to the length of the beam. If the load from the beam is unusually large, the dimension B may become excessive. For such a condition, one or more shallow-depth I-beams, placed parallel to the wall length, may be used instead of a plate. The dimensions B and N are usually in even inches, and a great variety of thicknesses is available.

The thickness of the plate is determined by considering the projection n (Fig. 5-10*b*) as an inverted cantilever; the uniform

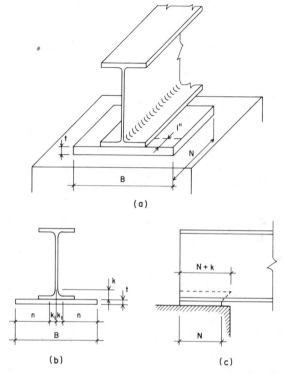

(a)

(b)   (c)

FIGURE 5-10. Reference dimensions for beam end bearing plates.

bearing pressure on the bottom of the plate tends to curl it upward about the beam flange. The required thickness may be computed readily by the following formula, which does not involve direct computation of bending moment and section modulus:

$$t = \sqrt{\frac{3f_p n^2}{F_b}}$$

where t = thickness of the plate in inches,
f_p = *actual* bearing pressure of the plate on the masonry, in psi or ksi,
F_b = allowable bending stress in the plate (the AISC Specifi-

cation gives the value of F_b as 0.75 F_y; for A36 steel $F_y = 36$ ksi; therefore $F_b = 0.75 \times 36 = 27$ ksi),

$n = \dfrac{B}{2} - k_1$, in inches (see Fig. 5-10b),

k_1 = the distance from the center of the web to the toe of the fillet; values of k_1 for various beam sizes may be found in the tables in the AISC Manual.

The foregoing formula is derived by considering a strip of plate 1 in. wide (Fig. 5-10a) and t in. thick, with a projecting length of n inches, as a cantilever. Because the upward pressure on the steel strip is f_p, the bending moment at distance n from the edge of the plate is

$$M = f_p n \times \frac{n}{2} = \frac{f_p n^2}{2}$$

TABLE 5-11. Allowable Bearing Pressure on Masonry and Concrete

Type of Material and Conditions	Allowable Unit Stress in Bearing, F_p (psi)	(kPa)
Solid brick, unreinforced, type S mortar		
$f'_m = 1500$ psi	170	1200
$f'_m = 4500$ psi	338	2300
Hollow unit masonry, unreinforced, type S mortar,	225	1500
$f'_m = 1500$ psi		
(on net area of masonry)		
Concrete*		
(1) Bearing on full area of support		
$f'_c = 2000$ psi	500	3500
$f'_c = 3000$ psi	750	5000
(2) Bearing on 1/3 or less of support area		
$f'_c = 2000$ psi	750	5000
$f'_c = 3000$ psi	1125	7500

* Stresses for areas between these limits may be determined by direct proportion.

For this strip with rectangular cross section,

$$\frac{I}{c} = \frac{bd^2}{6}$$ (Section 4-5)

and because $b = 1$ in. and $d = t$ in.,

$$\frac{I}{c} = \frac{1 \times t^2}{6} = \frac{t^2}{6}$$

Then from the beam formula,

$$\frac{M}{F_b} = \frac{I}{c}$$ (Section 4-2)

Substituting the values of M and I/c determined above,

$$\frac{f_p n^2}{2} \times \frac{1}{F_b} = \frac{t^2}{6}$$

and

$$t^2 = \frac{6f_p n^2}{2F_b} \quad \text{or} \quad t = \sqrt{\frac{3f_p n^2}{F_b}}$$

When the dimensions of the bearing plate are determined, the beam should be investigated for web crippling on the length $(N + k)$ shown in Fig. 5-10c. This is explained in Section 5-24.

Example 1. A W 21 × 57 of A36 steel transfers an end reaction of 44 kips [196 kN] to a wall built of solid brick by means of a bearing plate of A36 steel. Assume type S mortar and a brick with $f'_m = 1500$ psi. The N dimension of the plate (see Fig. 5-10) is 10 in. [254 mm]. Design the bearing plate.

Solution: In the AISC Manual we find that k_1 for the beam is 0.875 in. [22 mm]. From Table 5-11 the allowable bearing pressure F_p for this wall is 170 psi [1200 kPa]. The required area of the plate is then

$$A = \frac{R}{F_p} = \frac{44,000}{170} = 259 \text{ in}^2$$

$$\left[A = \frac{196 \times 10^6}{1200} = 163,333 \text{ mm}^2 \right]$$

Then, because $N = 10$ in. [254 mm],

$$B = \frac{259}{10} = 25.9 \text{ in}^2$$

$$\left[= \frac{163,333}{254} = 643 \text{ mm} \right]$$

which is rounded off to 26 in. [650 mm].

With the true dimensions of the plate, we now compute the true bearing pressure:

$$f_p = \frac{R}{A} = \frac{44,000}{10 \times 26} = 169 \text{ psi}$$

$$\left[f_p = \frac{196 \times 10^6}{254 \times 650} = 1187 \text{ kPa} \right]$$

To find the thickness, we first determine the value of n.

$$n = \frac{B}{2} - k_1 = \frac{26}{2} - 0.875 = 12.125 \text{ in.}$$

$$\left[n = \frac{650}{2} - 22 = 303 \text{ mm} \right]$$

Then

$$t = \sqrt{\frac{3f_p n^2}{F_b}} = \sqrt{\frac{3 \times 169 \times (12.125)^2}{27,000}} = \sqrt{2.760} = 1.66 \text{ in.}$$

$$\left[t = \sqrt{\frac{3 \times 1187 \times (303)^2}{186,000}} = \sqrt{1758} = 42 \text{ mm} \right]$$

The complete design for this problem would include a check of the web crippling in the beam. This has been done already as Example 1 in Section 5-23.

Problem 5-24-A*.   A W 14 × 30 with a reaction of 20 kips [89 kN] rests on a brick wall with brick of $f'_m = 1500$ psi and type S mortar. The beam has a bearing length of 8 in. [203 mm] parallel to the length of the beam. If the bearing plate is A36 steel, determine its dimensions. ($k_1 = 0.625$ in.)

Problem 5-24-B.   A wall of brick with f'_m of 1500 psi and type S mortar supports a W 18 × 50 of A36 steel. The beam reaction is 25 kips [111 kN], and the bearing length N is 9 in. [229 mm]. Design the beam bearing plate. ($k_1 = 0.8125$ in.)

5-25 Open-Web Steel Joists

Open-web steel joists are shop-fabricated, parallel chord trusses used for direct support of roof and floor decks. (See Fig. 5-11.) They are produced in a wide range of sizes with various fabrication details by a number of manufacturers. Although specific data should be obtained from individual suppliers, design and fabrication usually comply with the requirements of the standard specifications of the Steel Joist Institute.

Joists are produced in two series, designated J and H. J series joists are fabricated typically from A36 steel, whereas H series joists use a steel with a yield stress of 50 ksi [350 MPa]. In addition to the basic J and H series joists, there are two series of large joists designated LJ and LH (for long span) and DLJ and DLH (for deep, long span).

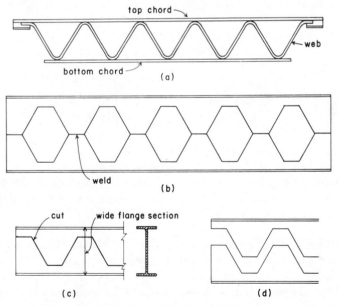

FIGURE 5-11. Open-web steel joists.

Table 5-12 is adapted from the standard tables of the Steel Joist Institute. This table lists the range of joist sizes available in the basic H series. (*Note:* A few of the heavier sizes have been omitted to shorten the table.) Joists are identified by a three-unit designation. The first number indicates the overall depth of the joist, the letter tells the series, and the second number gives the class of size of the members used; the higher the number, the heavier the joist.

Table 5-12 can be used to select the proper joist for a determined load and span condition. There are usually two entries in the table for each span; the first number represents the total load capacity of the joist, and the number in parentheses identifies the load that will produce a deflection of $\frac{1}{360}$ of the span. The following examples illustrate the use of the table data for some typical design situations.

Example 1.   Open-web steel joists are to be used to support a roof with a unit live load of 20 psf and a unit dead load of 15 psf (not including the weight of the joists) on a span of 40 ft. Joists are spaced at 6 ft center to center. Select the lightest joist if deflection under live load is limited to $\frac{1}{360}$ of the span.
Solution: We first determine the load per ft on the joist:

Live load	$6 \times 20 =$	120 lb/ft
Dead load	$6 \times 15 =$	90 lb/ft
Total load		210 lb/ft

We then scan the entries in Table 5-12 for the joists that will just carry these loads, noting that the joist weight must be deducted from the entry for total capacity. The possible choices for this example are summarized in Table 5-13. Although the joist weights are all very close, the 24H7 is the lightest choice.

Example 2.   Open-web steel joists are to be used for a floor with a unit live load of 75 psf and a unit dead load of 40 psf (not including the joists) on a span of 30 ft. Joists are 2 ft center to center, and deflection is limited to $\frac{1}{360}$ of the span under live load only and to $\frac{1}{240}$ of the span under total load. Determine (1) the

TABLE 5-12. Load Table for Open Web Steel Joists

Joist Designation	8H3	10H3	10H4	12H3	12H4	12H5	12H6	14H3	14H4	14H5	14H6	14H7	16H4	16H5	16H6	16H7	16H8
Joist Weight (lb/ft)	5.0	5.0	6.1	5.2	6.2	7.1	8.2	5.5	6.5	7.4	8.6	10.0	6.6	7.8	8.6	10.3	11.4
Span (ft)																	
8	600																
10	480 (460)	500	560														
12	400 (266)	417	467	467	533	600	650										
14	310 (167)	357 (270)	400 (334)	400 (393)	457	514	557	457	500	543	600	657					
16	232 (112)	302 (181)	350 (223)	350 (264)	400 (345)	450 (404)	488 (480)	400 (366)	438	475	525	575	475	538	575	613	650
18		239 (127)	305 (157)	288 (185)	356 (242)	400 (284)	433 (337)	340 (257)	389 (336)	422 (393)	467	511	422 (413)	478	511	544	578
20		193 (92)	247 (114)	233 (135)	300 (177)	360 (207)	390 (246)	275 (187)	350 (245)	380 (287)	420 (342)	460 (403)	368 (301)	430 (370)	460 (437)	490	520
22				193 (101)	248 (133)	306 (155)	355 (185)	227 (141)	292 (184)	345 (215)	382 (257)	418 (302)	304 (226)	391 (278)	418 (328)	445 (395)	473 (454)
24				162 (78)	208 (102)	257 (120)	301 (142)	191 (108)	245 (142)	300 (166)	350 (198)	383 (233)	256 (174)	334 (214)	383 (253)	408 (304)	433 (350)
26								163 (85)	209 (111)	255 (131)	303 (156)	354 (183)	218 (137)	285 (169)	339 (199)	377 (239)	400 (275)
28								140 (68)	180 (89)	220 (104)	261 (125)	314 (147)	188 (110)	246 (135)	293 (159)	350 (192)	371 (220)
30													164 (89)	214 (110)	255 (129)	306 (155)	347 (179)
32													144 (74)	188 (90)	224 (107)	269 (128)	311 (148)

TABLE 5-12. (Continued)

Joist Designation	18H5	18H6	18H7	18H8	18H9	18H10	20H5	20H6	20H7	20H8	20H9	20H10	22H6	22H7	22H8	22H9	22H10	22H11
Joist Weight (lb/ft)	8.0	9.2	10.4	11.6	12.6	14.0	8.4	9.6	10.7	12.2	13.2	14.6	9.7	10.7	12.0	13.8	15.2	16.9
Span (ft)																		
18	500	533	578	600	621	629	480	510	540	560	640	636						
20	450	480	520	540	590	629	436	464	491	509	582	636						
22	409 (356)	436 (420)	473	491	536	600	400 (335)	425 (382)	450	467	533	583	491	509	527	609	626	648
24	375 (274)	400 (324)	433 (388)	450 (444)	492 (484)	550 (546)	360 (263)	392 (300)	415 (365)	431	492 (476)	538	446 (446)	467	483	558	600	648
26	321 (216)	369 (255)	400 (305)	415 (349)	454 (380)	508 (429)	310 (211)	345 (240)	386 (292)	400 (352)	457 (381)	500 (431)	415 (351)	431 (426)	446	515	554	623
28	276 (173)	326 (204)	371 (244)	386 (280)	421 (305)	471 (344)	270 (171)	301 (195)	360 (238)	373 (286)	427 (310)	467 (350)	359 (281)	400 (341)	414	479 (468)	514	579
30	241 (140)	284 (166)	345 (199)	360 (227)	393 (248)	440 (280)	238 (141)	264 (161)	325 (196)	350 (236)	400 (255)	438 (288)	313 (228)	373 (277)	387 (343)	447 (381)	480 (428)	540 (487)
32	212 (116)	249 (137)	303 (164)	338 (187)	369 (204)	413 (230)	210 (118)	234 (134)	288 (163)	329 (196)	376 (213)	412 (240)	275 (188)	342 (228)	363 (282)	419 (314)	450 (352)	506 (401)
34	187 (96)	221 (114)	269 (136)	311 (156)	347 (170)	388 (192)	188 (99)	209 (113)	257 (137)	310 (166)	356 (179)	389 (203)	243 (157)	303 (190)	341 (235)	394 (261)	424 (294)	476 (335)
36	167 (81)	197 (96)	240 (115)	278 (132)	323 (143)	363 (162)	169 (84)	187 (96)	230 (117)	278 (141)	324 (153)	364 (172)	217 (132)	271 (160)	322 (198)	372 (220)	400 (247)	450 (282)
38							152 (72)	169 (82)	208 (100)	251 (121)	292 (131)	329 (148)	195 (112)	243 (136)	301 (169)	353 (187)	379 (210)	426 (240)
40													176 (96)	219 (117)	272 (145)	323 (161)	360 (180)	405 (205)
42													159 (83)	199 (101)	247 (125)	293 (139)	330 (156)	381 (177)
44													145 (72)	181 (88)	225 (109)	267 (121)	301 (136)	347 (154)

Joist Designation	30H11	30H10	30H9	30H8	28H11	28H10	28H9	28H8	26H11	26H10	26H9	26H8	24H11	24H10	24H9	24H8	24H7	24H6
Joist Weight (lb/ft)	18.8	17.3	15.4	14.2	18.3	16.8	15.2	13.5	17.9	16.2	14.8	12.8	17.5	15.5	14.0	12.7	11.5	10.3
Span (ft)																		
24													631	625	583	500	483	467
26													631	577	538	462	446	431
28					600	550	514	479	638	585	554	515	586	536	500	429	414 (406)	393 (336)
30	580	540	500	453	560	513	480	447	593	543	514	479	547	500	467 (457)	400	387 (330)	342 (273)
32	544	506	469	425	525	481	450	419	553	507	480	447	513 (482)	469 (423)	438 (376)	375 (339)	363 (272)	301 (225)
34	512	476	441	400	494	453	424	394 (370)	519	475	450 (445)	419 (380)	482 (402)	441 (353)	412 (314)	353 (283)	332 (227)	266 (188)
36	483	450	417	378 (359)	467	428 (410)	400 (364)	372 (311)	488 (476)	447 (417)	424 (371)	394 (317)	456 (339)	417 (297)	389 (264)	333 (238)	296 (191)	238 (158)
40	435 (395)	405 (345)	375 (306)	340 (262)	420 (342)	385 (299)	360 (266)	335 (227)	461 (401)	422 (352)	400 (312)	372 (267)	410 (247)	375 (217)	350 (193)	298 (174)	240 (139)	193 (115)
44	395 (297)	368 (259)	341 (230)	309 (196)	382 (257)	350 (225)	327 (200)	291 (171)	415 (292)	380 (256)	360 (228)	327 (194)	373 (186)	330 (163)	293 (145)	247 (130)	198 (105)	159 (87)
48	363 (229)	338 (200)	311 (177)	263 (151)	350 (198)	321 (173)	289 (154)	245 (131)	377 (220)	345 (193)	319 (171)	270 (146)	320 (143)	277 (125)	246 (111)	207 (100)	167 (81)	134 (67)
52	335 (180)	298 (157)	265 (139)	224 (119)	321 (156)	277 (136)	247 (121)	209 (103)	346 (169)	301 (148)	268 (132)	227 (112)						
56	297 (144)	257 (126)	229 (112)	193 (95)	276 (125)	239 (109)	213 (97)	180 (83)	297 (133)	256 (117)	228 (104)	193 (88)						
60	259 (117)	224 (102)	199 (91)	168 (77)														

Note: Loads in pounds per ft of joist span; first entry represents the total joist capacity; entry in parentheses is the load that produces a maximum deflection of $\frac{1}{360}$ of the span. Loads above the heavy line are governed by shear; loads below the dotted line are used for roofs only. Data adapted from more extensive tables published in the Standard Specifications, Load Tables, and Weight Tables for Steel Joists and Joist Girders, 1982 ed. (Ref. 6) with permission of the publishers, the Steel Joist Institute. The Steel Joist Institute publishes both Specifications and Load Tables; each of these contains standards which are to be used in conjunction with one another.

TABLE 5-13. Possible Choices for the Roof Joist

Load Condition	Load per Foot for the Indicated Joists			
	20H8	22H8	24H7	26H8
Total capacity (from Table 5-12)	251	272	240	327
Joist weight (from Table 5-12)	12.2	12.0	11.5	12.8
Net usable capacity	238.8	260	228.5	314.2
Load for 1/360 deflection (from Table 5-12)	121	145	139	194

lightest joist possible and (2) the joist with the least depth possible.

Solution: As in the preceding example, we first find the loads on the joist:

$$
\begin{array}{lll}
\text{Live load} & 2 \times 75 = & 150 \text{ lb/ft} \\
\text{Dead load} & 2 \times 40 = & \underline{80} \text{ lb/ft} \\
\text{Total load} & & 230 \text{ lb/ft}
\end{array}
$$

To satisfy the deflection criteria, we must find a table entry in parentheses of 150 lb/ft (for live load only) or $\frac{240}{360} \times 230 = 153$ lb/ft (for total load). The possible choices obtained from scanning Table 5-12 are summarized in Table 5-14, from which we observe

for (1), the lightest joist is the 20H5,

for (2), the shallowest joist is the 18H6.

Note that, although an entry is given for the 16H7, Table 5-12 recommends it for roofs only for this span.

Problem 5-25-A*.   Open-web steel joists are to be used for a roof with a live load of 25 psf and a dead load of 20 psf (not including joists) on a span of 48 ft. Joists are 4 ft center to center, and the deflection under live load is limited to $\frac{1}{360}$ of the span. Select the lightest joist possible.

TABLE 5-14. Possible Choices for the Floor Joist

Load Condition	Load per Foot for the Indicated Joists		
	16H7	18H6	20H5
Total capacity (from Table 5-12)	306	284	270
Joist weight (from Table 5-12)	10.3	9.2	8.4
Net usable capacity	295.7	274.8	261.6
Load for 1/360 deflection (from Table 5-12)	155	166	171

Problem 5-25-B. Open-web steel joists are to be used for a roof with a live load of 30 psf and a dead load of 18 psf (not including joists) on a span of 44 ft. Joists are 5 ft center to center, and deflection under live load is limited to $\frac{1}{360}$ of the span. Select the lightest joist.

Problem 5-25-C*. Open-web steel joists are to be used for a floor with a live load of 50 psf and a dead load of 45 psf (not including joists) on a span of 36 ft. Joists are 2 ft center to center, and deflection is limited to $\frac{1}{360}$ of the span under live load and to $\frac{1}{240}$ of the span under total load. Select (1) the lightest possible joist and (2) the shallowest possible joist.

Problem 5-25-D. Repeat Problem 5-25-C except that the live load is 100 psf, the dead load is 35 psf, and the span is 26 ft.

6

Steel Columns

|||||||||||||||||||||||||||||||||||||||||||||||||||||||||

6-1 Introduction

A column or strut is a compression member, the length of which is several times greater than its least lateral dimension. The term column denotes a relatively heavy vertical member, whereas the lighter vertical and inclined members, such as braces and the compression members of roof trusses, are called struts.

Under the discussion of direct stress in Section 1-3, it was pointed out that the unit compressive stress in the short block shown in Fig. 1-1b could be expressed by the direct stress formula $f_a = P/A$; but it was also stated that this relationship became invalid as the ratio of the length of the compression member to its least width increased. To pursue this further, consider a small block of steel 1 in. by 1 in. in cross section and 1 or 2 in. high. If the allowable compressive stress is 20 ksi, the block will safely support a load P (Fig. 6-1a) of 20 kips. If, however, we consider a bar of the same cross section with a length of, say, 30–40 in., we find that the value of P it will sustain is considerably less because of the tendency of this more slender bar to buckle or bend (Fig. 6-1b). Therefore, in columns the element of *slenderness* must be taken into account when we determine allowable loads. A short column or block fails by crushing, but long, slender columns fail by stresses that result from bending.

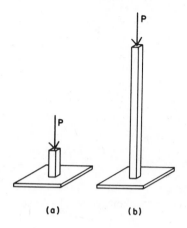

(a)   (b)

FIGURE 6-1.   Effect of column slenderness.

6-2   Column Shapes

Because of the tendency to bend, the safe load on a column depends not only on the number of square inches in the cross section but also on the manner in which the material is distributed with respect to the axes of the cross section; that is, the *shape* of the column section is an important factor. An axially loaded column tends to bend in a plane perpendicular to the axis of the cross section about which the moment of inertia is least. Since column cross sections are seldom symmetrical with respect to both major axes, the ideal section would be one in which the moment of inertia for each major axis is equal. Pipe columns and structural tubing meet this condition, but their use is somewhat limited because of difficulties in making beam connections.

Of the two major axes of a Standard I-beam, the moment of inertia about the axis parallel to the web is much the smaller; hence for the amount of material in the cross section, I-beams are not economical shapes when used as columns or struts. In former years built-up sections such as Fig. 6-2*c* and *d* were used extensively, but wide flange sections (Fig. 6-2*a*) are now rolled in a large variety of sizes and are used universally because they require a minimum of fabrication. They are sometimes called H-columns. For excessive loads or unusual conditions, plates are

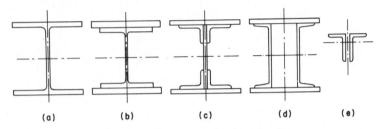

FIGURE 6-2. Typical steel column sections.

welded to the flanges of wide flange sections to give added strength (Fig. 6-2*b*). The compression members of steel trusses are often formed of two angles, as shown in Fig. 6-2*e*.

6-3 Slenderness Ratio

In the design of timber columns, the term *slenderness ratio* is defined as the unbraced length divided by the dimension of the least side, both in inches. For structural shapes such as those shown in Fig. 6-2, the least lateral dimension is not an accurate criterion; and the radius of gyration r, which relates more precisely to the stiffness of columns in general, is used in steel column design. As discussed in Section 4-9, $r = \sqrt{I/A}$. For rolled sections the value of the radius of gyration with respect to both major axes is given in the tables of properties for designing. For built-up sections it may be necessary to compute its value. The slenderness ratio of a steel column is then l/r, where l is the effective length of the column in inches and r is the *least* radius of gyration of the cross section, also in inches. The slenderness ratio for compression members should not exceed 200.

Example. A W 10 × 49 is used as a column whose effective length is 20 ft [6.10 m]. Compute the slenderness ratio.
Solution: Reference to Table 4-1 reveals that the radii of gyration for this section are $r_X = 4.35$ in. and $r_Y = 2.54$ in. Therefore the *least* radius of gyration is 2.54 in.

Because the effective length of the column is 20 ft [6.10 m], the slenderness ratio is

$$\frac{l}{r} = \frac{20 \times 12}{2.54} = 94.5 \left[\frac{6.10 \times 10^3}{64.5} = 94.6 \right]$$

It should be remembered that the tendency to bend due to buckling under the compression load is in a direction perpendicular to the axis about which the radius of gyration is least.

Problem 6-3-A*. The effective length of a W 8 × 31 used as a column is 16 ft [4.88 m]. Compute the slenderness ratio.

Problem 6-3-B. What is the slenderness ratio of a column whose section is a W 12 × 58 with an effective length of 30 ft [9.14 m]?

6-4 Effective Column Length

The AISC Specification requires that, in addition to the unbraced length of a column, the condition of the ends must be given consideration. The slenderness ratio is Kl/r, where K is a factor dependent on the restraint at the ends of a column and the means available to resist lateral motion. Figure 6-3 shows diagrammatically six idealized conditions in which joint rotation and joint translation are illustrated. The term K is the ratio of the effective column length to the actual unbraced length. For average conditions in building construction, the value of K is taken as 1; therefore the slenderness ratio Kl/r becomes simply l/r. (See Fig. 6-3d).

6-5 Column Formulas

The AISC Specification gives the following requirements for use in the design of compression members. The allowable unit stresses shall not exceed the following values:

1. On the gross section of axially loaded compression members, when Kl/r, the largest effective slenderness ratio of any unbraced segment, is less than C_c,

$$F_a = \frac{[1 - (Kl/r)^2/2C_c^2]F_y}{FS} \quad \text{(Formula 6-5-1)}$$

	(a)	(b)	(c)	(d)	(e)	(f)
Buckled shape of column is shown by dashed line	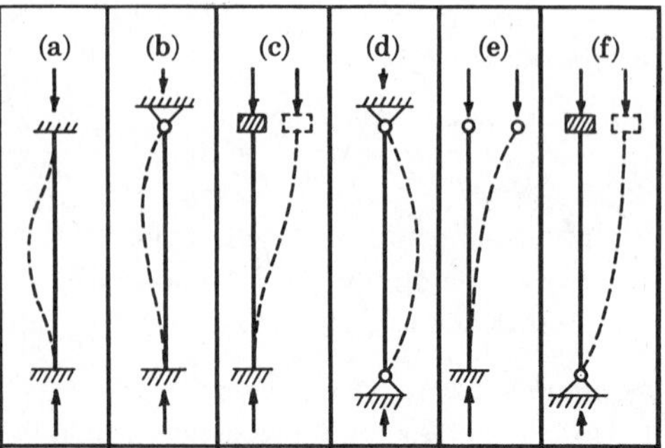					
Theoretical K value	0.5	0.7	1.0	1.0	2.0	2.0
Recommended design value when ideal conditions are approximated	0.65	0.80	1.2	1.0	2.10	2.0
End condition code		Rotation fixed and translation fixed Rotation free and translation fixed Rotation fixed and translation free Rotation free and translation free				

FIGURE 6-3.   Determination of effective column length. Reprinted from the *Manual of Steel Construction*, 8th ed., with permission of the publishers, American Institute of Steel Construction.

where

$$FS = \text{factor of safety} = \frac{5}{3} + \frac{3(Kl/r)}{8C_c} - \frac{(Kl/r)^3}{8C_c^3}$$

and

$$C_c = \sqrt{\frac{2\pi^2 E}{F_y}}$$

2.   On the gross section of axially loaded columns, when Kl/r exceeds C_c,

$$F_a = \frac{12\pi^2 E}{23(Kl/r)^2} \qquad \text{(Formula 6-5-2)}$$

3.   On the gross section of axially loaded bracing and secondary members, when l/r exceeds 120 (for this case, K is

taken as unity),

$$F_{as} = \frac{F_a(\text{by Formula 6-5-1 or 6-5-2})}{1.6 - l/200r} \quad \text{(Formula 6-5-3)}$$

In these formulas

F_a = axial compression stress permitted in the absence of bending stress,

K = effective length factor (see Section 6-4),

l = actual unbraced length,

r = governing radius of gyration (usually the least),

$C_c = \sqrt{2\pi^2 E/F_y}$; for A36 steel $C_c = 126.1$,

F_y = minimum yield point of the steel being used (for A36 steel $F_y = 36,000$),

FS = factor of safety (see above),

E = modulus of elasticity of structural steel, 29,000 ksi,

F_{as} = axial compressive stress permitted in the absence of bending stress for bracing and other secondary members.

To determine the allowable axial load that a main column will support, F_a, the allowable unit stress, is computed by Formula (1) or (2), and this stress is multiplied by the cross-sectional area of the column. If the column is a secondary member or is used for bracing, Formula (3) gives the allowable unit stress; these allowable unit stresses are somewhat greater than those permitted for main members. Table 6-1 gives allowable stresses computed in accordance with these formulas. It should be examined carefully because it will be of great assistance. Note particularly that this table is for use with A36 steel; tables based on other grades of steel are contained in the AISC Manual.

6-6   Allowable Column Loads

The allowable axial load that a steel column will support is found by multiplying the allowable unit stress by the cross-sectional area of the column. The value of Kl/r is first determined, and by referring to Table 6-1, we can establish the allowable unit stress.

TABLE 6-1. Allowable Unit Stresses for Columns of A36 Steel (in ksi)

Main and Secondary Members Kl/r not over 120						Main Members Kl/r 121 to 200				Secondary Members[a] l/r 121 to 200			
$\dfrac{Kl}{r}$	F_a (ksi)	$\dfrac{Kl}{r}$	F_a (ksi)	$\dfrac{Kl}{r}$	F_a (ksi)	$\dfrac{Kl}{r}$	F_a (ksi)	$\dfrac{Kl}{r}$	F_a (ksi)	$\dfrac{l}{r}$	F_{as} (ksi)	$\dfrac{l}{r}$	F_{as} (ksi)
1	21.56	41	19.11	81	15.24	121	10.14	161	5.76	121	10.19	161	7.25
2	21.52	42	19.03	82	15.13	122	9.99	162	5.69	122	10.09	162	7.20
3	21.48	43	18.95	83	15.02	123	9.85	163	5.62	123	10.00	163	7.16
4	21.44	44	18.86	84	14.90	124	9.70	164	5.55	124	9.90	164	7.12
5	21.39	45	18.78	85	14.79	125	9.55	165	5.49	125	9.80	165	7.08
6	21.35	46	18.70	86	14.67	126	9.41	166	5.42	126	9.70	166	7.04
7	21.30	47	18.61	87	14.56	127	9.26	167	5.35	127	9.59	167	7.00
8	21.25	48	18.53	88	14.44	128	9.11	168	5.29	128	9.49	168	6.96
9	21.21	49	18.44	89	14.32	129	8.97	169	5.23	129	9.40	169	6.93
10	21.16	50	18.35	90	14.20	130	8.84	170	5.17	130	9.30	170	6.89
11	21.10	51	18.26	91	14.09	131	8.70	171	5.11	131	9.21	171	6.85
12	21.05	52	18.17	92	13.97	132	8.57	172	5.05	132	9.12	172	6.82
13	21.00	53	18.08	93	13.84	133	8.44	173	4.99	133	9.03	173	6.79
14	20.95	54	17.99	94	13.72	134	8.32	174	4.93	134	8.94	174	6.76
15	20.89	55	17.90	95	13.60	135	8.19	175	4.88	135	8.86	175	6.73
16	20.83	56	17.81	96	13.48	136	8.07	176	4.82	136	8.78	176	6.70
17	20.78	57	17.71	97	13.35	137	7.96	177	4.77	137	8.70	177	6.67
18	20.72	58	17.62	98	13.23	138	7.84	178	4.71	138	8.62	178	6.64
19	20.66	59	17.53	99	13.10	139	7.73	179	4.66	139	8.54	179	6.61
20	20.60	60	17.43	100	12.98	140	7.62	180	4.61	140	8.47	180	6.58
21	20.54	61	17.33	101	12.85	141	7.51	181	4.56	141	8.39	181	6.56
22	20.48	62	17.24	102	12.72	142	7.41	182	4.51	142	8.32	182	6.53
23	20.41	63	17.14	103	12.59	143	7.30	183	4.46	143	8.25	183	6.51
24	20.35	64	17.04	104	12.47	144	7.20	184	4.41	144	8.18	184	6.49
25	20.28	65	16.94	105	12.33	145	7.10	185	4.36	145	8.12	185	6.46
26	20.22	66	16.84	106	12.20	146	7.01	186	4.32	146	8.05	186	6.44
27	20.15	67	16.74	107	12.07	147	6.91	187	4.27	147	7.99	187	6.42
28	20.08	68	16.64	108	11.94	148	6.82	188	4.23	148	7.93	188	6.40
29	20.01	69	16.53	109	11.81	149	6.73	189	4.18	149	7.87	189	6.38
30	19.94	70	16.43	110	11.67	150	6.64	190	4.14	150	7.81	190	6.36
31	19.87	71	16.33	111	11.54	151	6.55	191	4.09	151	7.75	191	6.35
32	19.80	72	16.22	112	11.40	152	6.46	192	4.05	152	7.69	192	6.33
33	19.73	73	16.12	113	11.26	153	6.38	193	4.01	153	7.64	193	6.31
34	19.65	74	16.01	114	11.13	154	6.30	194	3.97	154	7.59	194	6.30
35	19.58	75	15.90	115	10.99	155	6.22	195	3.93	155	7.53	195	6.28
36	19.50	76	15.79	116	10.85	156	6.14	196	3.89	156	7.48	196	6.27
37	19.42	77	15.69	117	10.71	157	6.06	197	3.85	157	7.43	197	6.26
38	19.35	78	15.58	118	10.57	158	5.98	198	3.81	158	7.39	198	6.24
39	19.27	79	15.47	119	10.43	159	5.91	199	3.77	159	7.34	199	6.23
40	19.19	80	15.36	120	10.28	160	5.83	200	3.73	160	7.29	200	6.22

Source: Reprinted from the *Manual of Steel Construction,* 8th ed., with permission of the publishers, American Institute of Steel Construction.

[a] K taken as 1.0 for secondary members.

Example 1.   A W 12 × 65 is used as a column with an unbraced length of 16 ft [4.88 m]. Compute the allowable load.

Solution: Referring to Table 4-1, we find that $A = 19.1$ in² [12,323 mm²], $r_x = 5.28$ in. [134 mm], and $r_Y = 3.02$ in. [76.7 mm]. Because the column is unbraced with respect to both axes, the least radius of gyration is used to determine the slenderness ratio. Also, with no qualifying conditions given, $K = 1.0$. The slenderness ratio is then

$$\frac{Kl}{r} = \frac{1 \times 16 \times 12}{3.02} = 63.6$$

$$\left[\frac{4.88 \times 10^3}{76.7} = 63.6\right]$$

In design work it is usually considered acceptable to round the slenderness ratio off in front of the decimal point because a typical lack of accuracy in the design data does not warrant greater precision. Therefore we consider the slenderness ratio to be 64; the allowable stress given in Table 6-1 is $F_a = 17.04$ ksi [117.5 MPa]. The allowable load on the column is then

$$P = A \times F_a = 19.1 \times 17.04 = 325.5 \text{ kips}$$

$$\left[P = \frac{12,323 \times 117.5}{10^3} = 1448 \text{ kN}\right]$$

Note: In the following problems assume A36 steel and a value of $K = 1$.

Problem 6-6-A*.   Compute the allowable axial load on a W 10 × 88 column with an unbraced height of 15 ft [4.57 m].

Problem 6-6-B.   A W 12 × 79 used as a column has an unbraced height of 22 ft [6.71 m]. Compute the allowable axial load.

6-7   Design of Steel Columns

In practice the design of steel columns is accomplished largely by the use of safe-load tables; if these tables are not available, design is carried out by the trial method. Data include the load and length of the column. The designer selects a trial cross section on the basis of experience and judgment and, by means of a column formula, computes the allowable load that it can support. If this load is less than the actual load the column will be required to support, the trial section is too small and another section is tested in a similar manner.

Table 6-2 lists allowable loads on a number of column sections. It has been compiled from more extensive tables in the AISC Manual, and the loads are computed in accordance with the formulas in Section 6-5. Note particularly that these allowable loads are for main members of A36 steel. The significance of the bending factors given at the extreme right of the table is considered in Section 6-11.

To illustrate the use of Table 6-2, refer to Example 1 of Section 6-6. This problem asked for the allowable load that could be supported by a W 12 × 65 of A36 steel with an unbraced height of 16 ft. Referring to Table 6-2, we see at a glance that the allowable axial load is 326 kips, which agrees closely with the value found by computation.

Although the designer may select the proper column section by merely referring to the safe-load tables, it is well to understand the application of the formulas by means of which the tables have been computed. To that end the *design procedure* is outlined below. When the design load and length have been ascertained, the following steps are taken:

Step 1: Assume a trial section and note from the table of properties the cross-sectional area and the least radius of gyration.

Step 2: Compute the slenderness ratio Kl/r, l being the unsupported length of the column in inches. For the value of K see Section 6-4.

Step 3: Compute F_a, the allowable unit stress, by using a column formula or Table 6-1.

Step 4: Multiply F_a found in Step (3) by the area of the column cross section. This gives the allowable load *on the trial column section*.

Step 5: Compare the allowable load found in Step (4) with the design load. If the allowable load on the trial section is less than the design load (or if it is so much greater that it makes use of the

TABLE 6-2. Allowable Column Loads for Selected W Shapes

Shape	Effective Length (KL) in Feet										Bending Factor	
	8	9	10	11	12	14	16	18	20	22	B_x	B_y
M 4 × 13	48	42	35	29	24	18					0.727	2.228
W 4 × 13	52	46	39	33	28	20	16				0.701	2.016
W 5 × 16	74	69	64	58	52	40	31	24	20		0.550	1.560
M 5 × 18.9	85	78	71	64	56	42	32	25			0.576	1.768
W 5 × 19	88	82	76	70	63	48	37	29	24		0.543	1.526
W 6 × 9	33	28	23	19	16	12					0.482	2.414
W 6 × 12	44	38	31	26	22	16					0.486	2.367
W 6 × 16	62	54	46	38	32	23	18				0.465	2.155
W 6 × 15	75	71	67	62	58	48	38	30	24	20	0.456	1.424
M 6 × 20	98	92	87	81	74	61	47	37	30	25	0.453	1.510
W 6 × 20	100	95	90	85	79	67	54	42	34	28	0.438	1.331
W 6 × 25	126	120	114	107	100	85	69	54	44	36	0.440	1.308
W 8 × 24	124	118	113	107	101	88	74	59	48	39	0.339	1.258
W 8 × 28	144	138	132	125	118	103	87	69	56	46	0.340	1.244
W 8 × 31	170	165	160	154	149	137	124	110	95	80	0.332	0.985
W 8 × 35	191	186	180	174	168	155	141	125	109	91	0.330	0.972
W 8 × 40	218	212	205	199	192	127	160	143	124	104	0.330	0.959
W 8 × 48	263	256	249	241	233	215	196	176	154	131	0.326	0.940
W 8 × 58	320	312	303	293	283	263	240	216	190	162	0.329	0.934
W 8 × 67	370	360	350	339	328	304	279	251	221	190	0.326	0.921
W 10 × 33	179	173	167	161	155	142	127	112	95	78	0.277	1.055
W 10 × 39	213	206	200	193	186	170	154	136	116	97	0.273	1.018
W 10 × 45	247	240	232	224	216	199	180	160	138	115	0.271	1.000
W 10 × 49	279	273	268	262	256	242	228	213	197	180	0.264	0.770
W 10 × 54	306	300	294	288	281	267	251	235	217	199	0.263	0.767
W 10 × 60	341	335	328	321	313	297	280	262	243	222	0.264	0.765
W 10 × 68	388	381	373	365	357	339	320	299	278	255	0.264	0.758
W 10 × 77	439	431	422	413	404	384	362	339	315	289	0.263	0.751
W 10 × 88	504	495	485	475	464	442	417	392	364	335	0.263	0.744
W 10 × 100	573	562	551	540	428	503	476	446	416	383	0.263	0.735
W 10 × 112	642	631	619	606	593	565	535	503	469	433	0.261	0.726
W 12 × 40	217	210	203	196	188	172	154	135	114	94	0.227	1.073
W 12 × 45	243	235	228	220	211	193	173	152	129	106	0.227	1.065
W 12 × 50	271	263	254	246	236	216	195	171	146	121	0.227	1.058
W 12 × 53	301	295	288	282	275	260	244	227	209	189	0.221	0.813
W 12 × 58	329	322	315	308	301	285	268	249	230	209	0.218	0.794
W 12 × 65	378	373	367	361	354	341	326	311	294	277	0.217	0.656
W 12 × 72	418	412	406	399	392	377	361	344	326	308	0.217	0.651
W 12 × 79	460	453	446	439	431	415	398	379	360	339	0.217	0.648
W 12 × 87	508	501	493	485	477	459	440	420	398	376	0.217	0.645
W 12 × 96	560	552	544	535	526	506	486	464	440	416	0.215	0.635
W 12 × 106	620	611	602	593	583	561	539	514	489	462	0.215	0.633

TABLE 6-2. (Continued)

Shape	8	10	12	14	16	18	20	22	24	26	B_x	B_y
W 12 × 120	702	692	660	636	611	584	555	525	493	460	0.217	0.630
W 12 × 136	795	772	747	721	693	662	630	597	561	524	0.215	0.621
W 12 × 152	891	866	839	810	778	745	710	673	633	592	0.214	0.614
W 12 × 170	998	970	940	908	873	837	798	757	714	668	0.213	0.608
W 12 × 190	1115	1084	1051	1016	978	937	894	849	802	752	0.212	0.600
W 12 × 210	1236	1202	1166	1127	1086	1042	995	946	894	840	0.212	0.594
W 12 × 230	1355	1319	1280	1238	1193	1145	1095	1041	985	927	0.211	0.589
W 12 × 252	1484	1445	1403	1358	1309	1258	1203	1146	1085	1022	0.210	0.583
W 12 × 279	1642	1600	1554	1505	1452	1396	1337	1275	1209	1141	0.208	0.573
W 12 × 305	1799	1753	1704	1651	1594	1534	1471	1404	1333	1260	0.206	0.564
W 12 × 336	1986	1937	1884	1827	1766	1701	1632	1560	1484	1404	0.205	0.558
W 14 × 43	230	215	199	181	161	140	117	96	81	69	0.201	1.115
W 14 × 48	258	242	224	204	182	159	133	110	93	79	0.201	1.102
W 14 × 53	286	268	248	226	202	177	149	123	104	88	0.201	1.091
W 14 × 61	345	330	314	297	278	258	237	214	190	165	0.194	0.833
W 14 × 68	385	369	351	332	311	289	266	241	214	186	0.194	0.826
W 14 × 74	421	403	384	363	341	317	292	265	236	206	0.195	0.820
W 14 × 82	465	446	425	402	377	351	323	293	261	227	0.196	0.823
W 14 × 90	536	524	511	497	482	466	449	432	413	394	0.185	0.531
W 14 × 99	589	575	561	546	529	512	494	475	454	433	0.185	0.527
W 14 × 109	647	633	618	601	583	564	544	523	501	478	0.185	0.523
W 14 × 120	714	699	682	663	644	623	601	578	554	528	0.186	0.523
W 14 × 132	786	768	750	730	708	686	662	637	610	583	0.186	0.521
W 14 × 145	869	851	832	812	790	767	743	718	691	663	0.184	0.489
W 14 × 159	950	931	911	889	865	840	814	786	758	727	0.184	0.485
W 14 × 176	1054	1034	1011	987	961	933	904	874	842	809	0.184	0.484
W 14 × 193	1157	1134	1110	1083	1055	1025	994	961	927	891	0.183	0.477
W 14 × 211	1263	1239	1212	1183	1153	1121	1087	1051	1014	975	0.183	0.477
W 14 × 233	1396	1370	1340	1309	1276	1241	1204	1165	1124	1081	0.183	0.472
W 14 × 257	1542	1513	1481	1447	1410	1372	1331	1289	1244	1198	0.182	0.470
W 14 × 283	1700	1668	1634	1597	1557	1515	1471	1425	1377	1326	0.181	0.465
W 14 × 311	1867	1832	1794	1754	1711	1666	1618	1568	1515	1460	0.181	0.459
W 14 × 342		2022	1985	1941	1894	1845	1793	1738	1681	1621	0.181	0.457
W 14 × 370		2181	2144	2097	2047	1995	1939	1881	1820	1756	0.180	0.452
W 14 × 398		2356	2304	2255	2202	2146	2087	2025	1961	1893	0.178	0.447
W 14 × 426		2515	2464	2411	2356	2296	2234	2169	2100	2029	0.177	0.442
W 14 × 455		2694	2644	2589	2430	2467	2401	2332	2260	2184	0.177	0.441
W 14 × 500		2952	2905	2845	2781	2714	2642	2568	2490	2409	0.175	0.434
W 14 × 550		3272	3206	3142	3073	3000	2923	2842	2758	2670	0.174	0.429
W 14 × 605		3591	3529	3459	3384	3306	3223	3136	3045	2951	0.171	0.421
W 14 × 665		3974	3892	3817	3737	3652	3563	3469	3372	3270	0.170	0.415
W 14 × 730		4355	4277	4196	4100	4019	3923	3823	3718	3609	0.168	0.408

Source: Adapted from data in the *Manual of Steel Construction*, 8th ed., with permission of the publisher, American Institute of Steel Construction.
Note: Loads in kips for shapes of steel with yield stress of 36 ksi [250 MPa].

section uneconomical), try another section and test it in the same manner. The reader should note that, except for assuming a trial section, these operations were carried out in Example 1 of Section 6-6.

Problem 6-7-A*.   Using Table 6-2, select a column section to support an axial load of 148 kips [658 kN] if the unbraced height is 12 ft [3.66 m]. A36 steel is to be used, and *K* is assumed to be 1.

Problem 6-7-B.   Same data as in Problem 6-7-A except that the load is 258 kips [1148 kN] and the unbraced height is 15 ft [4.57 m].

Problem 6-7-C.   Same data as in Problem 6-7-A except that the load is 355 kips [1579 kN] and the unbraced height is 20 ft [6.10 m].

6-8 Steel Pipe Columns

Round steel pipe columns are frequently installed in both steel and wood framed buildings. In routine work they are designed for simple axial load by the use of safe-load tables.

Table 6-3 gives allowable axial loads for standard weight steel pipe columns with a yield point of 36 ksi [250 MPa]. The outside diameters at the head of the table are *nominal* dimensions that designate the pipe sizes. True outside diameters are slightly larger and can be found from the Table 4-5. The AISC Manual contains additional tables that list allowable loads for the two heavier weight groups of steel pipe: extra strong and double-extra strong.

Example.   Using Table 6-3, select a steel pipe column to carry a load of 41 kips [182 kN] if the unbraced height is 12 ft [3.66 m]. Verify the value in the table by computing the allowable axial load.

Solution: Entering Table 6-3 with an effective length of 12 ft, we find that a load of 43 kips can be supported by a 4-in. column. From Table 4-5 we find that this section has $A = 3.17$ in^2 [2045 mm^2] and $r = 1.51$ in. [38.35 mm]. The slenderness ratio, with $K = 1$, is

$$\frac{KL}{r} = \frac{1 \times 12 \times 12}{1.51} = 95.4, \text{ say } 95$$

$$\left[\frac{KL}{r} = \frac{3658}{38.35} = 95.4 \right]$$

Nominal Dia.	12	10	8	6	5	4	3½	3
Wall Thickness	0.375	0.365	0.322	0.280	0.258	0.237	0.226	0.216
Weight per Foot	49.56	40.48	28.55	18.97	14.62	10.79	9.11	7.58
F_y	36 ksi							
0	315	257	181	121	93	68	58	48
6	303	246	171	110	83	59	48	38
7	301	243	168	108	81	57	46	36
8	299	241	166	106	78	54	44	34
9	296	238	163	103	76	52	41	31
10	293	235	161	101	73	49	38	28
11	291	232	158	98	71	46	35	25
12	288	229	155	95	68	43	32	22
13	285	226	152	92	65	40	29	19
14	282	223	149	89	61	36	25	16
15	278	220	145	86	58	33	22	14
16	275	216	142	82	55	29	19	12
17	272	213	138	79	51	26	17	11
18	268	209	135	75	47	23	15	10
19	265	205	131	71	43	21	14	9
20	261	201	127	67	39	19	12	
22	254	193	119	59	32	15	10	
24	246	185	111	51	27	13		
25	242	180	106	47	25	12		
26	238	176	102	43	23			
28	229	167	93	37	20			
30	220	158	83	32	17			
31	216	152	78	30	16			
32	211	148	73	29				
34	201	137	65	25				
36	192	127	58	23				
37	186	120	55	21				
38	181	115	52					
40	171	104	47					

Effective length in feet KL with respect to radius of gyration

Properties								
Area A (in.²)	14.6	11.9	8.40	5.58	4.30	3.17	2.68	2.23
I (in.⁴)	279	161	72.5	28.1	15.2	7.23	4.79	3.02
r (in.)	4.38	3.67	2.94	2.25	1.88	1.51	1.34	1.16
B } Bending factor	0.333	0.398	0.500	0.657	0.789	0.987	1.12	1.29
* a	41.7	23.9	10.8	4.21	2.26	1.08	0.717	0.447

* Tabulated values of a must be multiplied by 10^6.
Note: Heavy line indicates Kl/r of 200.

Source: Reprinted from the *Manual of Steel Construction*, 8th ed., with permission of the publishers, American Institute of Steel Construction.

Referring to Table 6-1, we find that the allowable unit stress F_a for a slenderness ratio of 95 is 13.60 ksi [93.8 MPa]. Thus the allowable axial load is

$$P = A \times F_a = 3.17 \times 13.60 = 43.1 \text{ kips}$$

$$\left[P = \frac{2045 \times 93.8}{10^3} = 192 \text{ kN} \right]$$

which agrees with the table value of 43 kips.

6-9 Structural Tubing Columns

Steel columns are fabricated from structural tubing in both square and rectangular shapes. Square tubing is available in sizes of 2–16 in. and rectangular sizes range from 3 × 2 to 20 × 12 in. Sections are produced with various wall thicknesses, thus allowing a considerable range of structural capacities. Although round pipe is specified by a nominal outside dimension, tubing is specified by its actual outside dimensions.

The AISC Manual contains safe-load tables for square and rectangular tubing based on $F_y = 46$ ksi [317 MPa]. Table 6-4 is a reproduction of one of these tables—for 3- and 4-in.-square tubing—and is presented to illustrate the form of the tables.

Example. Using the data given for the example on steel pipe columns in Section 6-8 ($P = 41$ kips and height = 12 ft), select a square structural tubing column from Table 6-4.

Solution: Entering Table 6-4 with an effective length of 12 ft, we find that a load of 43 kips can be supported by a square section $4 \times 4 \times \frac{3}{16}$.

Both pipe and tubing may be available in various steel strengths. We have used the properties in these examples because they appear in the AISC Manual. The choice between round pipe or rectangular tubing for a column is usually made for reasons other than simple structural efficiency. Freestanding columns are often round, but when built into wall construction, the rectangular shapes are often preferred.

Problem 6-9-A. A structural tubing column TS 4 × 4 × $\frac{3}{8}$, of steel with $F_y = 46$ ksi [317 MPa], is used on an effective length of 12 ft [3.66 m].

TABLE 6-4. Allowable Column Loads for Square Structural Steel Tubing with F_y of 46 ksi (in kips)

Nominal Size		4 x 4					3 x 3		
Thickness		½	⅜	5/16	¼	3/16	5/16	¼	3/16
Wt./ft.		21.63	17.27	14.83	12.21	9.42	10.58	8.81	6.87
F_y		46 ksi							
Effective length in feet KL with respect to radius of gyration	0	176	140	120	99	76	86	71	56
	2	168	134	115	95	73	80	67	53
	3	162	130	112	92	71	77	64	50
	4	156	126	108	89	69	73	61	48
	5	150	121	104	86	67	68	57	45
	6	143	115	100	83	64	63	53	42
	7	135	110	95	79	61	57	49	39
	8	126	103	90	75	58	51	44	35
	9	117	97	84	70	55	44	38	31
	10	108	89	78	65	51	37	33	27
	11	98	82	72	60	47	31	27	22
	12	87	74	65	55	43	26	23	19
	13	75	65	58	49	39	22	19	16
	14	65	57	51	43	35	19	17	14
	15	57	49	44	38	30	16	15	12
	16	50	43	39	33	27	14	13	11
	17	44	38	34	29	24	13	11	9
	18	39	34	31	26	21		10	8
	19	35	31	28	24	19			
	20	32	28	25	21	17			
	21	29	25	23	19	16			
	22	26	23	21	18	14			
	23	24	21	19	16	13			
	24		19	17	15	12			
	25				14	11			
Properties									
A (in.²)		6.36	5.08	4.36	3.59	2.77	3.11	2.59	2.02
I (in.⁴)		12.3	10.7	9.58	8.22	6.59	3.58	3.16	2.60
r (in.)		1.39	1.45	1.48	1.51	1.54	1.07	1.10	1.13
B {Bending factor		1.04	0.949	0.910	0.874	0.840	1.30	1.23	1.17
*a		1.83	1.59	1.43	1.22	0.983	0.533	0.470	0.387

* Tabulated values of a must be multiplied by 10^6.
Note: Heavy line indicates Kl/r of 200.

Source: Reprinted from the *Manual of Steel Construction,* 8th ed., with permission of the publishers, American Institute of Steel Construction.

200

Compute the allowable axial load, and compare it with the value in Table 6-4.

Problem 6-9-B*. Refer to Table 6-4 and select the lightest weight square tubing column with an effective length of 10 ft [3.05 m] that will support an axial load of 64 kips [285 kN].

6-10 Double-Angle Struts

Two angle sections, separated by the thickness of a connection plate at each end and fastened together at intervals by fillers and welds or bolts, are commonly used as compression members in roof trusses. (See Fig. 6-2e). These members, whether or not in a vertical position, are called struts; their size is determined in accordance with the requirements and formulas for columns in Section 6-5. To ensure that the angles act as a unit, the intermittent connections are made at intervals such that the slenderness ratio l/r of either angle between fasteners does not exceed the governing slenderness ratio of the built-up member. The least radius of gyration r is used in computing the slenderness ratio of each angle.

The AISC Manual contains safe-load tables for struts of two angles with $\frac{3}{8}$-in. separation back-to-back. Three series are given: equal-leg angles, unequal-leg angles with short legs back-to-back, and unequal-leg angles with long legs back-to-back. Table 6-5 has been abstracted from the latter series and lists allowable loads with respect to the X-X and Y-Y axes. The smaller (least) radius of gyration gives the smaller allowable load, and unless the member is braced with respect to the weaker axis, this is the tabular load to be used. The usual practice is to assume K equal to 1.0. The following example shows how the loads in the table are computed.

Example. Two $5 \times 3\frac{1}{2} \times \frac{1}{2}$-in. angle sections spaced with their long legs $\frac{3}{8}$ in. back-to-back are used as a compression member. If the member is A36 steel and has an effective length of 10 ft, compute the allowable axial load.

Solution: From Table 6-5 we find that the area of the two-angle member is 8.0 in^2 and that the radii of gyration are $r_x = 1.58$ in.

TABLE 6-5.   Allowable Axial Compression for Double Angle Struts

Size (in.)		8 × 6			6 × 4				5 × 3 1/2			5 × 3	
Thickness (in.)		3/4	1/2		5/8	1/2	3/8		1/2	3/8		1/2	3/8	5/16
Weight (lb/ft)		67.6	46.0		40.0	32.4	24.6		27.2	20.8		25.6	19.6	16.4
Area (in²)		19.9	13.5		11.7	9.50	7.22		8.00	6.09		7.50	5.72	4.80
r_x (in.)		2.53	2.56		1.90	1.91	1.93		1.58	1.60		1.59	1.61	1.61
r_y (in.)		2.48	2.44		1.67	1.64	1.62		1.49	1.46		1.25	1.23	1.22

Effective Length (KL) with Respect to Indicated Axis — **X-X Axis**

KL			KL				KL			KL			
0	430	266	0	253	205	142	0	173	129	0	162	121	94
10	370	231	8	214	174	122	4	159	119	4	149	112	88
12	353	222	10	200	163	115	6	150	113	6	141	106	83
14	334	211	12	185	151	107	8	139	105	8	130	98	77
16	315	200	14	168	137	99	10	126	96	10	119	90	71
20	271	175	16	150	123	89	12	113	86	12	106	81	64
24	222	148	20	110	90	69	14	97	75	14	92	70	57
28	168	117	24	76	62	48	16	81	63	16	76	59	49
32	129	90	28	56	46	36	20	52	40	20	49	38	32
36	102	71											

Effective Length (KL) with Respect to Indicated Axis — **Y-Y Axis**

KL			KL				KL			KL			
0	430	266	0	253	205	142	0	173	129	0	162	121	94
10	368	229	6	222	179	125	4	158	118	4	145	108	85
12	351	219	8	207	167	117	6	148	110	6	132	99	78
14	332	207	10	190	153	108	8	136	101	8	118	88	69
16	311	195	12	171	137	97	10	122	91	10	101	75	60
20	266	169	14	151	120	86	12	107	79	12	82	61	49
24	216	139	16	129	102	74	14	90	67	14	62	46	38
28	162	106	20	85	66	49	16	72	53	16	47	35	29
32	124	81	24	59	46	34	20	46	34	20	30	22	19
36	98	64											

and $r_y = 1.49$ in. Using the smaller r, the slenderness ratio is

$$\frac{Kl}{r} = 1 \times \frac{10 \times 12}{1.49} = 80.5, \text{ say } 81$$

Referring to Table 6-1, we find that $F_a = 15.24$ ksi, making the allowable load

$$P = A \times F_a = 8.0 \times 15.24 = 121.9 \text{ kips}$$

This value is, of course, readily verified by entering Table 6-5 under "*Y-Y Axis*," with an effective length of 10 ft and then finding the column of loads for the $5 \times 3\frac{1}{2} \times \frac{1}{2}$ angle, and proceeding down to the entry for an effective length of 10 ft.

4 × 3				3 1/2 × 2 1/2				3 × 2				2 1/2 × 2		
1/2	3/8	5/16		3/8	5/16	1/4		3/8	5/16	1/4		3/8	5/16	1/4
22.2	17.0	14.4		14.4	12.2	9.8		11.8	10.0	8.2		10.6	9.0	7.2
6.50	4.97	4.18		4.22	3.55	2.88		3.47	2.93	2.38		3.09	2.62	2.13
1.25	1.26	1.27		1.10	1.11	1.12		0.940	0.948	0.957		0.768	0.776	0.784
1.33	1.31	1.30		1.11	1.10	1.09		0.917	0.903	0.891		0.961	0.948	0.935

L	1/2	3/8	5/16	L	3/8	5/16	1/4	L	3/8	5/16	1/4	L	3/8	5/16	1/4
0	140	107	90	0	91	77	60	0	75	63	51	0	67	57	46
2	134	103	86	2	86	73	57	2	70	59	48	2	61	52	42
4	126	96	81	4	80	67	53	3	67	57	46	3	58	49	40
6	115	88	74	6	71	60	48	4	63	54	44	4	53	45	37
8	102	78	66	8	61	52	41	6	55	46	38	5	48	41	34
10	88	67	57	10	50	42	34	8	44	38	31	6	42	36	30
12	71	55	47	12	37	31	26	10	32	27	23	8	30	26	21
14	54	42	36	14	27	23	19	12	22	19	16	10	19	16	14
16	41	32	27	16	21	18	15	14	16	14	12	12	13	11	9
18	33	25	22	18	16	14	12								
20	26	20	17												

L	1/2	3/8	5/16	L	3/8	5/16	1/4	L	3/8	5/16	1/4	L	3/8	5/16	1/4
0	140	107	90	0	91	77	60	0	75	63	51	0	67	57	46
2	135	103	86	2	87	73	57	2	70	59	48	2	63	53	43
4	127	97	81	4	80	67	53	3	67	56	46	3	60	51	41
6	117	89	74	6	72	60	47	4	63	53	43	4	57	48	39
8	105	80	67	8	62	52	41	6	54	45	36	6	49	41	33
10	92	70	58	10	50	42	33	8	43	36	28	8	40	34	27
12	77	58	48	12	37	31	25	10	30	25	20	10	30	24	19
14	61	45	37	14	28	23	18	12	21	17	14	12	21	17	13
16	47	35	29	16	21	17	14	14	15	13	10	14	15	12	10
18	37	27	23	18	17	14	11								
20	30	22	18												

Source: Abstracted from data in the *Manual of Steel Construction*, 8th ed., with permission of the publishers, American Institute of Steel Corporation.

Notes: Loads in kips; $F_y = 36$ ksi [250 MPa]; long legs back-to-back with ⅜-in. separation.

The design of double-angle members for the compression elements in trusses is considered in Chapter 20.

When designing double angles or structural tees as compression members without the help of safe-load tables, we must consider the possibility that it may be necessary to reduce the allowable stress when these members have thin parts. This condition is indicated by the presence of a value for Q_s in the tabulated properties in the AISC Manual. When a value is given for Q_s, the safe axial load as normally calculated must be multiplied by this value for the true allowable load. Load values given in the safe-load tables in the AISC Manual have incorporated this requirement.

Problem 6-10-A*. A double-angle compression member 8 ft [2.44 m] long is composed of two angles 4 × 3 × ⅜ in. with the long legs ⅜ in. back-to-back. If the member is fabricated from A36 steel, compute the allowable concentric load.

Problem 6-10-B. Using Table 6-5, select a double-angle compression member that will support an axial load of 50 kips [222 kN] if the effective length is 10 ft [3.05 ml].

6-11 Eccentrically Loaded Columns

The columns previously discussed have been axially or concentrically loaded. It frequently happens, however, that in addition to the axial load, the column may be subjected to bending stresses that result from eccentric loads. Figure 6-4 shows a column with concentric and eccentric loads. The design of eccentrically loaded columns is accomplished by testing trial sections. As an aid to design, it is convenient to convert the axial and eccentric

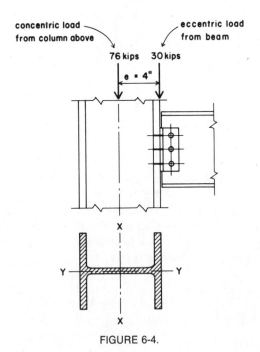

FIGURE 6-4.

loads into a single equivalent axial load. Having done this, we can use the safe-load tables to select the trial section.

The *bending factors* B_x and B_y are listed on the right-hand side of Table 6-2. The bending factor is the area of the cross section divided by its section modulus. Because there are two major section moduli, there are two bending factors: B_x for the X-X axis and B_y for the Y-Y axis. For example, the area of a W 10 × 49 is 14.4 sq in.; the section moduli with respect to the X-X and Y-Y axes are 54.6 and 18.7 in^3, respectively. Then

$$B_x = \frac{A}{S_x} = \frac{14.4}{54.6} = 0.264$$

and

$$B_y = \frac{A}{S_y} = \frac{14.4}{18.7} = 0.770$$

Note that these are the values given in Table 6-2.

Bending factors are used to convert the effect of eccentricity to an equivalent axial load. To accomplish this, we multiply the *bending moment* resulting from the eccentric load by the appropriate bending factor. Then *the total equivalent axial load (P') is equal to the sum of the axial load and the eccentric load plus the product of the bending moment due to the eccentric load and the appropriate bending factor.*

The trial section used in designing a column subjected to both axial and eccentric loads may be established by first finding an *approximate* equivalent axial load. This procedure is illustrated in the following example.

Example. An 8-in. wide flange column of A36 steel, with an unsupported height of 13 ft, supports an axial load of 76 kips and a load of 30 kips applied 4 in. from the X-X axis. The arrangement is shown in Fig. 6-4. Determine the trial column section.

Solution: (1) The bending moment produced by the eccentric load is

$$M = 30 \times 4 = 120 \text{ kip-in}$$

However, only the general dimensions of the section are known at this point; therefore we do not know the exact value of the bending factor.

(2) Referring to the 8-in. column sections in Table 6-2, tentatively select a bending factor of 0.333. This may be revised later. Then the bending moment multiplied by the bending factor is $120 \times 0.333 = 40$ kips, which is an equivalent axial load for the eccentric load.

(3) Now, in accordance with the principle previously stated, the approximate total equivalent load on the column is

$$P' = 76 + 30 + 40 = 146 \text{ kips}$$

Referring again to Table 6-2, we find that a W 8 × 35 with an effective length of 13 ft will carry 162 kips, and a W 8 × 31 will carry 143 kips.

It is necessary, of course, to verify the accuracy of our estimate for the bending factor. We assume a value of 0.333 and note from the tables that the W 8 × 35 has $B_x = 0.330$ and the W 8 × 31 has $B_x = 0.332$. These are sufficiently close to our estimate to make more work unnecessary. Had we been off by more than a few percent, it would have been necessary to compute a new value for P' from the true B_x values to verify our selections.

Because the foregoing procedure gives results that are approximate on the safe side, the W 8 × 35 could be the accepted section. However, if it is desired to determine the lightest weight section that can be used, the W 8 × 31 should be investigated more precisely for compliance with the AISC Specification requirements for the design of columns with combined loading. This is not a simple procedure, and the diversity of factors involved makes it inadvisable to include treatment of these complex requirements in a book of this scope. Reference to the AISC Manual is recommended for readers who wish to study the complete specification requirements that cover combined axial compression and bending in columns.

6-12 Column Base Plates

Steel columns generally transfer their loads to the supporting ground by means of reinforced concrete footings. As the allowable compressive strength of concrete is considerably less than the actual unit stresses in the steel column, it is necessary to provide a steel base plate under the column, to spread the load sufficiently to prevent overstressing of the concrete. The typical arrangement is shown in Fig. 6-5.

Rolled steel bearing plates used for column bases may be obtained in a great variety of sizes. The lengths and widths have dimensions usually in even inches, and for the sizes commonly used in building construction, the plates may be obtained in $\frac{1}{8}$-in. increments of thickness. For uniform distribution of the column load over the base plate, it is important that the column and plate be in absolute contact. Rolled steel bearing plates more than 2 in. thick, but not more than 4 in. thick, may be straightened by pressing or planing. Material more than 4 in. thick must be planed on the upper surface. The undersurface need not be planed because a full bearing contact is obtained by using a layer of cement grout on the concrete surface.

Steel columns are usually secured to the foundation by steel bolts that are embedded in the concrete and pass through the base plate and are secured to angles bolted or welded to the column

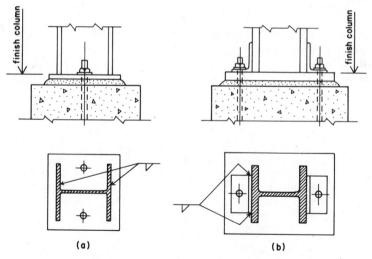

FIGURE 6-5. Typical column base details.

flange. (See Fig. 6-5b.) For light columns the angles are often omitted, and the base plate is secured to the column by fillet welds (Fig. 6-5a).

The first step in the design of a base plate is to determine its area. This is accomplished by use of the basic formula

$$A = \frac{P}{F_p}$$

where $A = B \times N$ = the area of the base plate in sq in. (see Fig. 6-6a),

P = column load in pounds,

F_p = allowable bearing value of the concrete in psi; the AISC Specification gives this stress as $0.25\,f'_c$ when the entire area of the concrete support is covered and $0.375\,f'_c$ when only one-third of the area is covered; a concrete commonly used has 3000 psi for the value of f'_c; hence F_p = 750 or 1125 psi.

The column load is assumed to be uniformly distributed over the dotted-line rectangle shown in Fig. 6-6a. In addition, the base plate is assumed to exert a uniform pressure on the concrete foundation.

After the minimum required area of the base plate has been found, B and N are established so that the projections m and n are approximately equal.

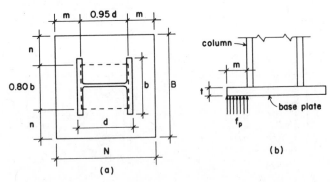

FIGURE 6-6.   Reference dimensions for design of column base plates.

The final step is to determine m and n in inches and to use the larger value in computing the thickness of the plate according to the formula

$$t = \sqrt{\frac{3f_p m^2}{F_b}} \qquad \text{or} \qquad t = \sqrt{\frac{3f_p n^2}{F_b}}$$

where t = thickness of the bearing plate, in inches,
f_p = *actual* bearing pressure on the foundation, in psi,
F_b = allowable bending stress in the base plate, in psi; the AISC Specification gives the value of F_b as 0.75 F_y, F_y being the yield point of the steel plate; thus for A36 steel F_y = 36,000 psi and F_b = 0.75 × 36,000, or F_b = 27,000 psi.

Review Problems

Note: In the following problems assume that the columns and base plates are A36 steel and that K, the effective length factor, is 1.0.

Problem 6-12-A.   A W 10 × 60 is used as a column with an effective length of 16 ft [4.9 m]. What safe load will it support?

Problem 6-12-B.   A W 6 × 25 is used as a column with an effective length of 12.5 ft [3.8 m]. Compute its allowable concentric load.

Problem 6-12-C.   The effective length of a W 8 × 40 used as a column is 22 ft [6.7 m]. Compute its allowable axial load.

Problem 6-12-D*.   What is the lightest weight wide flange section that can be used to support an axial compression load of 250 kips [1112 kN] if the effective length is 18 ft [5.5 m]?

Problem 6-12-E.   An S 10 × 25.4 is used as a column with an effective length of 8 ft [2.4 m]. Compute its allowable axial load.

Problem 6-12-F.   A W 14 × 283 has 20 × 1½-in. [508 × 38-mm] plates welded to its flanges. Determine the allowable axial load when the combined section is used as a column with an effective length of 16 ft [4.9 m].

Problem 6-12-G.   Two 12 × ½-in. [305 × 13-mm] plates are welded to the flanges of an S 12 × 31.8 to constitute a built-up member for use as a column. If its effective length is 13 ft [4 m], compute the allowable concentric load.

Problem 6-12-H*.   Design a column base plate for a W 8 × 31 column that is supported on concrete for which the allowable bearing capacity is 750 psi [5000 kPa]. The load on the column is 178 kips [792 kN].

Problem 6-12-I.   Design a column base plate for the W 8 × 31 in Problem 6-12-H if the bearing pressure allowed on the concrete is 1125 psi [7800 kPa].

7

Bolted Connections

||||||||||||||||||||||||||||||||||||||||||||||||||||

7-1 Introduction

Elements of structural steel are often connected by mating flat parts with common holes and inserting a pin-type device to hold them together. In times past the pin device was a rivet; today it is usually a bolt. A great number of types and sizes of bolt are available, as are many connections in which they are used. The material in this chapter deals with a few of the common bolting methods used in building structures.

7-2 Structural Bolts

The diagrams in Fig. 7-1 show a simple connection between two steel bars that functions to transfer a tension force from one bar to another. Although this is a tension-transfer connection, it is also referred to as a shear connection because of the manner in which the connecting device (the bolt) works in the connection. (See Fig. 7-1b.) If the bolt tension (due to tightening of the nut) is relatively low, the bolt serves primarily as a pin in the matched holes, bearing against the sides of the holes as shown in Fig. 7-1d. In addition to these functions, the bars develop tension stress that will be a maximum at the section through the bolt holes.

In the connection shown in Fig. 7-1, the failure of the bolt involves a slicing (shear) failure that is developed as a shear stress

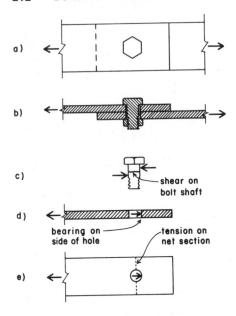

a)

b)

c) — shear on bolt shaft

d) bearing on side of hole — tension on net section

e)

FIGURE 7-1. Actions of bolted joints.

on the bolt cross section. The resistance of the bolt can be expressed as an allowable shear stress F_v times the area of the bolt cross section, or

$$R = F_v \times A$$

With the size of the bolt and the grade of steel known, it is a simple matter to establish this limit. In some types of connections, it may be necessary to slice the same bolt more than once to separate the connected parts. This is the case in the connection shown in Fig. 7-2, in which it may be observed that the bolt must be sliced twice to make the joint fail. When the bolt develops shear on only one section (Fig. 7-1), it is said to be in *single shear*; when it develops shear on two sections (Fig. 7-2), it is said to be in *double shear*.

When the bolt diameter is large or the bolt is made of strong steel, the connected parts must be sufficiently thick if they are to develop the full capacity of the bolts. The maximum bearing stress permitted for this situation by the AISC Specification is

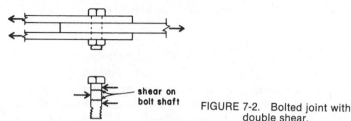

FIGURE 7-2. Bolted joint with double shear.

$F_p = 1.5F_u$, where F_u is the ultimate tensile strength of the steel in the part in which the hole occurs.

Bolts used for the connection of structural steel members come in two types. Bolts designated A307 and called *unfinished* have the lowest load capacity of the structural bolts. The nuts for these bolts are tightened just enough to secure a snug fit of the attached parts; because of this, plus the oversizing of the holes, there is some movement in the development of full resistance. These bolts are generally not used for major connections, especially when joint movement or loosening under vibration or repeated loading may be a problem.

Bolts designated A325 or A490 are called *high-strength bolts*. The nuts of these bolts are tightened to produce a considerable tension force, which results in a high degree of friction resistance between the attached parts. High-strength bolts are further designated as F, N, or X. The F designation denotes bolts for which the limiting resistance is that of friction. The N designation denotes bolts that function ultimately in bearing and shear but for which the threads are not excluded from the bolt shear planes. The X designation denotes bolts that function like the N bolts but for which the threads are excluded from the shear planes.

When bolts are loaded in tension, their capacities are based on the development of the ultimate resistance in tension stress at the reduced section through the threads. When loaded in shear, bolt capacities are based on the development of shear stress in the bolt shaft. The shear capacity of a single bolt is further designated as S for single shear (Fig. 7-1) or D for double shear (Fig. 7-2). The capacities of structural bolts in both tension and shear are given in Table 7-1. The size range given in the table— $\frac{5}{8}$–$1\frac{1}{2}$ in.—is that listed in the AISC Manual. However, the most commonly used sizes for structural steel framing are $\frac{3}{4}$ and $\frac{7}{8}$ in.

Bolts are ordinarily installed with a washer under both head and nut. Some manufactured high-strength bolts have specially formed heads or nuts that in effect have self-forming washers, eliminating the need for a separate, loose washer. When a washer is used, it is sometimes the limiting dimensional factor in detailing for bolt placement in tight locations, such as close to the fillet (inside radius) of angles or other rolled shapes.

For a given diameter of bolt, there is a minimum thickness required for the bolted parts in order to develop the full shear capacity of the bolt. This thickness is based on the bearing stress between the bolt and the side of the hole, which is limited to a maximum of $F_p = 1.5F_u$. The stress limit may be established by either the bolt steel or the steel of the bolted parts.

Steel rods are sometimes threaded for use as anchor bolts or tie rods. When they are loaded in tension, their capacities are usually limited by the stress on the reduced section at the threads. Tie rods are sometimes made with *upset ends,* which consist of larger diameter portions at the ends. When these enlarged ends are threaded, the net section at the thread is the same as the gross section in the remainder of the rods; the result is no loss of capacity for the rod.

7-3 Layout of Bolted Connections

Design of bolted connections generally involves a number of considerations in the dimensioned layout of the bolt-hole patterns for the attached structural members. Although we cannot develop all the points necessary for the production of structural steel construction and fabrication details, the material in this section presents basic factors that often must be included in the structural calculations.

Figure 7-3 shows the layout of a bolt pattern with bolts placed in two parallel rows. Two basic dimensions for this layout are limited by the size (nominal diameter) of the bolt. The first is the center-to-center spacing of the bolts, usually called the *pitch.* The

TABLE 7-1. Capacity of Structural Bolts (in kips)

ASTM Designation	Connection Type*	Loading Condition**	Nominal Diameter (in.)							
			5/8	3/4	7/8	1	1-1/8	1-1/4	1-3/8	1-1/2
			0.3068	0.4418	0.6013	0.7854	0.9940	1.227	1.485	1.767
			Area, Based on Nominal Diameter (in^2)							
A307		S	3.1	4.4	6.0	7.9	9.9	12.3	14.8	17.7
		D	6.1	8.8	12.0	15.7	19.9	24.5	29.7	35.3
		T	6.1	8.8	12.0	15.7	19.9	24.5	29.7	35.3
A325	F	S	5.4	7.7	10.5	13.7	17.4	21.5	26.0	30.9
		D	10.7	15.5	21.0	27.5	34.8	42.9	52.0	61.8
	N	S	6.4	9.3	12.6	16.5	20.9	25.8	31.2	37.1
		D	12.9	18.6	25.3	33.0	41.7	51.5	62.4	74.2
	X	S	9.2	13.3	18.0	23.6	29.8	36.8	44.5	53.0
		D	18.4	26.5	36.1	47.1	59.6	73.6	89.1	106.0
	All	T	13.5	19.4	26.5	34.6	43.7	54.0	65.3	77.7
A490	F	S	6.7	9.7	13.2	17.3	21.9	27.0	32.7	38.9
		D	13.5	19.4	26.5	34.6	43.7	54.0	65.3	77.7
	N	S	8.6	12.4	16.8	22.0	27.8	34.4	41.6	49.5
		D	17.2	24.7	33.7	44.0	55.7	68.7	83.2	99.0
	X	S	12.3	17.7	24.1	31.4	39.8	49.1	59.4	70.7
		D	24.5	35.3	48.1	62.8	79.5	98.2	119.0	141.0
	All	T	16.6	23.9	32.5	42.4	53.7	66.3	80.2	95.4

Source: Reproduced from data in the *Manual of Steel Construction*, 8th ed., with permission of the publishers, American Institute of Steel Construction.

*F = friction; N = bearing, threads not excluded; X = bearing, threads excluded.
**S = single shear; D = double shear; T = tension.

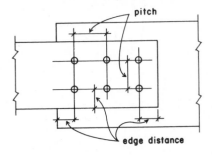

FIGURE 7-3. Pitch and edge distances for bolts.

AISC Specification limits this dimension to an absolute minimum of $2\frac{2}{3}$ times the bolt diameter. The preferred minimum, however, which is used in this book, is 3 times the diameter.

The second critical layout dimension is the *edge distance,* which is the distance from the center line of the bolt to the nearest edge. There is also a specified limit for this as a function of bolt size. This dimension may also be limited by edge tearing, which is discussed in Section 7-5.

Table 7-2 gives the recommended limits for pitch and edge distance for the bolt sizes used in ordinary steel construction.

In some cases bolts are staggered in parallel rows (Fig. 7-4). In this case the diagonal distance, labeled m in the illustration, must also be considered. For staggered bolts the spacing in the direction of the rows is usually referred to as the pitch; the spacing of the rows is called the gage. The reason for staggering the bolts is that sometimes the rows must be spaced closer (gage spacing) than the minimum spacing required for the bolts selected. Table 7-3 gives the pitch required for a given gage spacing to keep the diagonal spacing (m) within the recommended diameter limit.

Location of bolt lines is often related to the size and type of structural members being attached. This is especially true of bolts placed in the legs of angles or in the flanges of W, M, S, C, and structural tee shapes. Figure 7-5 shows the placement of bolts in the legs of angles. When a single row is placed in a leg, its recommended location is at the distance labeled g from the back of the angle. When two rows are used, the first row is placed at the distance g_1, and the second row is spaced a distance g_2 from the first. Table 7-4 gives the recommended values for these distances.

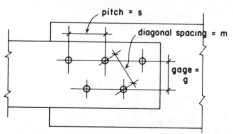

FIGURE 7-4. Standard reference dimensions for layout of bolted joints.

When placed at the recommended locations in rolled shapes, bolts will end up a certain distance from the edge of the part. Based on the recommended edge distance for rolled edges given in Table 7-2, it is thus possible to determine the maximum size of bolt that can be accommodated. For angles, the maximum fastener may be limited by the edge distance, especially when two rows are used: however, other factors may in some cases be more critical. The distance from the center of the bolts to the inside

TABLE 7-2. Pitch and Edge Distances for Bolts

Rivet or Bolt Diameter d (in.)	Minimum Edge Distance for Punched, Reamed, or Drilled Holes (in.)		Pitch, Center to Center (in.)	
	At Sheared Edges	At Rolled Edges of Plates, Shapes, or Bars, or Gas-Cut Edges[a]	Minimum Recommended	
			2–2/3 d	3d
5/8	1.125	0.875	1.67	1.875
3/4	1.25	1	2	2.25
7/8	1.5[b]	1.125	2.33	2.625
1	1.75[b]	1.25	2.67	3

Source: Reproduced from data in the *Manual of Steel Construction,* 8th ed., with permission of the publishers, American Institute of Steel Construction.

[a] May be reduced 1/8 in. when the hole is at a point where stress does not exceed 25% of the maximum allowed in the connected element.

[b] May be 1–1/4 in. at the ends of beam connection angles.

TABLE 7-3. Minimum Pitch to Maintain Three Diameters Center to Center of Holes

Diameter of Rivet	m	Distance, g (in.)								
		1	1 1/4	1 1/2	1 3/4	2	2 1/4	2 1/2	2 3/4	3
5/8	1 7/8	1 5/8	1 3/8	1 1/8	5/8	0				
3/4	2 1/4	2	1 7/8	1 5/8	1 3/8	1	0			
7/8	2 5/8	2 1/2	2 3/8	2 1/8	2	1 3/4	1 3/8	3/4	0	
1	3	2 7/8	2 3/4	2 5/8	2 1/2	2 1/4	2	1 3/8	1 1/8	0

Source: Reproduced from data in the *Manual of Steel Construction*, 8th ed., with permission of the publishers, American Institute of Steel Construction. (See Fig. 7-4.)

fillet of the angle may limit the use of a large washer where one is required. Another consideration may be the stress on the net section of the angle, especially if the member load is taken entirely by the attached leg. These problems are given some discussion in the design of the truss in Chapter 20.

Sections 1-16-4 and 1-16-5 of the AISC Specification provide additional criteria for minimum spacing and edge distances for fasteners as a function of the load per fastener and the thickness and the ultimate stress capacity of the connected parts.

7-4 Tension Connections

When tension members have reduced cross sections, two stress investigations must be considered. This is the case for members with holes for bolts or for bolts or rods with cut threads. For the member with a hole (Fig. 7-1*d*), the allowable tension stress at the

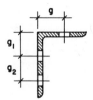

FIGURE 7-5. Gage distances for angles.

TABLE 7-4. Usual Gage Dimensions for Angles (in.)

Gage Dimension	Width of Angle Leg								
	8	7	6	5	4	3 1/2	3	2 1/2	2
g	4 1/2	4	3 1/2	3	2 1/2	2	1 3/4	1 3/8	1 1/8
g_1	3	2 1/2	2 1/4	2					
g_2	3	3	2 1/2	1 3/4					

Source: Reproduced from data in the *Manual of Steel Construction*, 8th ed., with permission of the publishers, American Institute of Steel Construction.

reduced cross section through the hole is $0.50F_u$, where F_u is the ultimate tensile strength of the steel. The total resistance at this reduced section (also called the net section) must be compared with the resistance at other, unreduced sections at which the allowable stress is $0.60 F_y$.

For threaded steel rods the maximum allowable tension stress at the threads is $0.33 F_u$. For steel bolts the allowable stress is specified as a value based on the type of bolt. The load capacity of various types and sizes of bolt is given in Table 7-1.

When tension elements consist of W, M, S, and tee shapes, the tension connection is usually not made in a manner that results in the attachment of all the parts of the section (e.g., both flanges plus the web for a W). In such cases the AISC Specification requires the determination of a reduced effective net area, A_e, that consists of

$$A_e = C_t A_n$$

where A_n = actual net area of the member,
C_t = reduction coefficient.

Unless a larger coefficient can be justified by tests, the following values are specified:

1. For W, M, or S shapes with flange widths not less than two-thirds the depth and structural tees cut from such shapes, when the connection is to the flanges and has at least three fasteners per line in the direction of stress, $C_t =$ 0.90.

2. For W, M, or S shapes not meeting the above conditions and for tees cut from such shapes, provided the connection has not fewer than three fasteners per line in the direction of stress, $C_t = 0.85$.

3. For all members with connections that have only two fasteners per line in the direction of stress, $C_t = 0.75$.

Angles used as tension members are often connected by only one leg. In a conservative design, the effective net area is only that of the connected leg, less the reduction caused by bolt holes. Rivet and bolt holes are punched larger in diameter than the nominal diameter of the fastener. The punching damages a small amount of the steel around the perimeter of the hole; consequently the diameter of the hole to be deducted in determining the net section is $1/8$ in. greater than the nominal diameter of the rivet.

When only one hole is involved, as in Fig. 7-1, or in a similar connection with a single row of fasteners along the line of stress, the net area of the cross section of one of the plates is found by multiplying the plate thickness by its net width (width of member minus diameter of hole).

When holes are staggered in two rows along the line of stress (Fig. 7-6), the net section is determined somewhat differently. The AISC Specification reads:

In the case of a chain of holes extending across a part in any diagonal or zigzag line, the net width of the part shall be obtained by deducting from the gross width the sum of the diameters of all the holes in the chain and adding, for each gage space in the chain, the quantity $s^2/4g$, where

s = longitudinal spacing (pitch) in inches of any two successive holes

and

g = transverse spacing (gage) in inches for the same two holes.

The critical net section of the part is obtained from that chain which gives the least net width.

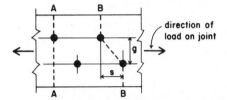

FIGURE 7-6.

The AISC Specification also provides that in no case shall the net section through a hole be considered as more than 85% of the corresponding gross section.

7-5 Tearing

One possible form of failure in a bolted connection is that of tearing out the edge of one of the attached members. The diagrams in Fig. 7-7 show this potentiality in a connection between two plates. The failure in this case involves a combination of shear and tension to produce the torn-out form shown. The total tearing force is computed as the sum required to cause both forms of failure. The allowable stress on the net tension area is specified as $0.50F_u$, where F_u is the maximum tensile strength of the steel. The allowable stress on the shear areas is specified as $0.30\,F_u$. With the edge distance, hole spacing, and diameter of the holes known, the net widths for tension and shear are determined and multiplied by the thickness of the part in which the tearing occurs. These areas are then multiplied by the appropriate stresses to find the total tearing force that can be resisted. If this force is

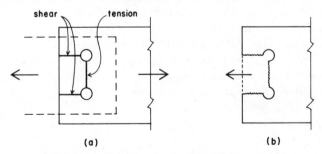

FIGURE 7-7. Tearing in a bolted tension joint.

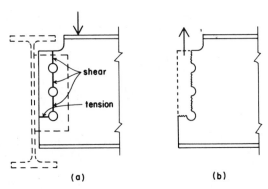

FIGURE 7-8.   Tearing in a bolted beam connection.

greater than the connection design load, the tearing problem is not critical.

Another case of potential tearing is shown in Fig. 7-8. This is the common situation for the end framing of a beam in which support is provided by another beam, whose top is aligned with that of the supported beam. The end portion of the top flange of the supported beam must be cut back to allow the beam web to extend to the side of the supporting beam. With the use of a bolted connection, the tearing condition shown is developed.

7-6   Design of a Bolted Tension Connection

The issues raised in several of the preceding sections are illustrated in the following design example. Before proceeding with the problem data, we should consider some of the general requirements for this joint.

If friction-type bolts are used, the surfaces of the connected parts must be cleaned and made reasonably true. If high-strength bolts are used, the determination to exclude threads from the shear failure planes must be established.

The AISC Specification has a number of general requirements for connections:

1.   Need for a minimum of two bolts per connection.
2.   Need for a minimum connection capacity of 6 kips.

3. Need for the connection to develop at least 50% of the full potential capacity of the member (for trusses only).

Although a part of the design problem may be the selection of the type of fastener or the required strength of steel for the attached parts, we provide this as given data in the example problem.

Example.   The connection shown in Fig. 7-9 consists of a pair of narrow plates that transfer a load of 100 kips [445 kN] in tension to a single 10-in. [254 mm] wide plate. The plates are A36 steel with F_u = 58 ksi [400 MPa] and are attached with $\frac{3}{4}$-in. A325F bolts placed in two rows. Determine the number of bolts required, the width and thickness of the narrow plates, the thickness of the wide plate, and the layout of the bolts.

Solution: From Table 7-1 we find the double shear (*D*) capacity for one bolt is 15.5 kips [69 kN]. The required number of bolts is thus

$$n = \frac{\text{connection load}}{\text{bolt capacity}} = \frac{100}{15.5} = 6.45$$

$$\left[n = \frac{445}{69} = 6.45 \right]$$

and the minimum number for a symmetrical connection is eight.

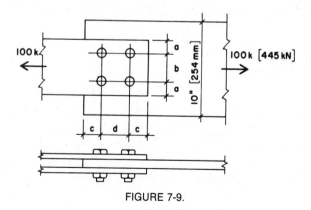

FIGURE 7-9.

With eight bolts used, the load on one bolt is

$$P = \frac{100}{8} = 12.5 \text{ kips}$$

$$\left[\frac{445}{8} = 55.6 \text{ kN} \right]$$

According to Table 7-2, the $\frac{3}{4}$-in. bolts require a minimum edge distance of 1.25 in. (at a sheared edge) and a recommended pitch of 2.25 in. The minimum width for the narrow plates is therefore (see Fig. 7-9)

$$w = b + 2(a)$$

$$w = 2.25 + 2(1.25) = 4.75 \text{ in. [121 mm]}$$

With no other constraining conditions given, we arbitrarily select a width of 6 in. [152.4 mm] for the narrow plates. Checking first for the requirement of a maximum tension stress of $0.60 \, F_y$ on the gross area, we find

$$F_t = 0.60F_y = 0.60(36) = 21.6 \text{ ksi [149 MPa]}$$

$$A_{req} = \frac{100}{21.6} = 4.63 \text{ in}^2$$

$$\left[\frac{445 \times 10^3}{149} = 2987 \text{ mm}^2 \right]$$

and the required thickness with the width selected is

$$t = \frac{4.63}{2(6)} = 0.386 \text{ in.}$$

$$\left[\frac{2987}{2(152.4)} = 9.80 \text{ mm} \right]$$

We therefore select a minimum thickness of $\frac{7}{16}$ in. (0.4375 in.) [11 mm]. The next step is to check the stress condition on the net section through the holes, for which the allowable stress is $0.50F_u$. For the computations, we assume a hole diameter $\frac{1}{8}$ in. [3.18 mm] larger than the bolt. Thus hole size = 0.875 in. [22.23 mm]

net width = 2{6 − (2 × 0.875)} = 8.5 in.

[2{152.4 − (2 × 22.23)} = 215.9 mm]

and the stress on the net section of the two plates is

$$f_t = \frac{100}{0.4375 \times 8.5} = 26.89 \text{ ksi}$$

$$\left[\frac{445 \times 10^3}{11 \times 215.9} = 187 \text{ MPa} \right]$$

This computed stress is compared with the specified allowable stress of

$$F_t = 0.50 F_u = 0.50 \times 58 = 29 \text{ ksi [200 MPa]}$$

Bearing stress is computed by dividing the load on a single bolt by the product of the bolt diameter and the plate thickness. Thus

$$f_p = \frac{12.5}{2 \times 0.75 \times 0.4375} = 19.05 \text{ ksi}$$

$$\left[f_p = \frac{55.6 \times 10^3}{2 \times 19.05 \times 10} = 146 \text{ MPa} \right]$$

This is compared with the allowable stress of

$$F_p = 1.5 F_u = 1.5 \times 58 = 87 \text{ ksi [600 MPa]}$$

For the middle plate the procedure is essentially the same except that, in this case, the plate width is given. As before, on the basis of stress on the unreduced section, we determine that the total area required is 4.63 in^2 [2987 mm^2]. Thus the thickness required is

$$t = \frac{4.63}{10} = 0.463 \text{ in}$$

$$\left[\frac{2987}{254} = 11.76 \text{ mm} \right]$$

We therefore select a minimum thickness of $\frac{1}{2}$ in. (0.50 in.) [13 mm]. We then proceed as before to check the stress on the net width. The net width through the two holes is

$$w = 10 - (2 \times 0.875) = 8.25 \text{ in.}$$

$$[w = 254 - (2 \times 22.23) = 209.5 \text{ mm}]$$

and the tension stress on this net cross section is

$$f_t = \frac{100}{8.25 \times 0.5} = 24.24 \text{ ksi}$$

$$\left[f_t = \frac{445 \times 10^3}{12 \times 209.5} = 177 \text{ MPa} \right]$$

which is less than the allowable stress of 29 ksi [200 MPa] determined previously.

The computed bearing stress on the wide plate is

$$f_p = \frac{12.5}{0.75 \times 0.50} = 33.3 \text{ ksi}$$

$$\left[\frac{55.6 \times 10^3}{19.05 \times 12} = 243 \text{ MPa} \right]$$

which is considerably less than the allowable determined before: $F_p = 87$ ksi [600 MPa].

In addition to the layout restrictions given in Section 7-3, the AISC Specification requires that the minimum spacing in the direction of the load be

$$\frac{2P}{F_u t} + \frac{d}{2} \qquad \text{(dimension } d \text{ in Fig. 7-9)}$$

and that the minimum edge distance in the direction of the load be

$$\frac{2P}{F_u t} \qquad \text{(dimension } c \text{ in Fig. 7-9)}$$

where P = force transmitted by one fastener to the critical connected part,

F_u = specified minimum (ultimate) tensile strength of the connected part,

t = thickness of the critical connected part.

For our case

$$\frac{2P}{F_u t} = \frac{2 \times 12.5}{58 \times 0.5} = 0.862 \text{ in.}$$

which is considerably less than the specified edge distance listed in Table 7-2 for a $\frac{3}{4}$-in. bolt at a sheared edge: 1.25 in.

For the spacing

$$\frac{2P}{F_u t} + \frac{d}{2} = 0.862 + 0.375 = 1.237 \text{ in.}$$

which is also not critical.

A final problem that must be considered is the potential of tearing out the two bolts at the ends of the plates. Because the combined thickness of the two outer plates is greater than that of the middle plate, the critical case in this connection is that of the middle plate. Figure 7-10 shows the condition for the tearing, which involves tension on the section labeled "1" and shear on the two sections labeled "2."

For the tension section

$$w_{(net)} = 3 - 0.875 = 2.125 \text{ in. [54.0 mm]}$$

$$F_t = 0.50 F_u = 29 \text{ ksi [200 MPa]}$$

For the shear sections

$$w_{(net)} = 2\left(1.25 - \frac{0.875}{2}\right) = 1.625 \text{ in. [41.3 mm]}$$

$$F_v = 0.30 F_u = 17.4 \text{ ksi [120 MPa]}$$

The total resistance to tearing is

$$T = (2.125 \times 0.5 \times 29) + (1.625 \times 0.5 \times 17.4) = 44.95 \text{ kips}$$

$$\left[T = \left(\frac{54 \times 13 \times 200}{10^3}\right) + \left(\frac{41.3 \times 13 \times 120}{10^3}\right) = 205 \text{ kN} \right]$$

Because this is greater than the combined load of 25 kips [111.2 kN] on the two bolts, the problem is not critical.

Connections that transfer compression between the joined parts are essentially the same with regard to the bolt stresses and

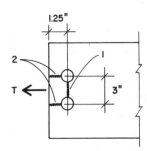

FIGURE 7-10.

bearing on the parts. Stress on the net section is less likely to be critical because the compression members will usually be designed for column action, with a considerably reduced value for the allowable compression stress.

Problem 7-6-A*. A bolted connection of the general form shown in Fig. 7-9 is to be used to transmit a tension force of 200 kips [890 kN] by using $\frac{7}{8}$-in. A490N bolts and plates of A36 steel. The outer plates are to be 8 in. [200 mm] wide, and the center plate is 12 in. [300 mm] wide. Find the required thicknesses of the plates and the number of bolts needed if the bolts are placed in two rows. Sketch the bolt layout with the necessary dimensions.

Problem 7-6-B. Design a connection for the data in Problem 7-6-A except that the bolts are 1-in. A325N, the outside plates are 9 in. wide, and the bolts are placed in three rows.

7-7 Framing Connections

The joining of structural steel members in a structural system generates a wide variety of situations, depending on the form of the connected parts, the type of connecting device used, and the nature and magnitude of the forces that must be transferred between the members. Figure 7-11 shows a number of common connections that are used to join steel columns and beams consisting of rolled shapes.

In the joint shown in Fig. 7-11a, a steel beam is connected to a supporting column by the simple means of resting it on top of a steel plate that is welded to the top of the column. The bolts in this case carry no computed loads if the force transfer is limited to

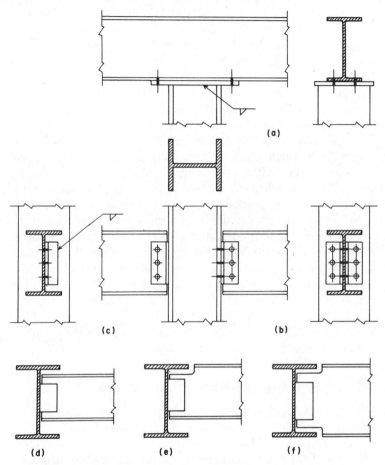

FIGURE 7-11. Typical bolted framing connections for steel structures.

that of the vertical end reaction of the beam. The only computed stress condition that is likely to be of concern in this situation is that of crippling the beam web (Section 5-23). This is a situation in which the use of unfinished bolts is indicated.

The remaining details in Fig. 7-11 illustrate situations in which the beam end reactions are transferred to the supports by attachment to the beam web. This is, in general, an appropriate form of force transfer because the vertical shear at the end of the beam is resisted primarily by the beam web. The most common form of connection is that which uses a pair of angles (Fig. 7-11b). The two most frequent examples of this type of connection are the joining of a steel beam to the side of a column (Fig. 7-11b) or to the side of another beam (Fig. 7-11d). A beam may also be joined to the web of a W shape column in this manner if the column depth provides enough space for the angles.

An alternative to this type of connection is shown in Fig. 7-11c, where a single plate is welded to the side of a column, and the beam web is bolted to one side of the plate. This is generally acceptable only when the magnitude of the load on the beam is low because the one-sided connection experiences some torsion.

When the two intersecting beams must have their tops at the same level, the supported beam must have its top flange cut back, as shown at Fig. 7-11e. This is to be avoided, if possible, because it represents an additional cost in the fabrication and also reduces the shear capacity of the beam. Even worse is the situation in which the two beams have the same depth and which requires cutting both flanges of the supported beam. (See Fig. 7-11f.) When these conditions produce critical shear in the beam web, it will be necessary to reinforce the beam end.

7-8 Conventional and Moment Connections

All of the beam-to-girder and beam-to-column connections discussed in this chapter come under the category of *simple* or *free-end* connections; that is, insofar as gravity loading is concerned, the ends of the beams and girders are connected for shear only and are relatively free to rotate under gravity load. We shall call connections of this nature *conventional connections;* they are used in Type 2 of the three types of steel construction recognized in the AISC Specification.

Type 1 construction, commonly designated as continuous or rigid frame, assumes that beam-to-column connections possess sufficient rigidity to prevent rotation of the beam ends as the member deflects under its load. This means that the connection must transmit some bending moment between beam and column.

Consequently it is called a moment-resisting connection or simply a *moment connection*. Type 3 construction, called partially restrained or semi-rigid framing, assumes that the connections possess a dependable and known moment-resisting capacity of a degree between the rigidity of Type 1 and the flexibility of Type 2.

Although Type 1 continuous framing can be achieved by the proper design of bolted or riveted connections, it is accomplished much more effectively by welded construction. This aspect of welded connections is considered briefly in Chapter 8. A fully continuous frame of Type 1 construction is statically indeterminate, and its analysis and design are beyond the scope of this book. Moment-resisting connections are used in multistory steel framed buildings to provide lateral stability against the effects of wind and earthquake forces.

In general, the design methods and procedures treated in this volume are applicable to Type 2 construction and follow the provisions of Section 1-12-1 of the AISC Specification, which states: "Beams, girders, and trusses shall ordinarily be designed on the basis of simple spans whose effective length is equal to the distance between centers of gravity of the members to which they deliver their end reactions."

8

Welded Connections

IIIIIIIIIIIIIIIIIIIIIIIIIIIIIIIIIIIIIIIIIIIIIIIIIIIIIIIII

8-1 Introduction

One of the distinguishing characteristics of welded construction is the facility with which one member may be attached directly to another without the use of additional plates or angles, which are necessary in bolted and riveted connections. A welded connection requires no holes for fasteners; therefore the gross, rather than the net, section may be considered when determining the effective cross-sectional area of members in tension.

As noted in Section 7-8, moment-resisting connections are readily achieved by welding; consequently, welded connections are customary in Type 1 construction, in order to develop continuity in the framing. Welding may also be used in Type 2 construction but care must be exercised in design, to ensure that a rigid connection is not provided where free-end conditions have been assumed in the design of the framing.

Welding is often used in combination with bolting in *shop-welded* and *field-bolted* construction. Here connection angles with holes in the outstanding legs may be welded to a beam in the fabricating shop and then bolted to a girder or column in the field.

8-2 Electric Arc Welding

Although there are many welding processes, electric arc welding is the one generally used in steel building construction. In this type of welding, an electric arc is formed between an electrode and the two pieces of metal that are to be joined. The intense heat melts a small portion of the members to be joined, as well as the end of the electrode or metallic wire. The term *penetration* is used to indicate the depth from the original surface of the base metal to the point at which fusion ceases. The globules of melted metal from the electrode flow into the molten seat and, when cool, are united with the members that are to be welded together. *Partial penetration* is the failure of the weld metal and base metal to fuse at the root of a weld. It may result from a number of items, and such incomplete fusion produces welds that are inferior to those of full penetration.

8-3 Welded Joints

When two members are to be joined, the ends may or may not be grooved in preparation for welding. In general, there are three classifications of joints: *butt joints, tee joints,* and *lap joints.* The selection of the type of weld to use depends on the magnitude of the load requirement, the manner in which it is applied, and the cost of preparation and welding. Several joints are shown in Fig. 8-1. The type of joint and preparation permit a number of variations. In addition, welding may be done from one or both sides. The scope of this book prevents a detailed discussion of the many joints and their uses and limitations.

The weld most commonly used for structural steel in building construction is the *fillet weld.* It is approximately triangular in cross section and is formed between the two intersecting surfaces of the joined members. (See Fig. 8-2a and b.) The *size* of a fillet weld is the leg length of the largest inscribed isosceles right triangle, AB or BC. (See Fig. 8-2a.) The *root* of the weld is the point at the bottom of the weld, point B in Fig. 8-2a. The *throat* of a fillet weld is the distance from the root to the hypotenuse of the largest isosceles right triangle that can be inscribed within the weld cross section, distance BC in Fig. 8-2a. The exposed surface of a weld is not the plane surface indicated in Fig. 8-2a but is usually somewhat convex, as shown in Fig. 8-2b. Therefore the actual throat may be greater than that shown in Fig. 8-2a. This additional mate-

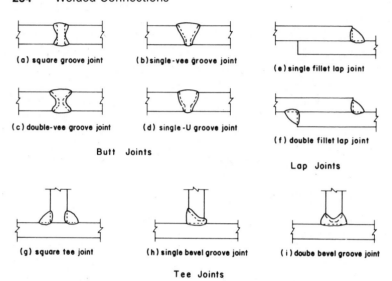

(a) square groove joint

(b) single-vee groove joint

(c) double-vee groove joint

(d) single-U groove joint

Butt Joints

(e) single fillet lap joint

(f) double fillet lap joint

Lap Joints

(g) square tee joint

(h) single bevel groove joint

(i) doube bevel groove joint

Tee Joints

FIGURE 8-1. Typical welded joints.

rial is called *reinforcement*. It is not included in determining the strength of a weld.

A single-vee groove weld between two members of unequal thickness is shown in Fig. 8-2c. The *size* of a butt weld is the thickness of the thinner part joined, with no allowance made for the weld reinforcement.

8-4 Stresses in Welds

If the dimension (size) of AB in Fig. 8-2a is one unit in length, $(AD)^2 + (BD)^2 = 1^2$. Because AD and BD are equal, $2(BD)^2 = 1^2$,

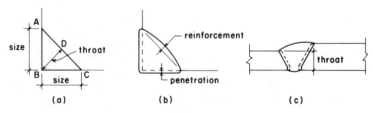

FIGURE 8-2. Properties of welded joints.

and $BD = \sqrt{0.5}$, or 0.707. Therefore the throat of a fillet weld is equal to the *size* of the weld multiplied by 0.707. As an example, consider a $\frac{1}{2}$-in. fillet weld. This would be a weld with dimensions AB or BC equal to $\frac{1}{2}$ in. In accordance with the above, the throat would be 0.5×0.707, or 0.3535 in. Then, if the allowable unit shearing stress on the throat is 21 ksi, the allowable working strength of a $\frac{1}{2}$-in. fillet weld is $0.3535 \times 21 = 7.42$ kips *per lin in. of weld.* If the allowable unit stress is 18 ksi, the allowable working strength is $0.3535 \times 18 = 6.36$ kips *per lin in. of weld.*

The permissible unit stresses used in the preceding paragraph are for welds made with E 70 XX- and E 60 XX-type electrodes on A36 steel. Particular attention is called to the fact that *the stress in a fillet weld is considered as shear on the throat, regardless of the direction of the applied load.* Neither plug nor slot welds shall be assigned any values in resistances other than shear. The allowable working strengths of fillet welds of various sizes are given in Table 8-1 with values rounded to $\frac{1}{10}$ kip.

The stresses allowed for the metal of the connected parts (known as the *base metal*) apply to complete penetration groove welds that are stressed in tension or compression parallel to the

TABLE 8-1. Allowable Working Strength of Fillet Welds

Size of Weld (in.)	Allowable Load (kips/in)		Allowable Load (kN/mm)		Size of Weld (mm)
	E 60 XX Electrodes $F_{vw} = 18$ (ksi)	E 70 XX Electrodes $F_{vw} = 21$ (ksi)	E 60 XX Electrodes $F_{vw} = 124$ (MPa)	E 70 XX Electrodes $F_{vw} = 145$ (Mpa)	
3/16	2.4	2.8	0.42	0.49	4.76
1/4	3.2	3.7	0.56	0.65	6.35
5/16	4.0	4.6	0.70	0.81	7.94
3/8	4.8	5.6	0.84	0.98	9.52
1/2	6.4	7.4	1.12	1.30	12.7
5/8	8.0	9.3	1.40	1.63	15.9
3/4	9.5	11.1	1.66	1.94	19.1

axis of the weld or are stressed in tension perpendicular to the effective throat. They apply also to complete or partial penetration groove welds stressed in compression normal to the effective throat and in shear on the effective throat. Consequently, allowable stresses for butt welds are the same as for the base metal.

The relation between the weld size and the maximum thickness of material in joints connected only by fillet welds is shown in Table 8-2. The maximum size of a fillet weld applied to the square edge of a plate or section that is $\frac{1}{4}$ in. or more in thickness should be $\frac{1}{16}$ in. less than the nominal thickness of the edge. Along edges of material less than $\frac{1}{4}$ in. thick, the maximum size may be equal to the thickness of the material.

The effective area of butt and fillet welds is considered to be the effective length of the weld multiplied by the effective throat thickness. The minimum effective length of a fillet weld should not be less than four times the weld size. For starting and stopping the arc, approximately $\frac{1}{4}$ in. should be added to the design length of fillet welds.

Figure 8-3a represents two plates connected by fillet welds. The welds marked A are longitudinal; B indicates a transverse weld. If a load is applied in the direction shown by the arrow, the stress distribution in the longitudinal weld is not uniform, and the stress in the transverse weld is approximately 30% higher per unit of length.

Added strength is given to a transverse fillet weld that terminates at the end of a member, as shown in Fig. 8-3b, if the weld is

TABLE 8-2.   Relation between Material Thickness and Minimum Size of Fillet Welds

Material Thickness of the Thicker Part Joined		Minimum Size of Fillet Weld	
(in.)	(mm)	(in.)	(mm)
To 1/4 inclusive	To 6.35 inclusive	1/8	3.18
Over 1/4 to 1/2	Over 6.35 to 12.7	3/16	4.76
Over 1/2 to 3/4	Over 12.7 to 19.1	1/4	6.35
Over 3/4	Over 19.1	5/16	7.94

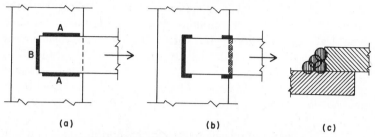

FIGURE 8-3. Welding of lapped plates.

returned around the corner for a distance not less than twice the weld size. These end returns, sometimes called *boxing,* afford considerable resistance to the tendency of tearing action on the weld.

The $\frac{1}{4}$-in. fillet weld is considered to be the minimum practical size, and a $\frac{5}{16}$-in. weld is probably the most economical size that can be obtained by one pass of the electrode. A small continuous weld is generally more economical than a larger discontinuous weld if both are made in one pass. Some specifications limit the singlepass fillet weld to $\frac{5}{16}$ in. Large fillet welds require two or more passes (multipass welds) of the electrode, as shown in Fig. 8-3c.

8-5   Design of Welded Joints

The most economical weld to use for a given condition depends on several factors. It should be borne in mind that members to be connected by welding must be firmly clamped or held rigidly in position during the welding process. When riveting a beam to a column, you must provide a seat angle as a support to keep the beam in position for riveting the connecting angles. The seat angle is not considered as adding strength to the connection. Similarly, seat angles are commonly used with welded connections. The designer must have in mind the actual conditions during erection and must provide for economy and ease in working the welds. Seat angles or similar members used to facilitate erection are *shop-welded* before the material is sent to the site. The welding done during erection is called *field-welding*. In preparing welding details, the designer indicates shop or field welds on the drawings. Conventional welding symbols are used to identify the type, size, and position of the various welds. Only engineers or architects experienced in the design of welded connections should design or supervise welded construction. It is apparent that a wide variety of connections is possible; experience is the best aid in determining the most economical and practical connection.

The following examples illustrate the basic principles on which welded connections are designed.

Example 1. A bar of A36 steel, $3 \times \frac{7}{16}$ in. [76.2 × 11 mm] in cross section, is to be welded with E 70 XX electrodes to the back of a channel so that the full tensile strength of the bar may be developed. What is the size of the weld? (See Fig. 8-4.)

Solution: The area of the bar is $3 \times 0.4375 = 1.313$ in² [76.2 × 11 = 838.2 mm²]. Because the allowable unit tensile stress of the steel is 22 ksi (Table 5-3), the tensile strength of the bar is $F_t \times A = 22 \times 1.313 = 28.9$ kips [152 × 838.2/10³ = 127 kN]. The weld must be of ample dimensions to resist a force of this magnitude.

A $\frac{3}{8}$-in. [9.52-mm] fillet weld will be used. Table 8-1 gives the allowable working strength as 5.6 kips/in. [0.98 kN/mm]. Hence the required length of weld to develop the strength of the bar is $28.9 \div 5.6 = 5.16$ in. [127 ÷ 0.98 = 130 mm]. The position of the weld with respect to the bar has several options, three of which are shown in Fig. 8-4a, c, and d.

Example 2. A $3\frac{1}{2} \times 3\frac{1}{2} \times \frac{5}{16}$-in. [89 × 89 × 7.94-mm] angle of A36 steel subjected to a tensile load is to be connected to a plate by fillet welds, using E 70 XX electrodes. What should the dimen-

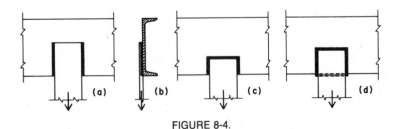

FIGURE 8-4.

sions of the welds be to develop the full tensile strength of the angle?

Solution: We shall use a $\frac{1}{4}$-in. fillet weld which has an allowable working strength of 3.7 kips/in. [0.65 kN/mm] (Table 8-1). From Table 4-4 the cross-sectional area of the angle is 2.09 in^2 [1348 mm^2]. By using the allowable tension stress of 22 ksi [152 MPa] for A36 steel (Table 5-3), the tensile strength of the angle is 22 × 2.09 = 46 kips [152 × 1348/10^3 = 205 kN]. Therefore the required total length of weld to develop the full strength of the angle is 46 ÷ 3.7 = 12.4 in. [205 ÷ 0.65 = 315 mm].

An angle is an unsymmetrical cross section, and the welds marked L_1 and L_2 in Fig. 8-5 are made unequal in length so that their individual resistance will be proportioned in accordance to the distributed area of the angle. From Table 4-4 we find that the centroid of the angle section is 0.99 in. [25 mm] from the back of the angle; hence the two welds are 0.99 in. [25 mm] and 2.51 in. [64 mm] from the centroidal axis, as shown in Fig. 8-5. The lengths of welds L_1 and L_2 are made inversely proportional to their distances from the axis, but the sum of their lengths is 12.4 in. [315 mm]. Therefore

$$L_1 = \frac{2.51}{3.5} \times 12.4 = 8.9 \text{ in.}$$

$$\left[\frac{64}{89} \times 315 = 227 \text{ mm} \right]$$

and

$$L_2 = \frac{0.99}{3.5} \times 12.4 = 3.5 \text{ in.}$$

$$\left[\frac{25}{89} \times 315 = 88 \text{ mm} \right]$$

These are the design lengths required, and as noted earlier, each weld would actually be made $\frac{1}{4}$ in. [6.4 mm] longer than its computed length.

When angle shapes are used as tension members and connected by fastening only one leg, it is questionable to assume a stress distribution of equal magnitude on the entire cross section.

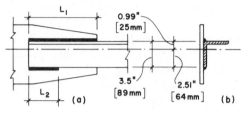

FIGURE 8-5.

Some designers therefore prefer to ignore the stress in the uncon-
nected leg and to limit the capacity of the member in tension to
the force obtained by multiplying the allowable stress by the area
of the connected leg only. If this is done, it is logical to use welds
of equal length on each side of the leg, as in Example 1.

Problem 8-5-A*. A 4 × 4 × $\frac{1}{2}$-in. angle of A36 steel is to be welded to a
plate with E 70 XX electrodes to develop the full tensile strength of the
angle. Using $\frac{3}{8}$-in. fillet welds, compute the design lengths L_1 and L_2, as
shown in Fig. 8-5, assuming the development of tension on the entire
cross section of the angle.

Problem 8-5-B. Redesign the welded connection in Problem 8-5-A assuming that the tension force is developed only by the connected leg of
the angle.

8-6 Plug and Slot Welds

One method of connecting two overlapping plates uses a weld in a
hole made in one of the two plates. (See Fig. 8-6.) Plug and slot
welds are those in which the entire area of the hole or slot receives weld metal. The maximum and minimum diameters of plug
and slot welds and the maximum length of slot welds are shown in
Fig. 8-6. If the plate containing the hole is not more than $\frac{5}{8}$ in.
thick, the hole should be filled with weld metal. If the plate is
more than $\frac{5}{8}$ in. thick, the weld metal should be at least one-half
the thickness of the material but not less than $\frac{5}{8}$ in.

The stress in a plug or slot weld is considered to be shear on
the area of the weld at the plane of contact of the two plates being
connected. The allowable unit shearing stress, when E 70 XX
electrodes are used, is 21 ksi [145 MPa].

D: minimum = t + $^5/_{16}$ in.   L: maximum = IO X t_1
    maximum = 2 $^1/_4$ X t_1

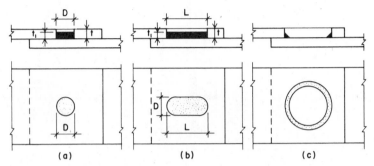

FIGURE 8-6.   Welds in holes: (a) plug weld, (b) slot weld, (c) fillet weld in
a large hole.

A somewhat similar weld consists of a continuous fillet weld at the circumference of a hole, as shown in Fig. 8-6c. This is not a plug or slot weld and is subject to the usual requirements for fillet welds discussed in Section 8-1.

8-7 Miscellaneous Welded Connections

Part 4 of the AISC Manual contains a series of tables that pertain to the design of welded connections. The tables cover free-end as well as moment-resisting connections. In addition, suggested framing details are shown for various situations.

A few common connections are shown in Fig. 8-7. As an aid to erection, certain parts are welded together in the shop before being sent to the site. Connection angles may be shop-welded to beams and the angles field-welded or field-bolted to girders or columns. The beam connection in Fig. 8-7a shows a beam supported on a seat that has been shop-welded to the column. A small connection plate is shop-welded to the lower flange of the beam, and the plate is bolted to the beam seat. After the beams have been erected and the frame plumbed, the beams are field-welded to the seat angles. This type of connection provides no degree of continuity in the form of moment transfer between the beam and column.

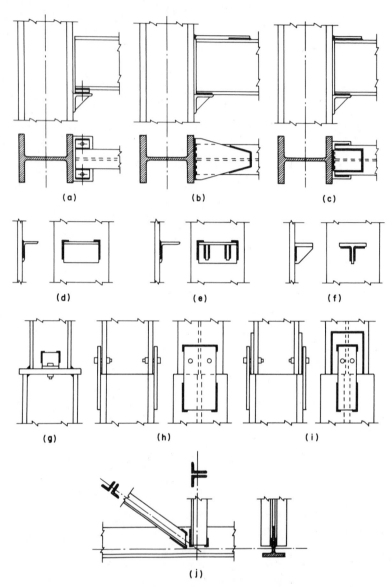

FIGURE 8-7. Welded framing connections.

The connections shown in Fig. 8-7*b* and *c* are designed to develop some moment transfer between the beam and its supporting column. Auxiliary plates are used to make the connection at the upper flanges.

Beam seats shop-welded to columns are shown in Fig. 8-7*d*, *e*, and *f*. A short length of angle welded to the column with no stiffeners is shown in Fig. 8-7*d*. Stiffeners consisting of triangular plates are welded to the legs of the angles shown in Fig. 8-7*e* and add materially to the strength of the seat. Another method of forming a seat, using a short piece of structural tee, is shown in Fig. 8-7*f*.

Various types of column splice are shown in Fig. 8-7*g*, *h*, and *i*. The auxiliary plates and angles are shop-welded to the columns and provide for bolted connections in the field before the permanent welds are made. Welded connections for column base plates are shown in Fig. 6-5.

Figure 8-7*j* shows a type of welded construction used in light trusses in which the lower chord consists of a structural tee. Truss web members consisting of pairs of angles are welded to the stem of the tee chord. Other truss connections are shown in the details in Chapter 20.

Some additional connection details are given in Fig. 8-8. The detail in Fig. 8-8*a* is an arrangement for framing a beam to a girder, in which welds are substituted for bolts or rivets. In this figure welds replace the fasteners that secure the connection angles to the web of the supported beam.

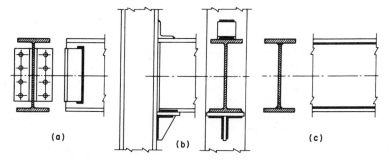

FIGURE 8-8. Some additional welded framing connections.

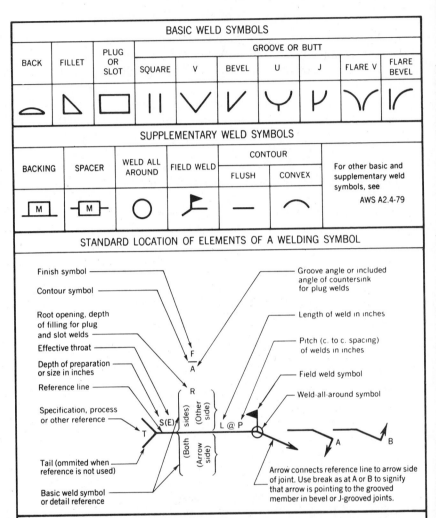

The image shows three sections: "BASIC WELD SYMBOLS", "SUPPLEMENTARY WELD SYMBOLS", and "STANDARD LOCATION OF ELEMENTS OF A WELDING SYMBOL".

BASIC WELD SYMBOLS

BACK	FILLET	PLUG OR SLOT	GROOVE OR BUTT						
			SQUARE	V	BEVEL	U	J	FLARE V	FLARE BEVEL

SUPPLEMENTARY WELD SYMBOLS

BACKING	SPACER	WELD ALL AROUND	FIELD WELD	CONTOUR		For other basic and supplementary weld symbols, see AWS A2.4-79
				FLUSH	CONVEX	

STANDARD LOCATION OF ELEMENTS OF A WELDING SYMBOL

Finish symbol

Contour symbol

Root opening, depth of filling for plug and slot welds

Effective throat

Depth of preparation or size in inches

Reference line

Specification, process or other reference

Tail (ommited when reference is not used)

Basic weld symbol or detail reference

Groove angle or included angle of countersink for plug welds

Length of weld in inches

Pitch (c. to c. spacing) of welds in inches

Field weld symbol

Weld-all-around symbol

Arrow connects reference line to arrow side of joint. Use break as at A or B to signify that arrow is pointing to the grooved member in bevel or J-grooved joints.

Note:
   Size, weld symbol, length of weld and spacing must read in that order from left to right along the reference line. Neither orientation of reference line nor location of the arrow alter this rule.
   The perpendicular leg of weld symbols must be at left.
   Arrow and Other Side welds are of the same size unless otherwise shown. Dimensions of fillet welds must be shown on both the Arrow Side and the Other Side Symbol.
   The point of the field weld symbol must point toward the tail.
   Symbols apply between abrupt changes in direction of welding unless governed by the "all around" symbol or otherwise dimensioned.
   These symbols do not explicitly provide for the case that frequently occurs in structural work, where duplicate material (such as stiffeners) occurs on the far side of a web or gusset plate. The fabricating industry has adopted this convention: that when the billing of the detail material discloses the existence of a member on the far side as well as on the near side, the welding shown for the near side shall be duplicated on the far side.

FIGURE 8-9. Standard weld symbols used on construction drawings. Reprinted from the *Manual of Steel Construction*, 8th ed., with permission of the publishers, American Institute of Steel Construction.

A welded connection for a stiffened seated beam connection to a column is shown in Fig. 8-8*b*. Figure 8-8*c* shows the simplicity of welding in connecting the upper and lower flanges of a plate girder to the web plate.

8-8 Symbols for Welds

Standard symbols are used in detail drawings of welded connections of structural elements. In addition to the type of weld, other information to be conveyed includes size, exact location, and finishes. Figure 8-9, reproduced from the AISC Manual, gives the standard symbols for welded joints. It will be noted that the symbol for a fillet weld is a triangle; this is drawn below the horizontal line if the weld is on the near side, above if it is on the far side; two triangles, one above and one below, are drawn for welds on both sides of the joint. The size of the weld is placed to the left of the vertical line of the triangle and the length to the right side of the hypotenuse.

9

Plastic Behavior and Strength Design

||||||||||||||||||||||||||||||||||||||||||||||||||||||||||

9-1   Plastic versus Elastic Behavior

Up to this point the discussions of the design of members in bending have been based on bending stresses well within the yield-point stress. In general, allowable stresses are based on the *theory of elastic behavior*. However, it has been found by tests that steel members can carry loads much higher than anticipated, even when the yield-point stress is reached at sections of maximum bending moment. This is particularly evident in continuous beams and in structures with rigid connections. An inherent property of structural steel is its ability to resist large deformations without failure. These large deformations occur chiefly in the *plastic range*, with no increase in the magnitude of the bending stress. Because of this phenomenon, the *plastic design theory*, sometimes called the *ultimate-strength design theory* (or more recently *strength design*), has been developed.

Figure 9-1 represents the typical form of a load-test response for a specimen of ductile steel. The graph shows that up to a stress f_y, the yield point, the deformations are directly proportional to the applied stresses and that beyond the yield point there is a deformation without an increase in stress. For A36 steel this

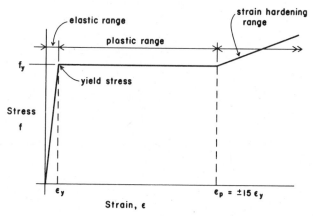

FIGURE 9-1. Form of the stress-strain response for ductile steel.

additional deformation, called the *plastic range,* is approximately
15 times that produced elastically. Note that beyond this range
strain hardening (loss of ductility) begins, when further deformation can occur only with an increase in stress.

For plastic behavior to be significant, the extent of the plastic
range of deformation must be several times the elastic deformation. As the yield point is increased in magnitude, this ratio of
deformations decreases, which is to say that higher strength
steels tend to be less ductile. At present, the theory of plastic
design is generally limited to steels with a yield point of not more
than 65 ksi [450 MPa].

9-2   Plastic Moment and the Plastic Hinge

Section 7-1 explains the design of members in bending in accordance with the theory of elasticity. When the extreme fiber stress
does not exceed the elastic limit, the bending stresses in the cross
section of a beam are directly proportional to their distances from
the neutral surface. In addition, the strains (deformations) in
these fibers are also proportional to their distances from the neutral surface. Both stresses and strains are zero at the neutral
surface, and both increase to maximum magnitudes at the fibers
farthest from the neutral surface.

The following example illustrates the analysis of a steel beam for bending, according to the theory of elastic behavior.

Example. A simple steel beam has a span of 16 ft [4.88 m] with a concentrated load of 18 kips [80 kN] at the center of the span. The section used is an S 12 × 31.8, the beam is adequately braced throughout its length, and the beam weight is ignored in the computations. Let us compute the maximum extreme fiber stress. (See Fig. 9-2 d.)

Solution: To do this we use the flexure formula

$$f = \frac{M}{S} \qquad \text{(Section 4-2)}$$

Then

$$M = \frac{PL}{4} = \frac{18 \times 16}{4} = 72 \text{ kip-ft}$$

$$\left[M = \frac{80 \times 4.88}{4} = 97.6 \text{ kN-m} \right]$$

which is the maximum bending moment. In Table 4-2 we find $S = 36.4 \text{ in}^3$ [$597 \times 10^3 \text{ mm}^3$]. Thus

$$f = \frac{M}{S} = \frac{72 \times 12}{36.4} = 24 \text{ ksi}$$

$$\left[f = \frac{97.6 \times 10^6}{597 \times 10^3} = 164 \text{ MPa} \right]$$

which is the stress on the fiber farthest from the neutral surface. (See Fig. 9-2d.)

Note that this stress occurs only at the beam section at the center of the span, where the bending moment has its maximum value. Figure 9-2e shows the deformations that accompany the stresses shown in Fig. 9-2d. Note that both stresses and deformations are directly proportional to their distances from the neutral surface in elastic analysis.

When a steel beam is loaded to produce an extreme fiber stress in excess of the yield point, the property of the material's ductility affects the distribution of the stresses in the beam cross sec-

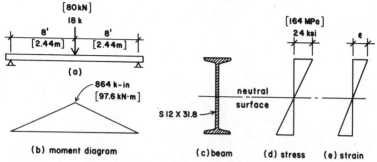

(a)

(b) moment diagram   (c) beam   (d) stress   (e) strain

FIGURE 9-2.   Elastic behavior of the beam.

tion. Elastic analysis does not suffice to explain this phenomenon because the beam will experience some plastic deformation.

Assume that the bending moment on a beam is of such magnitude that the extreme fiber stress is f_y, the yield stress. Then if M_y is the elastic bending moment at the yield stress, $M = M_y$, and the distribution of the stresses in the cross section is as shown in Fig. 9-3a; the maximum bending stress f_y is at the extreme fiber.

Next consider that the loading and the resulting bending moment have been increased; M is now greater than M_y. The stress on the extreme fiber is still f_y, but *the material has yielded* and a greater area of the cross section is also stressed to f_y. The stress distribution is shown in Fig. 9-3b.

Now imagine that the load is further increased. The stress on

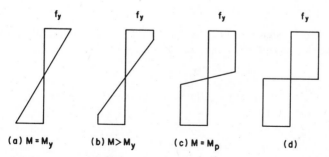

(a) $M = M_y$ (b) $M > M_y$ (c) $M = M_p$ (d)

FIGURE 9-3.   Progression of stress response—elastic to plastic.

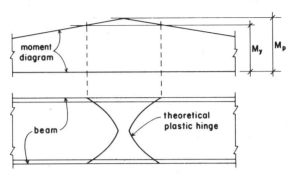

FIGURE 9-4. Development of the plastic hinge.

the extreme fiber is still f_y and, theoretically, *all fibers in the cross section are stressed to f_y*. This idealized plastic stress distribution is shown in Fig. 9-3d. The bending moment that produces this condition is M_p, the plastic bending moment. In reality about 10% of the central portion of the cross section continues to resist in an elastic manner, as indicated in Fig. 9-3c. This small resistance is quite negligible, and we assume that the stresses on all fibers of the cross section are f_y, as shown in Fig. 9-3d. The section is now said to be fully plastic, and any further increase in load will result in large deformations; the beam acts as if it were hinged at this section. We call this a plastic hinge, at which free rotation is permitted only after M_p has been attained. (See Fig. 9-4.) At sections of a beam in which this condition prevails, the bending resistance of the cross section has been exhausted.

9-3 Plastic Section Modulus

In elastic design the moment that produces the maximum allowable resisting moment may be found by the flexure formula

$$M = f \times S$$

where M = maximum allowable bending moment in inch-pounds,

f = maximum allowable bending stress in pounds per square inch,

S = section modulus in inches to the third power.

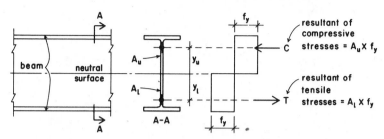

FIGURE 9-5. Development of the plastic resisting moment.

If the extreme fiber is stressed to the yield stress,

$$M_y = f_y \times S$$

where M_y = elastic bending moment at the yield stress,
  f_y = yield stress in pounds per square inch,
  S = section modulus in inches to the third power.

Now let us find a similar relation between the plastic moment and its plastic resisting moment. Refer to Fig. 9-5, which shows the cross section of a W or S section in which the bending stress f_y, the yield stress, is constant over the cross section. In the figure,

  A_u = upper area of the cross section above the neutral axis, in square inches,
  y_u = distance of the centroid of A_u from the neutral axis,
  A_l = lower area of the cross section below the neutral axis, in square inches,
  y_l = distance of the centroid of A_l from the neutral axis.

For equilibrium the algebraic sum of the horizontal forces must be zero. Then

$$\sum H = 0$$

or

$$[A_u \times (+f_y)] + [A_l \times (-f_y)] = 0$$

and

$$A_u = A_l$$

This shows that the neutral axis divides the cross section into equal areas, which is apparent in symmetrical sections, but it applies to unsymmetrical sections as well. Also the bending moment equals the sum of the moments of the stresses in the section. Thus for M_p, the plastic moment,

$$M_p = (A_u \times f_y \times y_u) + (A_l \times f_y \times y_l)$$

or

$$M_p = f_y[(A_u \times y_u) + (A_l \times y_l)]$$

and

$$M_p = f_y \times Z$$

The quantity $(A_u y_u + A_l y_l)$ is called the *plastic section modulus* of the cross section and is designated by the letter Z; because it is an area multiplied by a distance, it is in units to the third power. If the area is in units of square inches and the distance is in linear inches, Z, the section modulus, is in units of inches to the third power.

The plastic section modulus is always larger than the elastic section modulus.

It is important to note that in plastic design the neutral axis for unsymmetrical cross sections does not pass through the centroid of the section. In plastic design the neutral axis divides the cross section into *equal* areas.

9-4 Computation of the Plastic Section Modulus

The notation used in Section 9-3 is appropriate for both symmetrical and unsymmetrical sections. Consider now a symmetrical section such as a W or S shape, as shown in Fig. 9-5. $A_u = A_l$, $y_u = y_l$, and $A_u + A_l = A$, the total area of the cross section. Then

$$M_p = (A_u \times f_y \times y_u) + (A_l \times f_y \times y_l)$$

and

$$M_p = f_y \times A \times y \quad \text{or} \quad M_p = f_y \times Z$$

where f_y = yield stress,
 A = total area of the cross section,
 y = distance from the neutral axis to the centroid of the portion of the area on either side of the neutral axis,
 Z = plastic modulus of the section (in in^3 or mm^3).

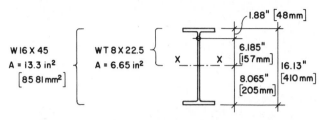

FIGURE 9-6.

Now, because $Z = A \times y$, we can readily compute the value of the plastic section modulus of a given cross section.

Consider a W 16 × 45. In Table 4-1 we find that its total depth is 16.13 in. [410 mm] and its cross-sectional area is 13.3 in^2 [8581 mm^2]. From the AISC Manual we find that a WT 8 × 22.5 (which is one-half a W 16 × 45) has its centroid located 1.88 in. [48 mm] from the outside of the flange. Therefore the distance from the centroid of either half of the W shape to the neutral axis is one-half the beam depth less the distance obtained for the tee. Thus the distance y is (see Fig. 9-6)

$$y = \left(\frac{16.13}{2}\right) - 1.88 = 6.185 \text{ in.}$$

$$\left[y = \left(\frac{410}{2}\right) - 48 = 157 \text{ mm}\right]$$

Then the plastic modulus of the W 16 × 45 is

$$Z = A \times y = 13.3 \times 6.185 = 82.26 \text{ in}^3$$

$$[Z = 8581 \times 157 = 1347 \times 10^3 \text{ mm}^3]$$

Use of the full value of the plastic hinge moment requires the shape to have limited values for the width–thickness ratio of the flanges and the depth–thickness ratio of the web. These requirements are given in Section 2-7 of the AISC Specification.

9-5 Load Factor

Consider a beam of A36 steel laterally supported throughout its length. Its span is 24 ft [7.315 m], and it carries a concentrated load of 42 kips [186.8 kN] at the center. Let us determine the size of the beam in accordance with the theory of elastic behavior.

The maximum bending moment for this beam is

$$M = \frac{PL}{4} = \frac{42 \times 24}{4} = 252 \text{ kip-ft}$$

$$\left[M = \frac{186.8 \times 7.315}{4} = 341.6 \text{ kN-m} \right]$$

and the required section modulus is

$$S = \frac{M}{f_b} = \frac{252 \times 12}{24} = 126 \text{ in}^3$$

$$\left[S = \frac{341.6 \times 10^6}{165} = 2070 \times 10^3 \text{ mm}^3 \right]$$

Table 4-1 shows that a W 21 × 62 has a section modulus of 127 in^3 and is acceptable.

Now let us compute the magnitude of the concentrated load at the center of the span that would produce a bending moment equal to the plastic resisting moment. In the AISC Manual, the plastic modulus for the W 21 × 62 is 144 in^3. Thus the plastic moment is

$$M_p = F_y \times Z_x = 36 \times 144 = 5184 \text{ kip-in, or } 432 \text{ kip-ft}$$

$$\left[M_p = 248 \times \frac{2360 \times 10^3}{10^6} = \ = 585.3 \text{ kN-m} \right]$$

Then the load that corresponds to this moment is

$$M_p = 432 = \frac{PL}{4} = \frac{P \times 24}{4}$$

and

$$P = \frac{4 \times 432}{24} = 72 \text{ kips}$$

$$\left[M_p = 585.3 = \frac{P \times 7.315}{4}, P = 320 \text{ kN} \right]$$

This load would produce a plastic hinge at the center of the span and a slight increase in load would result in failure.

The term *load factor* is given to the ratio of the ultimate load to the design load. In this example it is 72/42 = 1.714.

In elastic design the allowable bending stress F_b is decreased to a fraction of F_y, the yield stress. For compact sections, $F_b = 0.66$ F_y. Therefore the implied factor of safety against yielding is 1/0.66, or 1.5. This factor is higher, of course, if the beam's limiting capacity is taken as the plastic moment.

In plastic design the concept of allowable stress is not used, and computations are based strictly on the limit of the yield stress. Safety is produced by the use of the load factor, by which the beam is literally designed to fail but at a load larger than that it actually must sustain. For simple and continuous beams, the load factor is specified as 1.7; for rigid frames it is 1.85.

9-6 Design of a Simple Beam

The design of simple beams by the elastic or plastic theory will usually result in the same size beam, as illustrated in the following examples.

Example 1. A simple beam of A36 steel has a span of 20 ft [6.1 m] and supports a uniformly distributed load of 4.8 kips/ft [70 kN/m], including its own weight. Design this beam in accordance with the elastic theory, assuming that it is laterally supported throughout its length.

Solution: The maximum bending moment is $wL^2/8$. Then

$$M = \frac{4.8 \times (20)^2}{8} = 240 \text{ kip-ft}$$

$$\left[\frac{70 \times (6.1)^2}{8} = 325.6 \text{ kN-m} \right]$$

By use of the allowable bending stress of 24 ksi [165 MPa] for a compact section, the required section modulus is

$$S = \frac{M}{F_b} = \frac{240 \times 12}{24} = 120 \text{ in}^3$$

$$\left[S = \frac{325.6 \times 10^6}{165} = 1973 \times 10^3 \text{ mm}^3 \right]$$

In Table 4-1 we find a W 21 × 62 with $S_x = 127 \text{ in}^3$.

Example 2.   Design the same beam in accordance with the plastic theory.

Solution: We adjust the load with the load factor, which is given in Section 9-5 as 1.7. Thus

$$w_p = 4.8 \times 1.7 = 8.16 \text{ kips/ft}$$

$$[w_p = 70 \times 1.7 = 119 \text{ kN/m}]$$

and the maximum bending moment is

$$M_p = \frac{w_p L^2}{8} = \frac{8.16 \times (20)^2}{8} = 408 \text{ kip-ft}$$

$$\left[M_p = \frac{119 \times (6.1)^2}{8} = 553.5 \text{ kN-m} \right]$$

and

$$M_p = F_y \times Z, \, Z = \frac{M_p}{F_y} = \frac{408 \times 12}{36} = 136 \text{ in}^3$$

$$\left[Z = \frac{553.5 \times 10^6}{248} = 2232 \times 10^3 \text{ mm}^3 \right]$$

which is the minimum plastic section modulus. From the AISC Manual select a W 21 × 62 for which $Z_x = 144$ in³. Note that the elastic and plastic theories have yielded the same result in this example, which is common for simple beams.

The reader who wishes to pursue the subject of plastic design is referred to *Commentary on Plastic Design in Steel* by the American Society of Civil Engineers or *Plastic Design in Steel* by the American Institute of Steel Construction.

WOOD CONSTRUCTION

||||||||||||||||||||||||||||||||||||||||||||||||||||||||||||||||||

10

Wood Beams

IIIIIIIIIIIIIIIIIIIIIIIIIIIIIIIIIIIIIIIIIIIIIIIIIIIIIIIIIIIII

10-1 Structural Lumber

Unlike the metals, wood is not a processed material but an organic material generally used in its natural state. Aside from the natural properties of the species, the most important factors that influence its strength are density, natural defects (knots, checks, slope of grain, etc.), and moisture content. Because the effects of natural defects on the strength of lumber vary with the type of loading to which an individual piece is subjected, structural lumber is classified according to its *size and use*. The three major classifications are:

1. *Joists and planks*. Rectangular cross sections with nominal dimensions 2–4 in. thick and 4 or more in. wide, graded primarily for strength in bending edgewise or flatwise.

2. *Beams and stringers*. Rectangular cross sections with nominal dimensions 5×8 in. and larger, graded for strength in bending when loaded on the narrow face.

3. *Posts and timbers*. Square or nearly square cross sections with nominal dimensions 5×5 in. and larger, graded primarily for use as posts or columns but adapted to other uses where bending strength is not especially important.

The two groups of trees used for building purposes are the *softwoods* and the *hardwoods*. Softwoods such as the pines or cypress are coniferous or cone bearing, whereas hardwoods have broad leaves, as exemplified by the oaks and maples. The terms softwood and hardwood are not accurate indications of the degree of hardness of the various species of trees. Certain softwoods are as hard as the medium density hardwoods, while some species of hardwoods have softer wood than some of the softwood species. The two species of trees used most extensively in the United States for structural members are Douglas Fir and Southern Pine, both of which are classified among the softwoods.

10-2   Nominal and Dressed Sizes

Note: For sake of brevity, SI units have been omitted from the text discussion and most tabular data. However, example computations and exercise problems are presented with data in both U.S. and SI Units.

An individual piece of structural lumber is designated by its *nominal* cross-sectional dimensions; the size is indicated by the breadth and depth of the cross section in inches. As an example, we speak of a "6 by 12" (written 6 × 12), and by this we mean a timber with a nominal breadth of 6 in. and depth of 12 in.; the length is variable. However, after being dressed or surfaced on four sides (S4S), the actual dimensions of this piece are $5\frac{1}{2} \times 11\frac{1}{2}$ in. Since lumber used in structural design is almost exclusively dressed lumber, the sectional properties (A, I, and S) given in Table 4-7 are for standard dressed sizes conforming to those established by the industry.

10-3   Allowable Stresses for Structural Lumber

Many factors are taken into account in determining the allowable unit stresses for structural lumber. Numerous tests by the Forest Products Laboratory of the U.S. Department of Agriculture made on material free from defects have resulted in a tabulation known as *clear wood strength values*. To obtain allowable design stresses, we must reduce the clear wood values by factors that take into consideration the loss of strength from defects, size and position of knots, size of member, degree of density, and condition of seasoning. These adjustments are made in accordance with various industry grading standards. Grading is necessary to identify lumber quality. Individual grades are given a commercial designation, such as No. 1, No. 2, select structural, dense No. 2, etc., to which a schedule of allowable unit stresses is assigned.

Table 10-1, which has been compiled from more extensive data given in the 1982 edition of *Design Values for Wood Construction,* lists some of the most commonly used stress-grade woods and their allowable working stresses. The working stresses tabulated therein are for normal loading conditions and are applicable to lumber that will be used under continuously dry conditions, such as exist in most covered structures. Where wet conditions exist, and in situations where a member is fully stressed to the maximum allowable unit stress for many years (full load permanently applied), the allowable stress values in Table 10-1 are subject to adjustments. Methods for making such adjustments are given in the reference cited above. The stresses given are: extreme fiber in bending, F_b; tension parallel to grain, F_t; horizontal shear, F_v; compression perpendicular to grain, $F_{c\perp}$; compression parallel to grain, F_c; and modulus of elasticity, E. In addition, there is a column on size classification, which relates to the three major classifications described in Section 10-1.

It will be noted that two sets of values are given in the table for F_b, the extreme fiber stress in bending. The values listed for single-member uses apply where an individual beam or girder carries its full design load; the values given for repetitive-member uses are intended for design of members in bending, such as joists, rafters, or similar members, that are spaced not more than 24 in., are three or more in number, and are joined by floor, roof, and other load-distributing construction adequate to support the design load.

The notes in parentheses following species designations indicate whether the lumber is surfaced (dressed) when dry or green and the maximum moisture content at which it is intended to be used. Dry lumber is defined as lumber that has been seasoned to a

TABLE 10-1. Allowable Unit Stresses for Structural Lumber—Visual Grading
(allowable unit stresses listed are for normal loading conditions)

Species and Commercial Grade	Size Classification	Design Values in Pounds Per Square Inch						
		Extreme Fiber in Bending F_b		Tension Parallel to Grain F_t	Horizontal Shear F_v	Compression Perpendicular to Grain $F_{c\perp}$	Compression Parallel to Grain F_c	Modulus of Elasticity E
		Single-Member Uses	Repetitive-Member Uses					
Douglas Fir–Larch (Surfaced dry or surfaced green. Used at 19% max moisture content)								
Dense Select Structural	2–4 in. thick, 5 in. and wider	2100	2400	1400	95	730	1650	1,900,000
Select Structural		1800	2050	1200	95	625	1400	1,800,000
Dense No. 1		1800	2050	1200	95	730	1450	1,900,000
No. 1		1500	1750	1000	95	625	1250	1,800,000
Dense No. 2		1450	1700	775	95	730	1250	1,700,000
No. 2		1250	1450	650	95	625	1050	1,700,000
No. 3		725	850	375	95	625	675	1,500,000
Dense Select Structural	beams and stringers	1900	—	1100	85	730	1300	1,700,000
Select Structural		1600	—	950	85	625	1100	1,600,000
Dense No. 1		1550	—	775	85	730	1100	1,700,000
No. 1		1300	—	675	85	625	925	1,600,000
Dense Select Structural	posts and	1750	—	1150	85	730	1350	1,700,000

Grade	Size							
Select Structural	timbers	1500	—	1000	85	625	1150	1,600,000
Dense No. 1		1400	—	950	85	730	1200	1,700,000
No. 1		1200	—	825	85	625	1000	1,600,000
Southern Pine (Surfaced dry. Used at 19% max moisture content)								
Select Structural	2–4 in. thick,	1750	2000	1150	90	565	1350	1,700,000
Dense Select Structural	5 in. and wider	2050	2350	1300	90	660	1600	1,800,000
No. 1		1450	1700	975	90	565	1250	1,700,000
No. 1 Dense		1700	2000	1150	90	660	1450	1,800,000
No. 2		1200	1400	625	90	565	1000	1,600,000
No. 2 Dense		1400	1650	725	90	660	1200	1,600,000
No. 3		700	800	350	90	565	625	1,400,000
No. 3 Dense		825	925	425	90	660	725	1,500,000
Southern Pine (Surfaced green. Used in any condition.)								
Dense Structural 86	2.5 in. and thicker	2100	2400	1400	145	440	1300	1,600,000
Dense Structural 72		1750	2050	1200	120	440	1100	1,600,000
Dense Structural 65		1600	1800	1050	110	440	1000	1,600,000
No. 1 SR	5 in. and thicker	1350	—	875	110	375	775	1,500,000
No. 1 Dense SR		1550	—	1050	110	440	925	1,600,000
No. 2 SR		1000	—	725	95	375	625	1,400,000
No. 2 Dense SR		1250	—	850	95	440	725	1,400,000

Source: Adapted from more extensive tables in *Design Values for Wood Construction*, 1982 ed., with permission of the publishers, National Forest Products Association.

moisture content of 19% or less; green lumber has a moisture content in excess of 19%.

The allowable unit stresses to be used in actual design practice must, of course, conform to the requirements of the local building code. As noted earlier, many municipal codes are revised only infrequently and, consequently, may not be in agreement with current editions of industry-recommended allowable stresses. However, the allowable stresses for wood construction used throughout this book are those given in the National Design Specification and recommended by the National Forest Products Association.

10-4 Design for Bending

The design of a wood beam for strength in bending is accomplished by use of the flexure formula (Section 4-7). The form of this equation used in design is

$$S = \frac{M}{F_b}$$

in which M = maximum bending moment,
F_b = allowable extreme fiber (bending) stress,
S = required section modulus.

Although section moduli for standard rectangular wood beam sizes are given in Table 4-7, it is sometimes convenient to use the formula $S = bd^2/6$, which was developed in Section 4-6.

To determine the dimensions of a wood beam as governed by bending, first compute the maximum bending moment. Next, refer to a table such as Table 10-1, select the species and grade of lumber that is to be used, and note the corresponding allowable extreme fiber stress, F_b. Substitute these values in the flexure formula, and solve for the required section modulus. The proper beam size may be determined by referring to Table 4-7, which lists S for standard dressed sizes of structural lumber. Obviously, a number of different sections may be acceptable.

Example. A simple beam has a span of 16 ft [4.88 m] and supports a load, including its own weight, of 6500 lb [28.9 kN]. If the wood to be used is Douglas fir–larch, select structural grade, determine the size of the beam with the least cross-sectional area on the basis of limiting bending stress.

Solution: The maximum bending moment for the simple beam, Case 2 of Fig. 3-19, is

$$M = \frac{WL}{8} = \frac{6500 \times 16}{8} = 13{,}000 \text{ ft-lb } [17.63 \text{ kN-m}]$$

Referring to Table 10-1, we find under Douglas fir–larch, beams and stringers, select structural grade, that the limiting bending stress is 1600 psi [11.03 MPa]. Then, substituting in the beam formula and converting the bending moment to inch-pounds, we can calculate the required section modulus

$$S = \frac{M}{F_b} = \frac{13{,}000 \times 12}{1600} = 97.5 \text{ in}^3 [1.60 \times 10^6 \text{ mm}^3]$$

From Table 4-7 we find the section with the least area to be a 4×14 with $S = 102.411$ in^3 [1.68×10^6 mm^3]. Actually this size section falls into the higher stress category, for sections 2–4 in. thick and 5 in. and wider. Thus the allowable stress is 1800 psi [12.41 MPa], the required S drops to 86.67 in^3 [1.42×10^6 mm^3], and a 3×16 is the lightest choice.

A complete design would also require consideration of shear and deflection, as explained in Sections 10-5 and 10-6.

Problem 10-4-A*. The No. 1 grade of Douglas fir–larch is to be used for a series of floor beams spanning 14 ft [4.27 m]. If the total uniformly distributed load on each beam, including its own weight, is 3200 lb [14.23 kN], select the beam with the least cross-sectional area based on bending stress.

Problem 10-4-B. A simple beam of Douglas fir–larch, select structural grade, has a span of 18 ft [5.49 m] with two concentrated loads of 3 kips [13.34 kN] each placed at the third points of the span. Neglecting its own weight, determine the size of the beam with the least cross-sectional area based on bending stress.

Problem 10-4-C. A Southern pine beam of No. 1 SR grade has a span of 15 ft [4.57 m] with a single concentrated load of 6 kips [26.69 kN] placed 5 ft [1.52 m] from one support. Based on bending stress, select the beam with the least cross-sectional area. Include the effect of the beam weight as a uniformly distributed load, using a density of 35 pcf [561 kg/m³] for the wood.

Problem 10-4-D. A cantilever beam projects 6 ft [1.83 m] from the face of a masonry wall and supports a uniformly distributed load of 2 kips [8.90 kN], including its own weight. Determine the size of the required beam with the least cross-sectional area based on bending stress if Southern pine, No. 2 SR grade, is to be used.

10-5 Horizontal Shear

As discussed in Section 3-1 and illustrated in Figs. 3-1*b* and *d*, a beam has a tendency to fail in shear by the fibers sliding past each other both vertically and horizontally. Also, at any point in a beam, the intensity of the vertical and horizontal shearing stresses are equal. The vertical shear strength of wood beams is seldom of concern because the shear resistance of wood *across* the grain is much larger than it is *parallel* to the grain, where the horizontal shear forces develop.

The horizontal shearing stresses are not uniformly distributed over the cross section of a beam but are greatest at the neutral surface. The maximum horizontal unit shearing stress for rectangular sections is $\frac{3}{2}$ times the average vertical unit shearing stress. This is expressed by the formula

$$v = \frac{3}{2} \times \frac{V}{bd}$$

in which v = maximum unit horizontal shearing stress in psi,
   V = total vertical shear in pounds,
   b = width of cross section in inches,
   d = depth of cross section in inches.

It should be noted that the depth of the cross section is called h in Table 4-7. Notation usage is not entirely consistent in wood structural design, but the variations are of minor consequence only.

This formula applies only to rectangular cross sections. Timber is relatively weak in resistance to horizontal shear, and short spans with large loads should always be tested for this shearing tendency. Frequently a beam large enough to resist bending stresses must be made larger in order to resist horizontal shear.

Table 10-1 gives allowable horizontal unit shearing stresses for several stress grades of lumber under the column headed "Horizontal Shear, F_v."

Example. A 6 × 10 beam of Southern pine, No. 2 dense SR grade, has a total uniformly distributed load of 6000 lb [26.7 kN]. Investigate the beam for structural shear.

Solution: Since the beam is symmetrically loaded, $R_1 = R_2 = 6000/2 = 3000$ lb [13.35 kN]; this is also the value of the maximum shear force. Reference to Table 4.7 shows that the dressed dimensions of a 6 × 10 are 5.5 × 9.5 in. [140 × 240 mm]. Then

$$v = \frac{3}{2} \times \frac{V}{bd} = \frac{3}{2} \times \frac{3000}{5.5 \times 9.5} = 86.1 \text{ psi } [0.594 \text{ MPa}]$$

Referring to Table 10-1, under the size classification of 5 in. and thicker for the species and grade specified, we find that the allowable shear stress $F_v = 95$ psi [0.655 MPa]. The beam is therefore acceptable on the basis of consideration of shear stress.

For beams that are supported by end bearing, the code permits a computation for critical shear stress at a distance of the beam depth from the support. When the computed shear stress at the support is just slightly in excess of the allowable, this more precise computation may determine that the beam is acceptable. However, for simplicity in the procedure, we will ordinarily use the maximum shear at the support and make the computation directed by the code only when this value is not acceptable.

Note: In solving the following problems, use Tables 4-7 and 10-1 and neglect the beam weight.

Problem 10-5-A*. A 10 × 10 beam of Douglas fir–larch, select structural grade, supports a single concentrated load of 10 kips [44.5 kN] at the center of the span. Investigate the beam for shear stress.

Problem 10-5-B. A 10 × 14 beam of Douglas fir–larch, dense select structural grade, is loaded symmetrically with three concentrated loads of 4300 lb [19.13 kN] each, placed at the quarter points of the span. Is the beam safe in shear?

Problem 10-5-C. A 10 × 12 beam of Southern pine, No. 2 dense SR grade, is 8 ft [2.44 m] long and has a concentrated load of 8 kips [35.58 kN] located 3 ft [0.914 m] from one end. Investigate the beam for shear.

Problem 10-5-D. What should be the nominal cross-sectional dimensions for the beam of least weight that supports a total uniformly distributed load of 12 kips [53.4 kN] on a simple span and consists of Southern pine, No. 1 SR grade? Consider only the condition of limiting shear stress.

Problem 10-5-E. A 6 × 10 beam of Douglas fir–larch, dense select structural grade, is 18 ft [5.49 m] long. It supports a uniformly distributed load of 300 lb per linear ft [4.38 kN/m] over its entire length. Is the beam safe with respect to shear stress?

10-6 Deflection

Excessive vertical deflection of beams may be unsatisfactory because of a number of concerns. Some of these are visible sag, resulting cracks in plastered ceilings, bounciness of floors, and possible bearing of the beam on construction underneath the beam. In design work, deflection limitations are often expressed in terms of a certain percent of the span, specified in fraction form such as $\frac{1}{360}$, $\frac{1}{240}$, and so on, of the beam span length. These limitations sometimes are specified by building codes or are simply rules of thumb used commonly in practice.

There are two types of deflection that may be critical. The first is that caused by the live loads after the construction is completed. The most common limitation for this deflection is $\frac{1}{360}$ of the span—supposedly related to the curvature that may cause cracking of plaster ceiling surfaces. The second concern is for the total deflection caused by all the loads, both dead and live. This may be limited in some actual dimension by the details of the construction; if not, the most common restriction is for $\frac{1}{240}$ of the span.

For ordinary loadings, deflections may be computed using the formulas given in Fig. 3-19. When complex loadings occur, an approximate value for the maximum deflection can be obtained by using an equivalent uniformly distributed load (ETL as described in Section 5-14) and employing the formula for a simple beam with uniformly distributed load.

Example. Investigate the deflection of a 10 × 14 beam of Douglas fir–larch, dense select structural grade, 15 ft long and carrying a total uniformly distributed load of 16 kips, including its own weight. Deflection under total load is limited to $\frac{1}{240}$ of the beam span.

Solution: The deflection formula for this loading (Fig. 3-19, Case 2) is

$$D = \frac{5}{384} \times \frac{WL^3}{EI}$$

Referring to Table 4-7, we find that the moment of inertia of the 10×14 (actually 9.5×13.5 in.) is 1948 in^4 (rounded off from the tabular value). For the formula the length of the span must be used in inches; thus $L = 15 \times 12 = 180$ in. Substituting in the deflection formula,

$$D = \frac{5}{384} \times \frac{16,000 \times 180^3}{1,700,000 \times 1948} = 0.36 \text{ in.}$$

The limit for deflection is

$$D = \frac{L}{240} = \frac{180}{240} = 0.75 \text{ in.}$$

and we therefore observe that the deflection is not critical.

Note: In solving the following problems, use Tables 4-7 and 10-1 and Fig. 3-19. Neglect the beam weight, and consider the deflection to be limited to $\frac{1}{240}$ of the span.

Problem 10-6-A*. A 6×14 Southern pine beam, No. 1 SR grade, is 16 ft [4.88 m] long and supports a total uniformly distributed load of 6000 lb [26.7 kN]. Investigate the deflection.

Problem 10-6-B. An 8×12 beam of Douglas fir–larch, dense No. 1 grade, is 12 ft [3.66 m] in length and has a concentrated load of 5 kips [22.2 kN] at the center of the span. Investigate the deflection.

Problem 10-6-C. Two concentrated loads of 3500 lb [15.6 kN] each are located at the third points of a 15-ft [4.57-m] beam. The 10×14 beam is Douglas fir–larch, select structural grade. Investigate the deflection.

Problem 10-6-D*. An 8×14 beam of Douglas fir–larch, select structural grade, has a span of 16 ft [4.88 m] and a total uniformly distributed load of 8 kips [35.6 kN]. Investigate the deflection.

Problem 10-6-E. Find the least weight nominal section that can be used for a simple beam with an 18-ft [5.49-m] span carrying a total uniformly distributed load of 10 kips [44.5 kN]. The wood is Southern pine, No. 1 SR grade.

10-7 Beam Design Procedure

In general, three steps are necessary for the proper design of wood beams.

Step 1: Compute the required section modulus using the flexure formula $S = M/F_b$, as explained in Section 10-4. Beam sizes may be chosen from tables, such as Table 4-7, in which sectional properties are given for the available nominal sizes. Lacking other considerations, the section with the least cross-sectional area will ordinarily be chosen. When the section depth exceeds the width, the code requires some form of lateral bracing. For very slender sections with long, unbraced lengths, the code provides a reduction of the allowable bending stress on the basis of lateral buckling. For a complete presentation of such requirements, the reader is referred to Article 3-3-3 of the *National Design Specification for Wood Construction* published by the National Forest Products Association.

Step 2: Investigate the size selected under Step (1) for shear, as explained in Section 10-5, and increase the dimensions if necessary. In general, shear will be most critical for beams with heavy loadings on short spans or with concentrated loads.

Step 3: Investigate the beam for deflection, as explained in Section 10-6. For ordinary loadings, formulas for computation of deflection may be obtained from Fig. 3-19.

The following examples demonstrate the use of these procedures.

Example 1. Design beam A of the floor framing shown in Fig. 10-1. Wood is Southern pine, No. 2 dense SR grade. A 2-in. [51-mm] wood plank floor is used on the 5-ft [1.52-m] spans between beams, and there is a $\frac{7}{8}$-in. [22-mm] hardwood finish flooring laid over the planking. Live load is 90 psf [4.31 kN/m²], and deflection is limited to $\frac{1}{240}$ of the span under total load.

Solution: For the specified wood, we obtain the following from Table 10-1:

$$F_b = 1250 \text{ psi } [8.62 \text{ MPa}], \ F_v = 95 \text{ psi } [0.655 \text{ MPa}]$$

$$E = 1,400,000 \text{ psi } [9.65 \text{ GPa}]$$

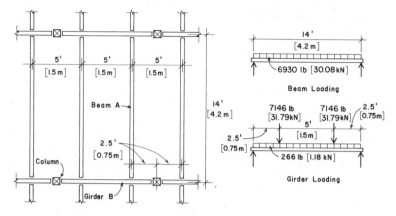

FIGURE 10-1.

The dead load is determined as follows, using data from Table 5-7.

> Plank: 1.5-in. actual thickness—4 psf
> Flooring: 0.875 in. at 5/lb/in.— <u>5 psf</u>
> Total superimposed dead load— 9 psf

The total superimposed load on the beams is thus 99 psf [4.74 kN/m²], and the load per linear ft is 99 × 5 = 495 lb/ft [7.22 kN/m]. The total load on the beam, ignoring its own weight, is 495 × 14 = 6930 lb [30.8 kN]. Then

$$M = \frac{WL}{8} = \frac{6930 \times 14}{8} = 12{,}127 \text{ ft-lb } [16.44 \text{ kN-m}]$$

and the required section modulus is

$$S = \frac{M}{F_b} = \frac{12{,}127 \times 12}{1250} = 116.4 \text{ in}^3 \; [1908 \times 10^3 \text{ mm}^3]$$

Referring to Table 4-7, we find that a 6 × 12 beam has a section modulus of 121.3 in³ (rounded off) and weighs approximately 15.4 lb/ft. To account for the beam weight, we therefore add 15.4 × 14 = 216 lb to the superimposed load, making a total of 7146 lb. This increased load will require a section modulus of 120.1 in³. Thus the selected size is still adequate.

For the uniformly loaded simple span beam, the reactions and the maximum shear will be one-half the total load, or 3573 lb [15.89 kN], and the maximum shear stress at the end of the span is computed as

$$v = \frac{3}{2} \times \frac{V}{bd} = \frac{3}{2} \times \frac{3573}{5.5 \times 11.5} = 84.7 \text{ psi } [0.584 \text{ MPa}]$$

which is less than the allowable stress, so the section is acceptable.

To investigate the deflection, we use the formula for Case 2 in Fig. 3-19 and obtain the value of $I = 697.1$ in^4 for the 6 × 12 section from Table 4.7. Then

$$D = \frac{5}{384} \times \frac{WL^3}{EI} = \frac{5}{384} \times \frac{7146 \times 168^3}{1,400,000 \times 697.1} = 0.45 \text{ in.}$$

$$[11.5 \text{ mm}]$$

Since this value is less than the allowable deflection (168/240 = 0.70 in. [17.8 mm]), the 6 × 12 beam meets all the requirements.

Example 2. Design girder B in Fig. 10-1, using the same data given for beam A.

Solution: The loading diagram for the girder is shown in Fig. 10-1. The reaction from beam A is 3573 lb [15.89 kN], but because similar beams frame into the girder on each side, the concentrated loads are 2 × 3573 = 7146 lb [31.79 kN]. The uniform load indicated on the diagram represents the weight of the girder; this is yet to be determined.

The maximum bending moment due to both the concentrated and distributed loadings will occur at midspan. For the concentrated loads, $R_1 = R_2 = 7146$ lb [31.79 kN], and the maximum moment is

$$M = (7146 \times 5) - (7146 \times 2.5) = 17,865 \text{ ft-lb } [24.22 \text{ kN-m}]$$

This bending moment requires a section modulus of

$$S = \frac{M}{F_b} = \frac{17,865 \times 12}{1250} = 172 \text{ in}^3 \text{ } [2.82 \times 10^6 \text{ mm}^3]$$

From Table 4-7 we find that a 10 × 12 has $S = 209.4$ in^3 [3.43 ×

10^6 mm³] and weighs 26.6 lb/ft [388 N/m]. The beam weight as a distributed load produces a maximum moment of

$$M = \frac{WL}{8} = \frac{(26.6 \times 10) \times 10}{8} = 333 \text{ ft-lb } [0.45 \text{ kN-m}]$$

which makes a total maximum moment of $17,865 + 333 = 18,198$ ft-lb [24.68 kN-m]. The revised required section modulus is

$$S = \frac{M}{F_b} = \frac{18,198 \times 12}{1250} = 175 \text{ in}^3 \, [2.87 \times 10^6 \text{ mm}^3]$$

which is still less than that of the 10 × 12.

The maximum shear force is the same as the reaction, which is equal to the concentrated load plus one-half the beam weight, or $7146 + 133 = 7279$ lb [32.38 kN]. Then

$$v = \frac{3}{2} \times \frac{V}{bd} = \frac{3}{2} \times \frac{7279}{9.5 \times 11.5} = 100 \text{ psi } [0.690 \text{ MPa}]$$

which exceeds the limit, so the 10 × 12 is not acceptable, and a 10 × 14 is required.

The total beam deflection under the combined loads can be found by computing the separate deflections for the two loads (Cases 2 and 3 from Fig. 3-19) and adding them. When the concentrated loading is not of a type shown in Fig. 3-19, an approximate deflection can be found by using the equivalent uniformly distributed load method. This consists of finding the load W that will produce the same maximum moment as that found for the combined loads and then using the deflection formula for uniform loading (Case 2, Fig. 3-19). Thus

$$M = \frac{WL}{8} = 18,198 \text{ ft-lb}$$

$$W = \frac{8M}{L} = \frac{8 \times 18,198}{10} = 14,558 \text{ lb}$$

$$D = \frac{5}{384} \frac{WL^3}{EI} = \frac{5}{384} \frac{14,558 \times (120)^3}{1,400,000 \times 1204} = 0.19 \text{ in.}$$

This is considerably less than the limit of 120/240 = 0.50 in., so a more accurate deflection computation is not required.

The 10×12 section is thus shown to be acceptable for bending and shear stress and for deflection. In practice it should be determined that the beam and girder are adequately braced against buckling. The end connections for both members should provide some resistance to rotation. Lateral buckling of the beam is most likely not critical, since the plank deck will probably be attached continuously in a manner to provide lateral bracing. If the girder is braced only by the beams, it has a lateral unsupported length of 5 ft at midspan, and an investigation should be made as described in Step (1) of Section 10-7 to see if the allowable bending stress must be significantly reduced.

Problem 10-7-A. Design the beam for the floor system shown in Fig. 10-2a. Floor construction consists of 3-in. nominal plank deck and $\frac{7}{8}$-in. hardwood flooring. Live load is 120 psf [5.75 kN/m²], and Southern pine, No. 1 SR grade, is to be used. Deflection is limited to $\frac{1}{360}$ of the span under live load only and to $\frac{1}{240}$ of the span under total load.

Problem 10-7-B. Design the girder shown in Fig. 10-2a, using the same data given for Problem 10-7-A.

Problem 10-7-C. For the floor system shown in Fig. 10-2b, the live load is 60 psf [2.87 kN/m²], and the floor consists of 4-in. nominal plank deck and $\frac{7}{8}$-in. hardwood flooring. Design the beam using Douglas fir–larch, dense No. 1 grade. Deflection is limited as for Problem 10-7-A.

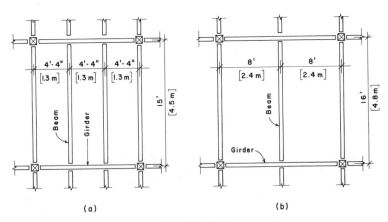

FIGURE 10-2

Problem 10-7-D.   Design the girder for the system shown in Fig. 10-2*b*, using the same data given for Problem 10-7-C.

10-8   Bearing on Supports

Beam bearings must have ample dimensions so that compressive stresses perpendicular to the grain do not exceed the allowable values of F_c given in Table 10-1. The allowable stresses in the table apply to bearings of any length at the ends of beams and to all bearings 6 in. or more in length at any other location.

For bearings less than 6 in. in length and not nearer than 3 in. to the end of the member, the National Design Specification provides that the allowable stress in compression perpendicular to the grain may be increased by the factor $(L_b + 0.375)/L_b$, in which L_b is the length of bearing measured along the grain of the wood.

Example 1.   An 8 × 14 Southern pine beam, No. 1 SR grade, has a bearing length of 6 in. [152 mm] at its supports. If the end reaction is 7400 lb [32.9 kN], is the beam adequate for bearing?
Solution: The developed bearing stress is equal to the end reaction divided by the product of the bearing length and the beam width; 7400/(7.5 × 6) = 164 psi [1.13 MPa]. This is compared to the allowable stress from Table 10-1, 375 psi [2.59 MPa]. The beam is therefore adequate for bearing.

Example 2.   A 2 × 10 roof joist cantilevers over and is supported by the top plate of a 2 × 4 stud wall. The load from the joist is 800 lb [3.56 kN]. If both the joist and the plate are Douglas fir–larch, No. 2 grade, is the situation adequate for bearing?
Solution: The developed bearing area is the product of the width of the joist (1.5 in. [38 mm]) times the width of the flat 2 × 4 plate (3.5 in [89 mm]). The bearing stress is thus

$$F = \frac{800}{1.5 \times 3.5} = 152 \text{ psi } [1.05 \text{ MPa}]$$

This is considerably less than the allowable stress from Table 10-1: 625 psi [4.31 MPa]. Bearing stress is therefore not critical.

Example 3.   A two-span 3 × 12 beam of Douglas fir–larch, No. 1 grade, bears on a 3 × 14 supporting beam at its center support. If the reaction force at the center support is 4200 lb [18.7 kN], is the situation critical for bearing?

Solution: If we assume the beam to bear at right angles to its supporting member, the developed bearing area is the product of the beam width times the support beam width: 2.5 × 2.5 = 6.25 in² [4033 mm²]. The developed bearing stress is thus

$$F = \frac{4200}{6.25} = 672 \text{ psi } [4.63 \text{ MPa}]$$

This exceeds the allowable stress of 625 psi [4.31 MPa] from Table 10-1. However, the situation qualifies for the increase factor for bearing lengths less than 6 in. The modified allowable stress is thus

$$F_{c\perp} = \frac{L_b + 0.375}{L_b} \times 625 = \frac{2.875}{2.5} \times 625 = 719 \text{ psi } [4.96 \text{ MPa}]$$

and the condition is not critical.

Problem 10-8-A. A 6 × 12 beam of Douglas fir–larch, No. 1 grade, has 3 in. of end bearing to develop an end reaction force of 5000 lb [22.2 kN]. Is the situation adequate for bearing?

Problem 10-8-B. A 3 × 16 roof joist cantilevers over a 3 × 14 support beam. If both beams are Southern pine, No. 1 grade, is the situation adequate for bearing? The joist load on the support beam is 3000 lb [13.3 kN].

10-9 Floor Joists

Joists are comparatively small, closely spaced beams. The sizes most commonly used are those of 2-in. nominal thickness, from 2 × 4 to 2 × 12. Joists of 3-in. nominal thickness are used when conditions require greater width for nailing decks, greater depth (up to 16 in. nominal), or simply a larger cross section for stress conditions.

Spacing of joists is usually determined by the choice of decking or—when ceilings exist—choice of ceiling construction. Most commonly used are spacings of 12, 16, and 24 in., based on the use of plywood decking or ceiling paneling in 4 ft by 8 ft sheets.

Cross-bridging or blocking is often required for joists as lateral bracing for the thin beams. Bridging and blocking also serve to provide for distribution of loads to adjacent joists. Such load sharing is the basis for permitting the use of higher bending stress values—those associated with the category labelled "repetitive-member uses" in Table 10-1 and discussed in Section 10-3. Bridging consists of rows of crisscrossed members of wood or metal. Blocking consists of rows of short pieces of the same size lumber as used for the joists; these are tightly fitted and nailed between the joists. Blocking is usually used where provision must be made for nailing the edges of plywood sheets that are perpendicular to the joists.

The design of joists consists of determining the load to be supported and then applying the procedures for beam design as explained in Section 10-7. However, to facilitate the selection of joists carrying uniformly distributed loads (by far the most common loading), many tables have been prepared that give maximum safe spans for joists of various sizes and spacings under different loadings per square foot. Table 10-2 is representative of such tables and has been reproduced from the 1982 edition of the *Uniform Building Code*. Examining the table, we note that spans are computed on the basis of modulus of elasticity, with the required bending stress F_b listed below each span; that is, both stiffness (deflection) and bending strength have been taken into account. Maximum safe clear spans in feet and inches are tabulated for three joist spacings and selected values of E. As stated in the table notes, deflection is limited to $\frac{1}{360}$ of the span due to live load only, whereas both live load and dead load allowance are used in determining the required bending stress.

The use of Table 10-2 is illustrated in the following example.

Example. Using Table 10-2, select joists to carry a live load of 40 psf on a span of 15 ft 6 in. if the spacing is 16 in. on centers. *Solution:* Referring to Table 10-2, we find that 2×10 joists with an E of 1,400,000 psi and F_b of 1150 psi may be used on a span of 15 ft 8 in.

Turning to Table 10-1, the reader should observe that among the species and grades listed, the following selections would be satisfactory: Southern pine No. 2 and Douglas fir–larch No. 2.

TABLE 10-2.   Allowable Spans in Feet and Inches for Floor Joists

Joist Size	Spacing (in)	Modulus of Elasticity, E, in 1,000,000 psi													
		0.8	0.9	1.0	1.1	1.2	1.3	1.4	1.5	1.6	1.7	1.8	1.9	2.0	2.2
2x6	12.0	8-6 / 720	8-10 / 780	9-2 / 830	9-6 / 890	9-9 / 940	10-0 / 990	10-3 / 1040	10-6 / 1090	10-9 / 1140	10-11 / 1190	11-2 / 1230	11-4 / 1280	11-7 / 1320	11-11 / 1410
	16.0	7-9 / 790	8-0 / 860	8-4 / 920	8-7 / 980	8-10 / 1040	9-1 / 1090	9-4 / 1150	9-6 / 1200	9-9 / 1250	9-11 / 1310	10-2 / 1360	10-4 / 1410	10-6 / 1460	10-10 / 1550
	24.0	6-9 / 900	7-0 / 980	7-3 / 1050	7-6 / 1120	7-9 / 1190	7-11 / 1250	8-2 / 1310	8-4 / 1380	8-6 / 1440	8-8 / 1500	8-10 / 1550	9-0 / 1610	9-2 / 1670	9-6 / 1780
2x8	12.0	11-3 / 720	11-8 / 780	12-1 / 830	12-6 / 890	12-10 / 940	13-2 / 990	13-6 / 1040	13-10 / 1090	14-2 / 1140	14-5 / 1190	14-8 / 1230	15-0 / 1280	15-3 / 1320	15-9 / 1410
	16.0	10-2 / 790	10-7 / 850	11-0 / 920	11-4 / 980	11-8 / 1040	12-0 / 1090	12-3 / 1150	12-7 / 1200	12-10 / 1250	13-1 / 1310	13-4 / 1360	13-7 / 1410	13-10 / 1460	14-3 / 1550
	24.0	8-11 / 900	9-3 / 980	9-7 / 1050	9-11 / 1120	10-2 / 1190	10-6 / 1250	10-9 / 1310	11-0 / 1380	11-3 / 1440	11-5 / 1500	11-8 / 1550	11-11 / 1610	12-1 / 1670	12-6 / 1780
2x10	12.0	14-4 / 720	14-11 / 780	15-5 / 830	15-11 / 890	16-5 / 940	16-10 / 990	17-3 / 1040	17-8 / 1090	18-0 / 1140	18-5 / 1190	18-9 / 1230	19-1 / 1280	19-5 / 1320	20-1 / 1410
	16.0	13-0 / 790	13-6 / 850	14-0 / 920	14-6 / 980	14-11 / 1040	15-3 / 1090	15-8 / 1150	16-0 / 1200	16-5 / 1250	16-9 / 1310	17-0 / 1360	17-4 / 1410	17-8 / 1460	18-3 / 1550
	24.0	11-4 / 900	11-10 / 980	12-3 / 1050	12-8 / 1120	13-0 / 1190	13-4 / 1250	13-8 / 1310	14-0 / 1380	14-4 / 1440	14-7 / 1500	14-11 / 1550	15-2 / 1610	15-5 / 1670	15-11 / 1780
2x12	12.0	17-5 / 720	18-1 / 780	18-9 / 830	19-4 / 890	19-11 / 940	20-6 / 990	21-0 / 1040	21-6 / 1090	21-11 / 1140	22-5 / 1190	22-10 / 1230	23-3 / 1280	23-7 / 1320	24-5 / 1410
	16.0	15-10 / 790	16-5 / 860	17-0 / 920	17-7 / 980	18-1 / 1040	18-7 / 1090	19-1 / 1150	19-6 / 1200	19-11 / 1250	20-4 / 1310	20-9 / 1360	21-1 / 1410	21-6 / 1460	22-2 / 1550
	24.0	13-10 / 900	14-4 / 980	14-11 / 1050	15-4 / 1120	15-10 / 1190	16-3 / 1250	16-8 / 1310	17-0 / 1380	17-5 / 1440	17-9 / 1500	18-1 / 1550	18-5 / 1610	18-9 / 1670	19-4 / 1780

Source: Reproduced from the *Uniform Building Code*, 1982 ed., with permission of the publishers, International Conference of Building Officials.

Notes: Criteria: 40 psf live load; 10 psf dead load; live load deflection limited to 1/360 of the span. Number indicated below each span is the required allowable bending stress in psi.

Remember that the values for F_b in the column headed "repetitive member uses" generally applies to joists.

As with all safe-load tabulations, caution must be exercised with respect to the criteria used in preparation of the table data. The 10 psf allowance provided in Table 10-2 is adequate to cover the weight of the joists, wood deck and flooring, and a gypsum dry-wall ceiling. However, if a heavier type of flooring or a plastered ceiling is used, this additional weight must be accounted for. This is readily accomplished, of course, if the usual beam design procedure is followed, with the table used only for a preliminary estimate of the required joist.

Problem 10-9-A*-B-C-D.   Using Douglas fir–larch, No. 2 grade, pick the joist size required for the stated conditions. Live load is 40 psf; dead load is 10 psf; deflection is limited to $\frac{1}{360}$ of the span under live load only.

	Joist Spacing (in.)	Joist Span (ft)
A	16	14
B	12	14
C	16	18
D	12	22

10-10   Rafters

Rafters are the comparatively small, closely spaced beams used to support the load on roofs. As with floor joists, the most common sizes are the available range of 2-in. and 3-in. nominal thickness lumber, and the most used spacings are 12, 16, and 24 in. The terms *rafter* and *roof joist* are frequently used interchangeably, the former mostly for sloping roof surfaces and the latter mostly for horizontal roof surfaces.

For sloping rafters it is common practice to consider the span to be the horizontal projection, as indicated in Fig. 10-3. This applies only to consideration for gravity loads. Wind forces are considered as applied perpendicular to the roof surface, and the span must thus be the actual rafter length.

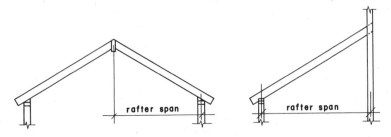

FIGURE 10-3.   Determination of rafter span.

Design of rafters is generally accomplished by the use of safe-load tables. Table 10-3 is representative of such tables and has been reproduced from the 1982 edition of the *Uniform Building Code*. The table gives maximum safe spans for rafters in feet and inches for three spacings and selected values of the allowable bending stress F_b. The modulus of elasticity E, required to maintain the stated deflection limit, is listed below each span.

The live load value provided for in Table 10-3 is 20 psf. This is the usual minimum value required by codes; where snow accumulation is possible, a higher value is usually specified. For high-slope rafters, codes usually provide for some reduction of the live load, based on the unlikelihood of accumulation of anything on the sloping surface. Magnitudes of wind loads and the manner of their application to roof surfaces are a regional matter, and the prevailing code must be used. Except for very steep slopes or exceptionally light construction, wind loads are usually critical only in areas with histories of frequent windstorm conditions.

The following example illustrates the use of the data in Table 10-3.

Example.   Rafters are to be used on 24-in. centers for a roof span of 16 ft. Live load is 20 psf; total dead load is 15 psf; live load deflection is limited to $\frac{1}{240}$ of the span. Find the rafter size required for Douglas fir–larch of (1) No. 1 grade and (2) No. 2 grade.

Solution: From Table 10-1 we find the design values for No. 1 grade to be $E = 1,800,000$ psi and $F_b = 1750$ psi. Although Table 10.3 does not have a column for $F_b = 1750$ psi, it is apparent that

TABLE 10-3. Allowable Spans in Feet and Inches for Low- or High-Slope Rafters

RAFTER SIZE (IN)	SPACING (IN)	Allowable Extreme Fiber Stress In Bending F_b (psi)														
		500	600	700	800	900	1000	1100	1200	1300	1400	1500	1600	1700	1800	1900
2x6	12.0	8-6 / 0.26	9-4 / 0.35	10-0 / 0.44	10-9 / 0.54	11-5 / 0.64	12-0 / 0.75	12-7 / 0.86	13-2 / 0.98	13-8 / 1.11	14-2 / 1.24	14-8 / 1.37	15-2 / 1.51	15-8 / 1.66	16-1 / 1.81	16-7 / 1.96
	16.0	7-4 / 0.23	8-1 / 0.30	8-8 / 0.38	9-4 / 0.46	9-10 / 0.55	10-5 / 0.65	10-11 / 0.75	11-5 / 0.85	11-10 / 0.97	12-4 / 1.07	12-9 / 1.19	13-2 / 1.31	13-7 / 1.44	13-11 / 1.56	14-4 / 1.70
	24.0	6-0 / 0.19	6-7 / 0.25	7-1 / 0.31	7-7 / 0.38	8-1 / 0.45	8-6 / 0.53	8-11 / 0.61	9-4 / 0.70	9-8 / 0.78	10-0 / 0.88	10-5 / 0.97	10-9 / 1.07	11-1 / 1.17	11-5 / 1.28	11-8 / 1.39
2x8	12.0	11-2 / 0.26	12-3 / 0.35	13-3 / 0.44	14-2 / 0.54	15-0 / 0.64	15-10 / 0.75	16-7 / 0.86	17-4 / 0.98	18-0 / 1.11	18-9 / 1.24	19-5 / 1.37	20-0 / 1.51	20-8 / 1.66	21-3 / 1.81	21-10 / 1.96
	16.0	9-8 / 0.23	10-7 / 0.30	11-6 / 0.38	12-3 / 0.46	13-0 / 0.55	13-8 / 0.65	14-4 / 0.75	15-0 / 0.85	15-7 / 0.96	16-3 / 1.07	16-9 / 1.19	17-4 / 1.31	17-10 / 1.44	18-5 / 1.56	18-11 / 1.70
	24.0	7-11 / 0.19	8-8 / 0.25	9-4 / 0.31	10-0 / 0.38	10-7 / 0.45	11-2 / 0.53	11-9 / 0.61	12-3 / 0.70	12-9 / 0.78	13-3 / 0.88	13-8 / 0.97	14-2 / 1.07	14-7 / 1.17	15-0 / 1.28	15-5 / 1.39
2x10	12.0	14-3 / 0.26	15-8 / 0.35	16-11 / 0.44	18-1 / 0.54	19-2 / 0.64	20-2 / 0.75	21-2 / 0.86	22-1 / 0.98	23-0 / 1.11	23-11 / 1.24	24-9 / 1.37	25-6 / 1.51	26-4 / 1.66	27-1 / 1.81	27-10 / 1.96
	16.0	12-4 / 0.23	13-6 / 0.30	14-8 / 0.38	15-8 / 0.46	16-7 / 0.55	17-6 / 0.65	18-4 / 0.75	19-2 / 0.85	19-11 / 0.96	20-8 / 1.07	21-5 / 1.19	22-1 / 1.31	22-10 / 1.44	23-5 / 1.56	24-1 / 1.70
	24.0	10-1 / 0.19	11-1 / 0.25	11-11 / 0.31	12-9 / 0.38	13-6 / 0.45	14-3 / 0.53	15-0 / 0.61	15-8 / 0.70	16-3 / 0.78	16-11 / 0.88	17-6 / 0.97	18-1 / 1.07	18-7 / 1.17	19-2 / 1.28	19-8 / 1.39
2x12	12.0	17-4 / 0.26	19-0 / 0.35	20-6 / 0.44	21-11 / 0.54	23-3 / 0.64	24-7 / 0.75	25-9 / 0.86	26-11 / 0.98	28-0 / 1.11	29-1 / 1.24	30-1 / 1.37	31-1 / 1.51	32-0 / 1.66	32-11 / 1.81	33-10 / 1.96
	16.0	15-0 / 0.23	16-6 / 0.30	17-9 / 0.38	19-0 / 0.46	20-2 / 0.55	21-3 / 0.65	22-4 / 0.75	23-3 / 0.85	24-3 / 0.97	25-2 / 1.07	26-0 / 1.19	26-11 / 1.31	27-9 / 1.44	28-6 / 1.56	29-4 / 1.70
	24.0	12-3 / 0.19	13-5 / 0.25	14-6 / 0.31	15-6 / 0.38	16-6 / 0.45	17-4 / 0.53	18-2 / 0.61	19-0 / 0.70	19-10 / 0.78	20-6 / 0.88	21-3 / 0.97	21-11 / 1.07	22-8 / 1.17	23-3 / 1.28	23-11 / 1.39

Source: Reproduced from the *Uniform Building Code*, 1982 ed., with permission of the publishers, International Conference of Building Officials.

Notes: Criteria: 20 psf live load; 15 psf dead load; live load deflection limited to 1/360 of the span.
Number indicated below each span is the required minimum value for the modulus of elasticity E; E in psi equals the listed value times 1,000,000.

the size choice is for a 2 × 10 rafter. This observation is made by comparing the listed data for F_b = 1700 and 1800 for both 2 × 8 and 2 × 10 rafters. Thus a 2 × 8 will span between 14 ft 7 in. and 15 ft, while a 2 × 10 will span between 18 ft 7 in. and 19 ft 2 in. It should also be apparent that E will not be critical.

From Table 10-1 we find design values of E = 1,700,000 psi and F_b = 1450 psi for No. 2 grade. Observing the values listed in Table 10.3 for F_b = 1400 and 1500 psi, it should be apparent that the 2 × 10 is again the choice and that E is not critical.

Problem 10-10-A*-B-C-D. Select the smallest joist from Table 10-3 for the conditions stated. Wood is Southern pine, No. 2 grade; live load is 20 psf; dead load is 15 psf; live load deflection is limited to $\frac{1}{240}$ of the span.

	Rafter Spacing (in.)	Rafter Span (ft)
A	16	12
B	24	12
C	16	18
D	24	18

10-11 Plank Decking

The most widely used material for roof and floor decking is Douglas fir plywood. When the construction is exposed to view on its underside, and there is concern for its appearance, a popular alternative is tongue-and-groove planking of 2-in. or greater nominal thickness. Another reason for selecting this decking in some situations is its increased fire resistance, owing to its greater thickness.

Figure 10-4 shows two common forms for plank decking: solid and glued laminated. Although available in various thicknesses,

FIGURE 10-4. Common forms of plank decking.

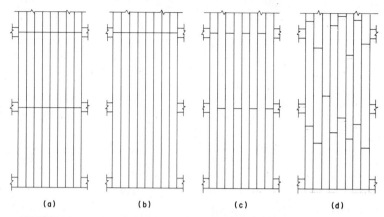

FIGURE 10-5. Plank deck installation and span conditions: (a) simple span, (b) two-span continuous, (c) combination simple and two-span continuous, (d) controlled random length.

the most widely used is the 2-in. nominal decking. The span continuity conditions for the decking are of major concern in determining its load-carrying capacity. Figure 10-5 shows four possibilities for this, based on the location of end-to-end joints in the decking. Although the first three lay-up arrangements will result in a cleaner exposed appearance (no exposed end joints), the controlled random lay-up (Fig. 10-5d) is usually more economical.

Table 10-4 presents data for 2-in. planking used for either roof or floor decking. This material is adapted from a more extensive table in the reference cited. Although code limits must be acknowledged, it is best to obtain information from material suppliers as to the products available locally.

10-12 Glued Laminated Products

In addition to sheets of plywood, there are a number of other products used for wood structures that are fabricated by gluing together pieces of wood into a solid form. Girders, framed bents, and arch ribs of large size are produced by assembling standard 2-in. nominal lumber (2 × 6, etc.). The resulting thickness of such elements is essentially the width of the standard lumber used,

TABLE 10-4. Allowable Spans for Two-Inch Tongue-and-Groove Decking

Allowable Span (ft)	Live Load (psf)	Live Load Deflection Limit	Minimum Properties Allowable Bending Stress F_b (psi)	Modulus of Elasticity E (psi)
For Roof Decks				
4	20	1/240	160	170,000
		1/360	160	256,000
	30	1/240	210	256,000
		1/360	210	384,000
5	20	1/240	250	332,000
		1/360	250	500,000
	30	1/240	330	495,000
		1/360	330	742,000
6	20	1/240	360	575,000
		1/360	360	862,000
	30	1/240	480	862,000
		1/360	480	1,295,000
7	20	1/240	490	910,000
		1/360	490	1,360,000
	30	1/240	650	1,370,000
		1/360	650	2,000,000
8	20	1/240	640	1,360,000
		1/360	640	2,040,000
	30	1/240	850	2,040,000
For Floor Decks				
4	40	1/360	840	1,000,000
4.5	40	1/360	950	1,300,000
5	40	1/360	1060	1,600,000

Source: Adapted from data in the *Uniform Building Code*, 1982 ed., with permission of the publishers, International Conference of Building Officials.

Notes: Based on simple span or controlled random lay-up (Fig. 10-5a or b). Provision for 10-psf dead load and alternate 300-lb concentrated live load. Actual lumber thickness 1.5 in.

with a small dimensional loss due to finishing. The depth is a multiple of the lumber thickness of 1.5 in.

Availability of glued laminated products should be investigated on a regional basis. Information can be obtained from local suppliers. Fabricators and suppliers also commonly provide some assistance in engineering design when their products are specified for a building. The National Design Specification and most local building codes provide some design requirements and guides for the design of ordinary glued laminated elements fabricated from standard lumber.

11

Wood Columns

||||||||||||||||||||||||||||||||||||||||||||||||||||||||||||||||||

11-1 Column Types

The wood column that is used most frequently is the *simple solid column*. It consists of a single piece of wood, square or rectangular in cross section. Solid columns of round cross section are also considered simple solid columns but are used less frequently. A *spaced column* is an assembly of two or more pieces separated at the ends and at intermediate points along their lengths by blocking. Two other types are *built-up columns,* consisting of multiple solid elements fastened together to form a solid mass, and *glued laminated columns*.

11-2 Slenderness Ratio

In wood construction the slenderness ratio of a freestanding simple solid column is the ratio of the unbraced (laterally unsupported) length to the dimension of its least side, or L/d. (Fig. 11-1a.) When members are braced so that the unsupported length with respect to one face is less than that with respect to the other, L is the distance between the points of support that prevent lateral movement in the direction along which the dimension of the section is measured. This is illustrated in Fig. 11-1b. If the section is not square or round, it may be necessary to investigate two L/d conditions for such a column, to determine which is the limiting

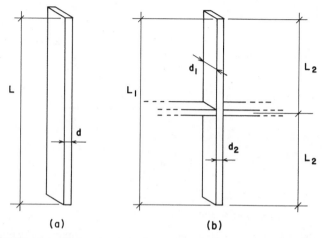

FIGURE 11-1. Determination of slenderness ratio for a wood column.

one. The slenderness ratio for simple solid columns is limited to $L/d = 50$; for spaced columns the limiting ratio is $L/d = 80$.

11-3 Capacity of Simple Solid Columns

Figure 11-2 illustrates the typical form of the relationship between axial compression capacity and slenderness for a linear compression member (column). The two limiting conditions are those of the very short member and the very long member. The short member (such as a block of wood) fails in crushing, which is limited by the mass of material and the stress limit in compression. The very long member (such as a yardstick) fails in elastic buckling, which is determined by the stiffness of the member; stiffness is determined by a combination of geometric property (shape of the cross section) and material stiffness property (modulus of elasticity). Between these two extremes—which is where most wood compression members fall—the behavior is indeterminate as the transition is made between the two distinctly different modes of behavior.

The National Design Specification currently provides for three separate compression stress calculations, corresponding to the

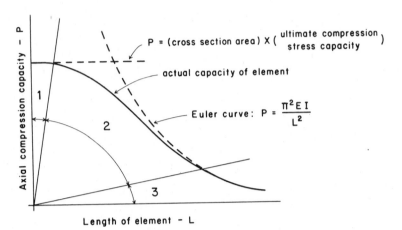

FIGURE 11-2.   Relation of member length to axial compression capacity.

three zones of behavior described in Fig. 11-2. The plot of these three stress formulas, for a specific example wood, is shown in Fig. 11-3. Typical analysis and design procedures for simple solid wood columns are illustrated in the following examples.

Example 1.   A wood compression member consists of a 3 × 6 of Douglas fir–larch, dense No. 1 grade. Find the allowable axial compression force for unbraced lengths of: (1) 2 ft [0.61 m], (2) 4 ft [1.22 m], (3) 8 ft [2.44 m].

Solution: We find from Table 10-1: $F_c = 1450$ psi [10.0 MPa] and $E = 1,900,000$ psi [13.1 GPa]. To establish the zone limits we compute the following:

$$11(d) = 11(2.5) = 27.5 \text{ in.}$$

$$50(d) = 50(2.5) = 125 \text{ in.}$$

and

$$K = 0.671 \sqrt{\frac{E}{F_c}} = 0.671 \sqrt{\frac{1,900,000}{1450}} = 24.29$$

Thus for (1), $L = 24$ in., which is in Zone 1; $F_c' = F_c = 1450$ psi; allowable $C = F_c' \times$ gross area $= 1450 \times 13.75 = 19,938$ lb [88.7 kN].

For (2), $L = 48$ in.; $L/d = \dfrac{48}{2.5} = 19.2$, which is in Zone 2.

$$F_c' = F_c\left\{1 - \frac{1}{3}\left(\frac{L/d}{K}\right)^4\right\}$$

$$= 1450\left\{1 - \frac{1}{3}\left(\frac{19.2}{24.29}\right)^4\right\}$$

$$= 1262 \text{ psi}$$

Allowable $C = 1262 \times (13.75) = 17,353$ lb [77.2 kN].

For (3), $L = 96$ in.; $L/d = \dfrac{96}{2.5} = 38.4$, which is in Zone 3.

$$F_c' = \frac{0.3(E)}{(L/d)^2} = \frac{0.3(1,900,000)}{(38.4)^2} = 387 \text{ psi}$$

Allowable compression $= 387 \times (13.75) = 5321$ lb [23.7 kN].

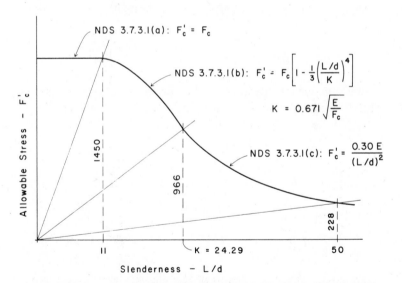

FIGURE 11-3. Allowable axial compression stress as a function of the slenderness ratio L/d. National Design Specification (NDS) requirements for Douglas fir–larch, Dense No. 1.

Example 2. Wood 2 × 4 elements are to be used as vertical compression members to form a wall (ordinary stud wall construction). If the wood is Douglas fir–larch, No. 3 grade, and the wall is 8.5 ft high, what is the column load capacity of a single stud?

Solution: In this case it must be assumed that the surfacing materials used for the wall (plywood, dry wall, plaster, etc.) will provide adequate bracing for the studs on their weak axis (the 1.5-in. [38-mm] direction). If not, the studs cannot be used, since the specified height of the wall is considerably in excess of the limit for L/d for a solid column (50). We therefore assume the direction of potential buckling to be that of the 3.5-in. [89-mm] dimension. Thus

$$\frac{L}{d} = \frac{8.5 \times 12}{3.5} = 29.1$$

In order to determine which column load formula must be used, we must find the value of K for this wood. From Table 10-1 we find $F_c = 675$ psi [4.65 MPa] and $E = 1,500,000$ psi [10.3 GPa]. Then

$$K = 0.671 \sqrt{\frac{E}{F_c}} = 0.671 \sqrt{\frac{1,500,000}{675}} = 31.63$$

We thus establish the condition for the stud as Zone 2 (Fig. 11-2), and the allowable compression stress is computed as

$$F'_c = F_c\left\{1 - \frac{1}{3}\left(\frac{L/d}{K}\right)^4\right\}$$

$$= 675\left\{1 - \frac{1}{3}\left(\frac{29.1}{31.63}\right)^4\right\} = 514 \text{ psi } [3.54 \text{ MPa}]$$

The allowable load for the stud is

$$P = F'_c \times \text{gross area} = 514 \times 5.25 = 2699 \text{ lb } [12.0 \text{ kN}]$$

Example 3. A wood column of Douglas fir–larch, dense No. 1 grade, must carry an axial load of 40 kips [178 kN]. Find the smallest section for unbraced lengths of: (1) 4 ft [1.22 m], (2) 8 ft [2.44 m], (3) 16 ft (4.88 m).

Solution: Since the size of the column is unknown, the values of L/d, F_c, and E cannot be predetermined. Therefore, without design aids (tables, graphs, or computer programs), the process becomes a cut-and-try approach, in which a specific value is assumed for d and the resulting values for L/d, F_c, E, and F_c' are determined. A required area is then determined and the sections with the assumed d compared with the requirement. If an acceptable member cannot be found, another try must be made with a different d. Although somewhat clumsy, the process is usually not all that laborious, since a limited number of available sizes are involved.

We first consider the possibility of a Zone 1 stress condition (Fig. 11-2), since this calculation is quite simple. If the maximum $L = 11(d)$, then the minimum $d = (4 \times 12)/11 = 4.36$ in. [111 mm]. This requires a nominal thickness of 6 in., which puts the size range into the "posts and timbers" category in Table 10-1, for which the allowable stress F_c is 1200 psi. The required area is thus

$$A = \frac{\text{load}}{F_c'} = \frac{40{,}000}{1200} = 33.3 \text{ in}^2 \ [21{,}485 \text{ mm}^2]$$

The smallest section is thus a 6×8, with an area of 41.25 in², since a 6×6 with 30.25 in² is not sufficient. (See Table 4-7.) If the rectangular-shape column is acceptable, this becomes the smallest member usable. If a square shape is desired, the smallest size would be an 8×8.

If the 6-in. nominal thickness is used for the 8-ft column, we determine that

$$\frac{L}{d} = \frac{8 \times 12}{5.5} = 17.45$$

Since this is greater than 11, the allowable stress is in the next Zone, for which

$$F_c' = F_c \left\{ 1 - \frac{1}{3} \left(\frac{L/d}{K} \right)^4 \right\}$$

$$= 1200 \left\{ 1 - \frac{1}{3} \left(\frac{17.45}{25.26} \right)^4 \right\}$$

$$= 1109 \text{ psi } [7.65 \text{ MPa}]$$

in which

$$F_c = 1200 \text{ psi and } E = 1,700,000 \text{ (from Table 10-1)}$$

and

$$K = 0.671 \sqrt{\frac{E}{F_c}} = 0.671 \sqrt{\frac{1,700,000}{1200}} = 25.26$$

The required area is thus

$$A = \frac{\text{load}}{F'_c} = \frac{40,000}{1109} = 36.07 \text{ in}^2 \text{ [23,272 mm}^2]$$

and the choices remain the same as for the 4-ft column.

If the 6-in. nominal thickness is used for the 16-ft column, we determine that

$$\frac{L}{d} = \frac{16 \times 12}{5.5} = 34.9$$

Since this is greater than the value of K, the stress condition is that of Zone 3 (Fig. 11-2), and the allowable stress is

$$F'_c = \frac{0.30E}{(L/d)^2} = \frac{(0.30)(1,700,000)}{34.9^2} = 419 \text{ psi [2.89 MPa]}$$

which requires an area for the column of

$$A = \frac{\text{load}}{F'_c} = \frac{40,000}{419} = 95.5 \text{ in}^2 \text{ [61,617 mm}^2]$$

This is greater than the area for the largest section with a nominal thickness of 6 in., as listed in Table 4-7. Although larger sections may be available in some areas, it is highly questionable to use a member with these proportions as a column. Therefore, we consider the next larger nominal thickness of 8 in. Then, if

$$\frac{L}{d} = \frac{16 \times 12}{7.5} = 25.6$$

we are still in the Zone 3 condition, and the allowable stress is

$$F'_c = \frac{0.30E}{(L/d)^2} = \frac{(0.30)(1,700,000)}{25.6^2} = 778 \text{ psi [5.36 MPa]}$$

which requires an area of

$$A = \frac{\text{load}}{F'_c} = \frac{40,000}{778} = 51.4 \text{ in}^2 \text{ [33,163 mm}^2\text{]}$$

The smallest member usable is thus an 8 × 8. It is interesting to note that the required square column remains the same for all the column lengths, even though the stress varies from 1200 psi to 778 psi. This is not uncommon and is simply due to the limited number of sizes available for the square column section.

Note: For the following problems use Douglas fir–larch, No. 1 grade.

Problems 11-3-A*-B-C-D. Find the allowable axial compression load for each of the following.

	Nominal Size (in.)	Unbraced Length (ft)	(m)
A	4 × 4	8	2.44
B	6 × 6	12	3.66
C	8 × 8	18	5.49
D	8 × 8	14	4.27

Problems 11-3-E*-F-G-H. Select the smallest square section for each of the following.

	Axial Load (kips)	(kN)	Unbraced Length (ft)	(m)
E	20	89	8	2.44
F	50	222	12	3.66
G	50	222	20	6.10
H	100	445	16	4.88

11-4 Spaced Columns

A type of structural element sometimes used in wood structures is the *spaced column*. This is an element in which two or more wood members are fastened together to share load as a single compression unit. The design of such elements is quite complex, owing to the numerous code requirements. The following example shows the general procedure for analysis of such an element, but the reader should refer to the applicable code for the various requirements for any design work.

Example.   A spaced column of the form shown in Fig. 11-4 consists of three 3 × 10 pieces of Douglas fir–larch, No. 1 dense grade. Dimension $L_1 = 15$ ft [4.57 m] and $x = 6$ in. [152 mm]. Find the axial compression capacity.

Solution: There are two separate conditions to be investigated for the spaced column. These relate to the effects of relative slenderness in the two directins, as designated by the x and y axes in Fig. 11-4. In the y-direction the column behaves simply as a set of solid wood columns. Thus the stress permitted is limited by the dimension d_2 and the ratio of L_2 to d_2, and F'_c for this condition is the same as that for a solid wood column. Thus, for the example

$$\frac{L_2}{d_2} = \frac{15 \times 12}{9.25} = 19.46$$

and F'_c is determined as usual for a solid section.

We determine that

$$K = 0.671 \sqrt{\frac{E}{F_c}} = 0.671 \sqrt{\frac{1,900,000}{1450}} = 24.29$$

using values for E and F_c from Table 10-1.

This establishes the stress condition as Zone 2 as shown in Fig. 11-2 (L/d between 11 and K). The allowable stress is thus

$$F'_c = F_c \left\{ 1 - \frac{1}{3} \left(\frac{L/d}{K} \right)^4 \right\}$$

$$= 1450 \left\{ 1 - \frac{1}{3} \left(\frac{19.46}{24.29} \right)^4 \right\} = 1251 \text{ psi [8.63 MPa]}$$

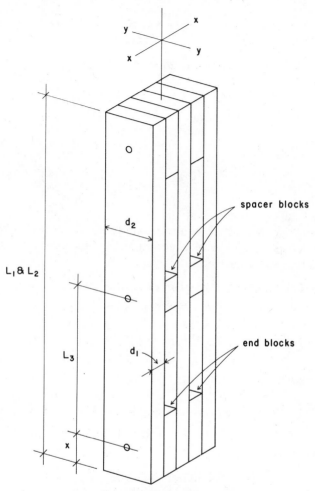

FIGURE 11-4. The spaced column.

For the condition of behavior with regard to the x-axis, we first check for conformance with two limitations:

1. Maximum value for $L_3/d_1 = 40$.
2. Maximum value for $L_1/d_1 = 80$.

Thus

$$\frac{84}{2.5} = 33.6 < 40$$

and

$$\frac{180}{2.5} = 72 < 80$$

so the limits are not exceeded.

The stress permitted for the x-axis condition depends on the value of L_1/d_1 and is one of three situations, similar to the requirements for solid columns:

1. For values of L_1/d_1 of 11 or less; $F'_c = F_c$.
2. For values of L_1/d_1 between 11 and K,

$$F'_c = F_c\left\{1 - \frac{1}{3}\left(\frac{L_1/d_1}{K}\right)^4\right\}$$

where $K = 0.671\sqrt{C_x\left(\frac{E}{F_c}\right)}$

3. For values of L_1/d_1 between K and 80,

$$F'_c = \frac{0.30(C_x)(E)}{(L_1/d_1)^2}$$

In the equations for conditions (2) and (3), the value for C_x is based on the conditions of the end blocks in the column. In the illustration in Fig. 11-4, the distance x indicates the distance from the end of the column to the centroid of the connectors that are used to bolt the end blocks into the column. Two values are given for C_x, based on the relation of the distance x to the column length—L_1 in the figure:

1. $C_x = 2.5$ when x is equal to or less than $L_1/20$.
2. $C_x = 3.0$ when x is between $L_1/20$ and $L_1/10$.

For our example, with $x = 6$ in., $L_1/20 = 180/20 = 9$. Thus $C_x = 2.5$ and

$$K = 0.671 \sqrt{C_x\left(\frac{E}{F_c}\right)} = 0.671 \sqrt{(2.5)\left(\frac{1,900,000}{1450}\right)} = 38.4$$

which indicates that the stress condition is condition (3), and

$$F'_c = \frac{0.30(C_x)(E)}{(L_1/d_1)^2} = \frac{(0.30)(2.5)(1,900,000)}{(72)^2}$$

$$= 275 \text{ psi } [1.90 \text{ MPa}]$$

We thus establish that the behavior with respect to the x-axis is critical for this column, that the stress is limited to 275 psi, and that the load permitted for the column is thus

$$\text{load} = \text{allowable stress} \times \text{gross area of column}$$

$$= 275 \times (3 \times 23.125)$$

$$= 19,078 \text{ lb } [84.9 \text{ kN}]$$

Problem 11-4-A. A spaced column of the form shown in Fig. 11-4 consists of two 2×8 pieces of Southern pine, No. 1 grade. Dimension $L_1 = 10$ ft [3.05 m] and $x = 5$ in. [127 mm]. Find the axial compression capacity for the column.

11-5 Use of Design Aids for Wood Columns

It should be apparent from these examples that the design of wood columns by these procedures is a laborious task. The working designer, therefore, typically utilizes some design aids in the form of graphs, tables, or computer-aided processes. One should exercise care in using such aids, however, to be sure that any specific values for E or F_c that are used correspond to the true conditions of the design work and that the aids are developed from criteria identical to those in any applicable code for the work.

Figure 11-5 consists of a graph from which the axial compression capacity of solid, square wood columns may be determined. Note that the graph curves are based on a specific species and

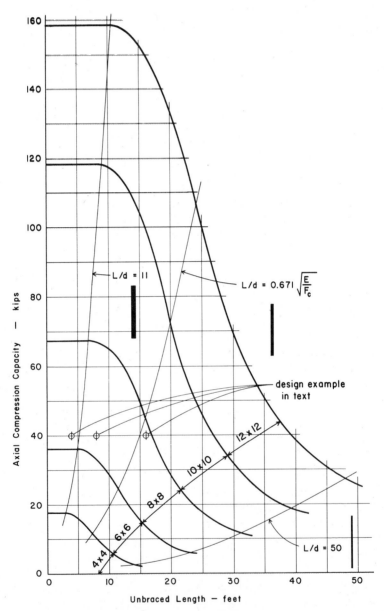

FIGURE 11-5. Axial compression load capacity for wood members of square cross section. Derived from *National Design Specification* requirements for Douglas fir–larch, dense No. 1.

grade of wood (Douglas fir–larch, dense No. 1 grade). The three circled points on the graph correspond to the design examples in Example 3 of Section 11-3.

Table 11-1 gives the axial compression capacity for a range of sizes and unbraced lengths of solid, rectangular wood sections. Note that the design values for elements with nominal thickness of 4 in. and less are different from those with nominal thickness of 6 in. and over, owing to the different size classifications as given in Table 10-1.

TABLE 11-1.   Axial Compression Capacity of Solid Wood Elements (kips)

Element Size		Unbraced Length (ft)							
Designation	Area of Section (in²)	6	8	10	12	14	16	18	20
2 × 3	3.375	0.8			*L/d* greater than 50				
2 × 4	5.25	1.3							
3 × 4	8.75	6.0	3.4	2.2					
3 × 6	13.75	9.4	5.3	3.4					
4 × 4	12.25	14.7	9.3	5.9	4.1	3.0			
4 × 6	19.25	23.1	14.6	9.3	6.5	4.8			
4 × 8	25.375	30.4	19.2	12.3	8.5	6.3			
6 × 6	30.25	35.4	33.5	29.6	22.5	16.6	12.7	10.0	8.1
6 × 8	41.25	48.3	45.7	40.3	30.6	22.6	17.3	13.6	11.0
6 × 10	52.25	61.2	57.9	51.1	38.8	28.6	21.9	17.2	14.0
6 × 12	63.25	74.1	70.1	61.8	47.0	34.7	26.5	20.9	17.0
8 × 8	56.25	67.5	66.0	63.9	60.0	53.6	43.8	34.6	28.0
8 × 10	71.25	85.5	83.6	80.9	75.9	67.9	55.4	46.9	35.5
8 × 12	86.25	103.5	101.2	98.0	91.9	82.2	67.1	53.0	42.9
8 × 14	101.25	121.5	118.9	115.0	107.9	96.5	78.8	62.3	50.4
10 × 10	90.25	108.3	108.3	106.0	103.6	99.6	93.5	84.5	72.1
10 × 12	109.25	131.1	131.1	128.4	125.4	120.6	113.2	102.4	87.3
10 × 14	128.25	153.9	153.9	150.7	147.2	141.6	132.9	120.2	102.5
10 × 16	147.25	176.7	176.7	173.0	169.0	162.5	152.5	138.0	117.7
12 × 12	132.25	158.7	158.7	158.7	155.5	152.7	148.6	142.5	134.1

Note: Wood used is dense No. 1 Douglas fir–larch under normal moisture and load conditions.

Problems 11-5-A-B-C-D. Select square column sections for the loading and lateral bracing conditions given for Problems 11-3-E-F-G-H, using data from Table 11-1. Note that in the problems in Section 11-3, the wood is No. 1 grade, while Table 11-1 uses dense No. 1 grade. It is possible, therefore, that the selections from the table may not agree with those made from the computations.

12

Wood Fastenings

IIIIIIIIIIIIIIIIIIIIIIIIIIIIIIIIIIIIIIIIIIIIIIIIIIIII

12-1 Introduction

Structures of wood typically consist of large numbers of separate pieces that must be fastened together for the structural action of the whole system. Fastening is sometimes achieved directly, that is, without an intermediate device. Such is the case for the tongue-and-groove, doweled, mortise-and-tenon, and rabbeted joints, used more often in cabinet and furniture work than in building construction. Fastening of wood elements for building structures is most often achieved by utilizing some steel device, common ones being nails, bolts, screws, sheet-metal fasteners, and split-ring connectors. Material in this chapter will explain the use of three devices: the bolt, common wire nail, and split-ring connector.

12-2 Bolted Joints in Wood Structures

When steel bolts are used to connect wood members, there are several design considerations. The principal concerns are the following:

1. Net Stress in Member.   Holes drilled for the placing of bolts reduce the member cross section. For this analysis the hole is assumed to have a diameter $\frac{1}{16}$ in. larger than

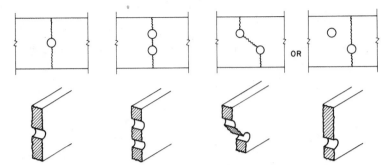

FIGURE 12-1.   Effect of bolt holes on reduction of cross section for tension members.

that of the bolt. The most common situations are those shown in Fig. 12-1. When bolts are staggered, it may be necessary to make two investigations, as shown in the illustration.

2. Bearing of the Bolt on the Wood and Bending in the Bolt.   When the members are thick and the bolt thin and long, the bending of the bolt will cause a concentration of stress at the edges of the members. The bearing on the wood is further limited by the angle of the load to the grain, since wood is much stronger in compression in the grain direction.

3. Number of Members Bolted.   The worst case, as shown in Fig. 12-2, is that of the two-member joint. In this case the lack of symmetry in the joint produces considerable twisting. This situation is referred to as single shear, since the bolt is subjected to shear on a single plane. When more members are joined, this twisting effect is reduced.

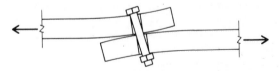

FIGURE 12-2.   Behavior of the single lapped joint with the bolt in single shear.

4. Ripping Out the Bolt When Too Close to an Edge. This problem, together with that of the minimum spacing of the bolts in multiple-bolt joints, is dealt with by using the criteria given in Fig. 12-3. Note that the limiting dimensions involve the consideration of: the bolt diameter D; the bolt length L; the type of force—tension or compression; and the angle of load to the grain of the wood.

The bolt design length is established on the basis of the number of members in the joint and the thickness of the wood pieces.

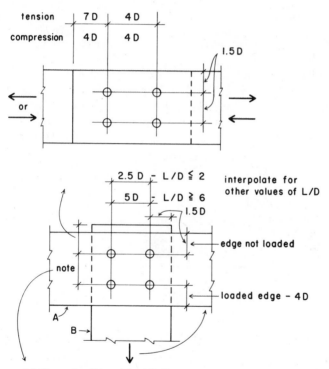

4 D if member B is critical limit.

Straight line proportion if design load is less than limit.

1.5 D minimum.

FIGURE 12-3. Edge, end, and spacing distances for bolts in wood structures.

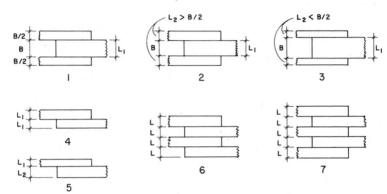

FIGURE 12-4. Various cases of lapped joints with relation to the determination of the critical bolt length.

There are many possible cases, but the most common are those shown in Fig. 12-4. The critical lengths for these cases are given in Table 12-1. Also given in the table is the factor for determining the allowable load on the bolt. Allowable loads ordinarily are tabulated for the three-member joint (Case 1 in Table 12-1), and the factors represent adjustments for other conditions.

Table 12-2 gives allowable loads for bolts with wood members of dense grades of Douglas fir–larch and Southern pine. The two loads given are that for a load parallel to the grain (P load) and that for a load perpendicular to the grain (Q load). Fig. 12-5

TABLE 12-1. Design Length for Bolts

Case[a]	Critical Length	Modification Factor
1	L_1	1.0
2	L_1	1.0
3	$2L_2$	1.0
4	L_1	0.5
5	Lesser of L_2 or $2L_1$	0.5
6	L	1.5
7	L	2.0

[a] See Figure 12-4.

TABLE 12-2. Bolt Design Values for Wood Joints for Douglas Fir–Larch and Southern Pine

		Design Values for One Bolt in Double Shear[a] (lb)			
		Parallel to Grain Load (*P*)		Perpendicular to Grain Load (*Q*)	
Design Length of Bolt (in.)	Diameter of Bolt (in.)	Dense Grades	Ordinary Grades	Dense Grades	Ordinary Grades
1.5	1/2	1100	940	500	430
	5/8	1380	1180	570	490
	3/4	1660	1420	630	540
	7/8	1940	1660	700	600
	1	2220	1890	760	650
2.5	1/2	1480	1260	840	720
	5/8	2140	1820	950	810
	3/4	2710	2310	1060	900
	7/8	3210	2740	1160	990
	1	3680	3150	1270	1080
3.0	1/2	1490	1270	1010	860
	5/8	2290	1960	1140	970
	3/4	3080	2630	1270	1080
	7/8	3770	3220	1390	1190
	1	4390	3750	1520	1300
3.5	1/2	1490	1270	1140	980
	5/8	2320	1980	1330	1130
	3/4	3280	2800	1480	1260
	7/8	4190	3580	1630	1390
	1	5000	4270	1770	1520
5.5	5/8	2330	1990	1650	1410
	3/4	3350	2860	2200	1880
	7/8	4570	3900	2550	2180
	1	5930	5070	2790	2380
	1 1/4	8940	7640	3260	2790
7.5	5/8	2330	1990	1480	1260
	3/4	3350	2860	2130	1820
	7/8	4560	3890	2840	2430
	1	5950	5080	3550	3030
	1 1/4	9310	7950	4450	3800

[a] See Table 12-1 for modification factors for other conditions.

305

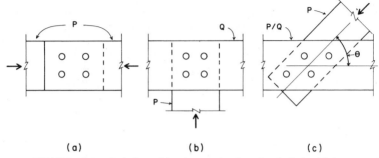

FIGURE 12-5. Relation of load to grain direction in bolted joints.

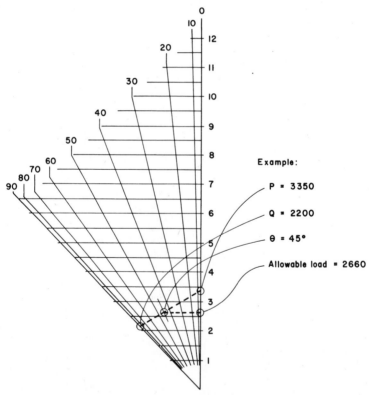

Example:

P = 3350

Q = 2200

θ = 45°

Allowable load = 2660

FIGURE 12-6. Hankinson graph for determination of load values with the loading at an angle to the wood grain.

illustrates these two loading conditions, together with the case of a load at some other angle (Θ). For such cases it is necessary to find the allowable load for the specific angle. Figure 12-6 is an adaptation of the Hankinson graph, which may be used to find values for loads at some angle to the grain.

The following examples illustrate the use of the data presented for the design of bolted joints.

Example 1. A three-member bolted joint is made with members of Douglas fir–larch, dense No. 1 grade. The members are loaded in a direction parallel to the grain (Fig. 12-5a), with a tension force of 10 kips. The middle member is a 3 × 12 and the outer members are each 2 × 12. If four ¾-in. bolts are used for the joint, is the joint capable of carrying the tension load?

Solution: The first step is to identify the critical bolt length (Fig. 12-4) and the load factor (Table 12-1). Since the outer members are greater than one-half the thickness of the middle member, the condition is that of Case 2 in Fig. 12-4, and the effective length is the thickness of the middle member: 2.5 in. The load factor from Table 12-1 is 1.0, indicating that the tabulated load may be used with no adjustment.

From Table 12-2 the allowable load per bolt is 2710 lb. (Bolt design length of 2.5 in.; bolt diameter of ¾ in.; P load; dense grade.) With the four bolts, the total capacity of the bolts is thus 4 × 2710 = 10,840 lb.

For tension stress in the wood, the critical condition is for the middle member, with the net section through the bolts being that shown in Fig. 12-1b. With the holes being considered as $\frac{1}{16}$ in. larger than the bolts, the net area for tension stress is thus

$$A = 2.5 \times \left[11.25 - \left(2 \times \frac{13}{16} \right) \right] = 24.06 \text{ in}^2$$

From Table 10-1 the allowable tension stress is 1200 psi. The maximum tension capacity of the member in tension at the net section is thus

$$T = \text{allowable stress} \times \text{net area} = 1200 \times 24.06 = 28{,}872 \text{ lb}$$

and the joint is adequate for the required load.

Example 2. A bolted two-member joint consists of two 2 × 10 members of Douglas fir–larch, dense No. 2 grade, attached at right angles to each other, as shown in Fig. 12-5b. If the joint is made with two ⅞-in. bolts, what is the maximum compression capacity of the joint?

Solution: This is Case 4 in Fig. 12-4, and the effective length is the member thickness of 1.5 in. The modification factor from Table 12-1 is 0.5, and the bolt capacity from Table 12-2 is 700 lb per bolt. (Bolt design length of 1.5 in.; bolt diameter of ⅞ in.; Q load; dense grade.) The total capacity of the bolts is thus

$$C = 2 \times 700 \times 0.5 = 700 \text{ lb.}$$

The net section is not a concern for the compression force, as the capacity of the members would be based on an analysis for the slenderness condition based on the L/d of the members.

Example 3. A three-member bolted joint consists of two outer members, each 2 × 10, and a middle member that is a 4 × 12. The outer members are arranged at an angle to the middle member, as shown in Fig. 12-5c, such that $\Theta = 60°$. Find the maximum compression force that can be transmitted through the joint by the outer members. Wood is Southern pine, No. 1 grade. The joint is made with two ¾-in. bolts.

Solution: In this case we must investigate both the outer and middle members. For the outer members the effective length is 2 × 1.5 = 3.0 in., and the modification factor is 1.0 (Case 3, Fig. 12-4 and Table 12-1). From Table 12-2 the bolt capacity based on the outer members is 2630 lb per bolt. (Bolt design length of 3.0 in.; bolt diameter of ¾-in.; P load; ordinary grade.)

For the middle member the effective bolt length is the member thickness of 3.5 in., and the unadjusted load per bolt from Table 12-2 is 2800 lb for the P condition and 1260 lb for the Q condition. If these values are used on the Hankinson graph in Fig. 12-6, the load per bolt for the 60° angle is found to be approximately 1700 lb. Since this value is lower than that found for the outer members, it represents the limit for the joint. The joint capacity based on the bolts is thus 2 × 1700 = 3400 lb.

Note: For all of the following problems, use Douglas fir–larch, dense No. 1 grade.

Problem 12-2-A*.   A three-member tension joint has 2 × 12 outer members and a 4 × 12 middle member. (Fig. 12-5*a*.) The joint is made with six ¾-in. bolts. Find the capacity of the joint as limited by the bolts and the tension stresses in the members.

Problem 12-2-B.   A two-member tension joint consists of 2 × 6 members bolted with two ⅞-in. bolts. (Fig. 12-5*a*.) What is the limit for the tension force?

Problem 12-2-C.   Two outer members, each 2 × 8, are bolted to a middle member consisting of a 3 × 12. The outer members form an angle of 45° with respect to the middle member (Fig. 12-5*c*; Θ = 45°). What is the maximum compression force that the outer members can transmit to the joint?

12-3   Nailed Joints

Nails are used in great variety in building construction. For structural fastening, the nail most commonly used is called—appropriately—the *common wire nail.* As shown in Fig. 12-7, the critical concerns for such nails are the following:

1. Nail Size.   Critical dimensions are the diameter and length. Sizes are specified in pennyweight units, designated as 4d, 6d, and so on, and referred to as four penny, six penny, and so on.

2. Load Direction.   Pull-out loading in the direction of the nail shaft is called *withdrawal;* shear loading perpendicular to the nail shaft is called *lateral load.*

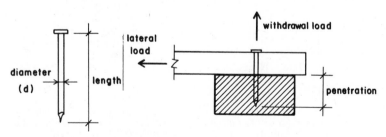

FIGURE 12-7.   Typical common wire nail and loading conditions.

3. Penetration. Nailing is typically done through one element and into another, and the load capacity is limited by the amount of length of embedment of the nail in the second member. The length of embedment is called the penetration.

4. Species and Grade of Wood. The harder, tougher, and heavier the wood, the more the load resistance capability.

Design of good nail joints requires a little engineering and a lot of good carpentry. Some obvious situations to avoid are those shown in Fig. 12-8. A little actual carpentry experience is highly desirable for anyone who designs nailed joints.

Withdrawal load capacities of common wire nails are given in Table 12-3. The capacities are given for both Douglas fir–larch and Southern pine. Note that the table values are given in units of capacity per inch of penetration and must be multiplied by the actual penetration length to obtain the load capacity in pounds. In general, it is best not to use structural joints that rely on withdrawal loading resistance.

Lateral load capacities for common wire nails are given in Table 12-4. These values also apply to Douglas fir–larch and Southern pine. Note that a penetration of at least 11 times the nail diameter is required for the development of the full capacity of the nail. A value of one-third of that in the table is permitted with a penetration of one-third of this length, which is the minimum penetration permitted. For actual penetration lengths between

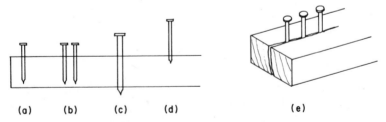

(a)   (b)   (c)   (d)   (e)

FIGURE 12-8. Poor nailing practices: (a) too close to edge, (b) nails too closely spaced, (c) nail too large for wood piece, (d) too little penetration in holding piece, (e) too many nails in a single row parallel to the wood grain.

TABLE 12-3. Withdrawal Load Capacity of Common Wire Nails (lb/in)

	Size of Nail				
Pennyweight	6	8	10	12	16
Diameter (in.)	0.113	0.131	0.148	0.148	0.162
Douglas Fir–Larch	29	34	38	38	42
Southern Pine	35	41	46	46	50

these limits, the load capacity may be determined by direct proportion. Orientation of the load to the direction of grain in the wood is not a concern when considering nails in terms of lateral loading.

The following example illustrates the design of a typical nailed joint for a wood truss.

Example. The truss heel joint shown in Fig. 12-9 is made with 2-in. nominal wood elements of Douglas fir–larch, dense No. 1 grade, and gusset plates of ½-in. plywood. Nails are 6d common, with the nail layout shown occurring on both sides of the joint. Find the tension force limit for the bottom chord (load 3 in the illustration).

TABLE 12-4. Lateral Load Capacity of Common Wire Nails (lb/nail)

	Size of Nail				
Pennyweight	6	8	10	12	16
Diameter (in.)	0.113	0.131	0.148	0.148	0.162
Length (in.)	2.0	2.5	3.0	3.25	3.5
Douglas Fir–Larch and Southern Pine	63	78	94	94	108
Penetration Required for 100% of Table Value[a] (in.)	1.24	1.44	1.63	1.63	1.78
Minimum Penetration[b] (in.)	0.42	0.48	0.54	0.54	0.59

[a] Eleven diameters; reduce by straight-line proportion for less penetration.
[b] One-third of that for full value; 11/3 diameters.

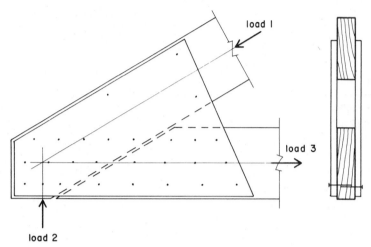

FIGURE 12-9.   Truss joint with nails and plywood gusset plates.

Solution: The two primary concerns are for the lateral capacity of the nails and the tension, tearing stress in the gussets. For the nails, we observe from Table 12-4 that

nail length is 2 in. [51 mm],
minimum penetration for full capacity is 1.24 in. [31.5 mm],
maximum capacity is 63 lb/nail [0.28 kN].

From inspection of the joint layout, we see that

actual penetration = nail length − plywood thickness
= 2.0 − 0.5 = 1.5 in. [38 mm]

Therefore, we may use the full table value for the nails. With 12 nails on each side of the member, the total capacity is thus

$$F_3 = (24)(63) = 1512 \text{ lb } [6.73 \text{ kN}]$$

If we consider the cross section of the plywood gussets only in the zone of the bottom chord member, the tension stress in the plywood will be approximately

$$f_t = \frac{1512}{(0.5)(2)(5 \text{ in. of width})} = 302 \text{ psi } [2.08 \text{ MPa}]$$

which is probably not a critical magnitude for the plywood.

A problem that must be considered in this type of joint is that of the pattern of placement of the nails (commonly called the layout of the nails). In order to accommodate the large number of nails required, they must be quite closely spaced, and since they are close to the ends of the wood pieces, the possibility of splitting the wood is a critical concern. The factors that determine this possibility include the size of the nail (essentially its diameter), the spacing of the nails, the distance of the nails from the end of the piece, and the tendency of the particular wood species to be split. There are no formal guidelines for this problem; it is largely a matter of good carpentry or some experimentation to establish the feasibility of a given layout.

One technique that can be used to reduce the possibility of splitting is to stagger the nails rather than to arrange them in single rows. Another technique is to use a single set of nails for both gusset plates, rather than to nail the plates independently, as shown in Fig. 12-10. The latter procedure consists simply of driving a nail of sufficient length so that its end protrudes from the gusset on the opposite side and then bending the end over—called clinching—so that the nail is anchored on both ends. A single nail may thus be utilized for twice its rated capacity for lateral load. This is similar to the development of a single bolt in double shear in a three-member joint.

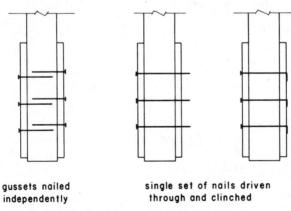

gussets nailed   single set of nails driven
independently   through and clinched

FIGURE 12-10. Nailing techniques for joints with plywood gusset plates.

It is also possible to glue the gusset plates to the wood pieces and to use the nails essentially to hold the plates in place only until the glue has set and hardened. The adequacy of such joints should be verified by load testing, and the nails should be capable of developing some significant percentage of the design load as a safety backup for the glue.

Problem 12-3-A. Two wood members of 3-in. nominal thickness are attached at right angles to each other by plywood gussets on both sides, to develop a tension force through the joint. The wood is Douglas fir–larch, No. 1 grade; the plywood gussets are $\frac{3}{4}$ in. thick; nails are common wire. Select the size of nail and find the number required if the tension force is 2000 lb [8.90 kN].

12-4   Split-Ring Connectors

Various devices are used to increase both the strength and the tightness of bolted joints in wood structures. One such device is the split-ring, shown in Fig. 12-11. Design considerations for the split-ring include the following:

1.   Size of the Ring.   Rings are available in the two sizes shown in the figure, with nominal diameters of 2.5 and 4 in.
2.   Stress on the Net Section of the Wood Member.   As

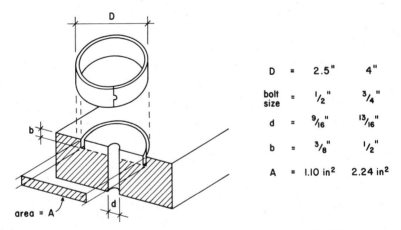

		D	=	2.5"	4"
bolt size	=	$\frac{1}{2}$"	$\frac{3}{4}$"
d	=	$\frac{9}{16}$"	$\frac{13}{16}$"
b	=	$\frac{3}{8}$"	$\frac{1}{2}$"
A	=	1.10 in²	2.24 in²

FIGURE 12-11.   Split-ring connectors for bolted wood joints.

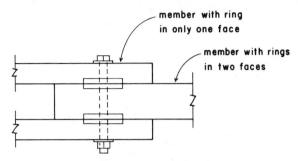

FIGURE 12-12. Determination of the number of faces of a member with split-ring connectors.

shown in Fig. 12-11, the cross section of the wood piece is reduced by the ring profile (*A* in the figure) and the bolt hole. If rings are placed on both sides of a wood piece, there will be two reductions for the ring profile.

3. Thickness of the Wood Piece. If the wood piece is too thin, the cut for the ring will bite excessively into the cross section. Rated load values reflect concern for this.

4. Number of Faces of the Wood Piece Having Rings. As shown in Fig. 12-12, the outside members in a joint will have rings on only one face, while the inside members will have rings on both faces. Thickness considerations, therefore, are more critical for the inside members.

5. Edge and End Distances. These must be sufficient to permit the placing of the rings and to prevent splitting out from the side of the wood piece when the joint is loaded. Concern is greatest for the edge in the direction of loading—called the loaded edge. (See Fig. 12-13.)

6. Spacing of Rings. Spacing must be sufficient to permit the placing of the rings and the full development of the ring capacity of the wood piece.

Figure 12-14 shows the four placement dimensions that must be considered. The limits for these dimensions are given in Table 12-5. In some cases, two limits are given. One limit is that required for the full development of the ring capacity (100% in the

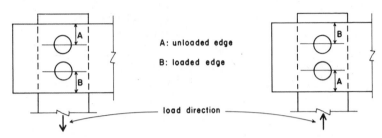

FIGURE 12-13.   Determination of the loaded edge condition.

table). The other limit is the minimum dimension permitted, for which some reduction factor is given for the ring capacity. Load capacities for dimensions between these limits can be directly proportioned.

Table 12-6 gives capacities for split-ring connections for both dense and regular grades of Douglas fir–larch and Southern pine. As with bolts, values are given for load directions both parallel to and perpendicular to the grain of the wood. Values for loadings at some angle to the grain can be determined with the use of the Hankinson graph shown in Fig. 12-6.

The following example illustrates the procedures for the analysis of a joint using split-ring connectors.

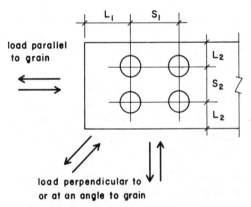

FIGURE 12-14.   Reference figure for the end, edge, and spacing distances for split-ring connectors. (See Table 12-5.)

TABLE 12-5. Spacing, Edge Distance, and End Distance for Split-Ring Connectors

Distances (in inches) and Corresponding Percentages of Design Values from Table 12-6

Load Direction with Respect to Grain	Parallel		Perpendicular or Angle	
Ring Size (in.)	2.5	4	2.5	4
L_1 tension	5.50 in., 100% 2.75 in. min, 62.5%	7 in., 100% 3.50 in. min, 62.5%	5.50 in., 100%	7 in., 100%
compression	4 in., 100% 2.50 in. min, 62.5%	5.50 in., 100% 3.25 in. min, 62.5%	2.75 in. min, 62.5%	3.25 in. min, 62.5%
L_2 unloaded	1.75 in. min	2.75 in. min	1.75 in. min	2.75 in. min
loaded[a]				
S_1	3.50 in. min, 50% 6.75 in., 100%	5 in. min, 50% 9 in., 100%	3.50 in. min	5 in. min
S_2	3.50 in. min	5 in. min	3.5 in. min, 50% 4.25 in., 100%	5 in. min, 50% 6 in., 100%

Note: See Fig. 12-14.

[a] See Table 12-6 and Fig. 12-13.

317

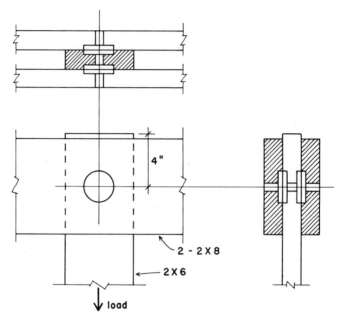

FIGURE 12-15.

Example. The joint shown in Fig. 12-15, using $2\frac{1}{2}$-in. split-rings and wood of Douglas fir–larch, dense No. 1 grade, sustains the load indicated. Find the limiting value for the load.
Solution: Separate investigations must be made for the members in this joint. For the 2 × 6,

load is parallel to the grain,

rings are in two faces,

critical dimensions are member thickness of 1.5 in. [38 mm] and end distance of 4 in.

From Table 12-5 we determine that the end distance required for use of the full capacity of the rings is 5.5 in. and that, if the minimum distance of 2.75 in. is used, the capacity must be reduced to 62.5% of the full value. The value to be used for the 4-in. end distance must be interpolated between these limits, as shown

TABLE 12-6. Design Values for Split-Ring Connectors

Ring Size (in.)	Bolt Diameter (in.)	Faces with Connectors[a]	Actual Thickness of Piece (in.)	Load Parallel to Grain Design Value/Connector (lb)		Distance to Loaded Edge[c] (in.)	Load Perpendicular to Grain Design Value/Connector (lb)	
				Group A Woods[b]	Group B Woods[b]		Group A Woods[b]	Group B Woods[b]
2.5	1/2	1	1 min	2630	2270	1.75 min	1580	1350
						2.75 or more	1900	1620
			1.5 or more	3160	2730	1.75 min	1900	1620
						2.75 or more	2280	1940
		2	1.5 min	2430	2100	1.75 min	1460	1250
						2.75 or more	1750	1500
			2 or more	3160	2730	1.75 min	1900	1620
						2.75 or more	2280	1940
4	3/4	1	1 min	4090	3510	2.75 min	2370	2030
						3.75 or more	2840	2440
			1.5 or more	6020	5160	2.75 min	3490	2990
						3.75 or more	4180	3590
		2	1.5 min	4110	3520	2.75 min	2480	2040
						3.75 or more	2980	2450
			2	4950	4250	2.75 min	2870	2470
						3.75 or more	3440	2960
			2.5	5830	5000	2.75 min	3380	2900
						3.75 or more	4050	3480
			3 or more	6140	5260	2.75 min	3560	3050
						3.75 or more	4270	3660

[a] See Fig. 12-12.
[b] Group A includes dense grades and Group B regular grades of Douglas fir–larch and Southern pine.
[c] See Fig. 12-13.

319

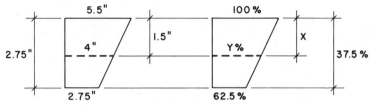

FIGURE 12-16.

in Fig. 12-16. Thus

$$\frac{1.5}{x} = \frac{2.75}{37.5} \qquad x = \frac{1.5}{2.75}(37.5) = 20.45\%$$

$$y = 100 - 20.45 = 79.55\%, \text{ or approximately } 80\%$$

From Table 12-6 we determine the full capacity to be 2430 lb per ring. Therefore, the usable capacity is

$$(0.80)(2430) = 1944 \text{ lb/ring } [8.65 \text{ kN/ring}]$$

For the 2 × 8,

load is perpendicular to the grain,
rings are in only one face,
loaded edge distance is one-half of 7.25 in., or 3.625 in. [92 mm].

For this situation the load value from Table 12-6 is 2880 lb [12.81 kN]. Therefore, the joint is limited by the conditions for

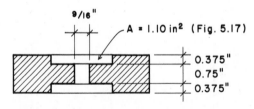

FIGURE 12-17. Determination of the net cross-sectional area.

the 2 × 6, and the capacity of the joint with the two rings is

$$T = (2)(1944) = 3888 \text{ lb } [17.3 \text{ kN}]$$

It should be verified that the 2 × 6 is capable of sustaining this load in tension stress on the net section at the joint. As shown in Fig. 12-17, the net area is

$$A = 8.25 - (2)(1.10) - \left(\frac{9}{16}\right)(0.75) = 5.63 \text{ in}^2 \text{ [3632 mm}^2]$$

From Table 10-1 the allowable tension stress is 1200 psi [8.27 MPa], and therefore the capacity of the 2 × 6 is

$$T = (1200)(5.63) = 6756 \text{ lb } [30.0 \text{ kN}]$$

and the member is not critical in tension stress.

Problem 12-4-A.   A joint similar to that shown in Fig. 12-15 is made with 4-in. split-rings and wood members of Southern pine, No. 1 grade. Find the limit for the tension load if the outer members are each 3 × 10 and the middle member is a 4 × 10.

IV

REINFORCED
CONCRETE

||||||||||||||||||||||||||||||||||||||||||||||||||||||||||||||||||

13

Stresses in Reinforced Concrete

||||||||||||||||||||||||||||||||||||||||||||||||||||||||||||

13-1 Introduction

Concrete is made by mixing a paste of cement and water with sand and crushed stone, gravel, or other inert material. The sand and other inert materials are called the *aggregate*. After this plastic mixture is placed in forms, a group of chemical reactions called hydration takes place and the mass hardens. Concrete, although strong in compression, is relatively weak in resisting tensile and shearing stresses which develop in structural members. To overcome this lack of resistance, steel bars are placed in the concrete at the proper positions; the result is *reinforced concrete*. In beams and slabs the principal function of the concrete is to resist compressive stresses, whereas the steel bars resist tensile stresses.

13-2 Design Methods

The design of reinforced concrete structural members may be accomplished by two different methods. The first, called *working stress design,* is the principal method used in this part of the book; the second method is known as *strength design.* At present the principal reference for the analysis and design of reinforced concrete members is *Building Code Requirements for Reinforced Concrete* (ACI 318-77), published by the American Concrete Institute, commonly referred to as the ACI Code. Most of the material in the present edition of the ACI Code (1977) is based on strength design methods. A limited treatment of working stress design is given in Appendix B of the code, under the title "Alternative Design Method." The last edition of the code to fully develop the working stress design method was the 1963 edition (ACI 318-63). The discussion of reinforced concrete design in Chapters 13 through 17 is keyed primarily to the 1963 code but also draws on the alternate method of the 1977 code. A brief introduction to strength design methods is presented in Chapter 18.

The design of reinforced concrete members must, of course, be carried out in compliance with the building code that has jurisdiction in your locality. The ACI Code is extensive, and those who desire more complete information should examine it in detail. In this elementary book, space permits discussions of only the basic structural members, and many of the items referred to in the code must necessarily be omitted.

13-3 Strength of Concrete

The designer of a reinforced concrete structure bases his or her computations on the use of concrete having a specified compressive strength (2500, 3000, 3500, etc., psi) at the end of a 28-day curing period. The symbol for this specified compressive strength is f'_c. Concretes of different strengths are produced by varying the proportions of cement, fine aggregate (sand), coarse aggregate, and water in the mix. The general theory in establishing the proportions of fine and coarse aggregates is that the voids in the coarse aggregate should be filled with the cement paste and fine aggregate. The proportioning of concrete mixes and the attendant procedures for strength verification are not discussed in this book. Readers interested in studying this aspect of concrete manufacture should consult *Recommended Practice for Selecting Proportions for Normal and Heavy Weight Concrete* (ACI 211.1-74) and Section 4-2 of the 1977 ACI Code.

Very little concrete is proportioned and mixed at the building site today. Central or ready-mixed concrete is used whenever it is available. The use of a concrete mixed under ideal controlled conditions at a central plant affords many advantages. It is delivered to the building site in a revolving mixer, the proportions of cement, aggregate, and water are maintained accurately, any desired strength may be ordered, and the concrete thus provided is uniform in quality.

Table 13-1 gives allowable stresses in flexure (bending), shear, and bearing for concretes of four different specified compressive strengths. Note that the individual values are functions of f'_c.

13-4 Water–Cement Ratio

An important factor affecting the strength of concrete is the *water–cement ratio*. This is expressed as the number of pounds of water per pound of cement used in the mix (or the number of gallons of water for each 94-lb bag of cement). The relationship is an inverse one, that is, lower values of the ratio produce higher strengths. However, in order to produce freshly mixed concrete that possesses *workability*—the property that controls the ease with which it can be placed in the forms and around the reinforcing bars—more water must be used than the amount required for hydration of the cement. The use of too much water may cause segregation of the mix components, thereby producing nonuniform concrete. To control this situation, building codes specify the maximum permissible water–cement ratios for concretes of specified design strengths.

13-5 Cement

The cement used most extensively in building construction is *portland cement*. Of the five types of standard portland cement generally available in the United States and for which the American Society for Testing and Materials has established specifications, two types account for most of the cement used in buildings. These are ASTM Type I, a general purpose cement for use in concrete designed to reach its required strength in about 28 days,

TABLE 13-1. Allowable Stresses in Concrete

Description (all values given for normal weight concrete, 145 lb/ft³)		Based on Specified Concrete Strength	Allowable Stresses (psi) For Strength of Concrete Shown Below, f'_c				Values in MPa Based on Specified Concrete Strength
			2500	3000	4000	5000	
Modulus of Elasticity	E_c	$57{,}000\sqrt{f'_c}$	2.85×10^6	3.12×10^6	3.60×10^6	4.03×10^6	$4730\sqrt{f'_c}$
Modular Ratio: $n = \dfrac{E_s}{E_c}$	n	$\dfrac{29{,}000}{57\sqrt{f'_c}}$	10.2	9.3	8.0	7.2	$\dfrac{200{,}000}{4730\sqrt{f'_c}}$
Flexure							
Extreme fiber stress in compression	f_c	$0.45 f'_c$	1125	1350	1800	2250	$0.45 f'_c$
Extreme fiber stress in tension, plain concrete walls and footings	f_c	$1.6\sqrt{f'_c}$	80	88	102	113	$0.13\sqrt{f'_c}$
Shear (carried by concrete)							
Beams, walls, one-way slabs	v_c	$1.1\sqrt{f'_c}$	55	60	70	78	$0.09\sqrt{f'_c}$
Joists	v_c	$1.2\sqrt{f'_c}$	61	66	77	86	$0.10\sqrt{f'_c}$
Two-way slabs and footings (peripheral shear)	v_c	$2\sqrt{f'_c}$	100	110	126	141	$0.17\sqrt{f'_c}$
Bearing							
On full area	f_c	$0.3 f'_c$	750	900	1200	1500	$0.3 f'_c$
When supporting surface A_2 is wider on all sides than loaded area A_1	f_c	$0.3 f'_c \times \sqrt{A_2/A_1}$ (but not more than $0.6 f'_c$)					

Source: Most tabulated stresses are those specified in Appendix B, "Alternate Design Method," *Building Code Requirements for Reinforced Concrete* (ACI 318-77), abstracted with permission of the publishers, American Concrete Institute. Extreme fiber stresses in tension in plain concrete walls and footings were abstracted from *Building Code Requirements for Reinforced Concrete* (ACI 318-63) with permission of the publishers, American Concrete Institute.

and ASTM Type III, a high-early-strength cement for use in concrete that attains its design strength in a period of a week or less. All portland cements set and harden by reacting with water, and this hydration process is accompanied by generation of heat. In massive concrete structures such as dams, the resulting temperature rise of the materials becomes a critical factor in both design and construction, but the problem is not usually significant in building construction. ASTM Type IV, a low-heat cement, is designed for use where the heat rise during hydration is a critical factor. It is, of course, essential that the cement actually used in construction correspond to that employed in designing the mix, to produce the specified compressive strength of the concrete.

13-6 Air-Entrained Concrete

Air-entrained concrete is produced by using special cement and by introducing an additive during mixing of the concrete. In addition to improving workability (mobility of the wet mix), entrainment permits lower water–cement ratios and significantly improves the durability of the concrete. Air-entraining agents produce billions of microscopic air cells throughout the concrete mass. These minute voids prevent accumulation of water in cracks and other large voids which, on freezing, would permit the water to expand and result in spalling away of the exposed surface of the concrete.

13-7 Steel Reinforcement

The steel used in reinforced concrete consists of round bars, mostly of the deformed type, with lugs or projections on their surfaces. The surface deformations help to develop a greater bond between the concrete and steel. The most common grades of reinforcing steel are Grade 60 and Grade 40, having yield strengths of 60,000 psi [414 MPa] and 40,000 psi [276 MPa], respectively. Table 13-2 gives the properties of standard deformed reinforcing bars.

Ample concrete protection, called *cover,* must be provided for the steel reinforcing. Cover is measured as the distance from the outside face of the concrete to the edge of a reinforcing bar. For

TABLE 13-2. Properties of Standard Deformed Reinforcing Bars

	Nominal Dimensions					
	U.S. Units			SI Units		
Bar Designation Number	Diameter in.	Cross-Sectional Area in²	Perimeter in.	Diameter mm	Cross-Sectional Area mm²	Perimeter mm
3	0.375	0.11	1.178	9.52	71	29.9
4	0.500	0.20	1.571	12.70	129	39.9
5	0.625	0.31	1.963	15.88	200	49.9
6	0.750	0.44	2.356	19.05	284	59.8
7	0.875	0.60	2.749	22.22	387	69.8
8	1.000	0.79	3.142	25.40	510	79.8
9	1.128	1.00	3.544	28.65	645	90.0
10	1.270	1.27	3.990	32.26	819	101.3
11	1.410	1.56	4.430	35.81	1006	112.5
14	1.693	2.25	5.32	43.00	1452	135.1
18	2.257	4.00	7.09	57.33	2581	180.1

reinforcement near surfaces not exposed to the ground or to weather, cover should be not less than $\frac{3}{4}$ in. [19 mm] for slabs, walls, and joists, and 1.5 in. [38 mm] for beams, girders, and columns. Where formed surfaces are exposed to earth or weather, the cover should be 1.5 in. [3.8 mm] for No. 5 bars and smaller and 2 in. [51 mm] for No. 6 through No. 18 bars. For foundation construction poured directly against ground without forms, cover should be 3 in. [76 mm].

13-8 Modulus of Elasticity

The modulus of elasticity E_c of hardened concrete is a measure of its resistance to deformation. The magnitude of E_c depends on w, the weight of the concrete, and on f'_c, its strength. Its value may be determined from the expression $E_c = w^{1.5}33\sqrt{f'_c}$ for values of w between 90 and 155 lb/ft³. For normal weight concrete (145 lb/ft³), E_c may be considered as equal to $57{,}000\sqrt{f'_c}$. $[E_c = w^{1.5}$

$0.043\sqrt{f_c'}$ for values of w between 1440 and 2480 kg/m^3. For normal weight concrete (2320 kg/m^3), E_c may be considered as equal to $4730\sqrt{f_c'}$.]

In the design of reinforced concrete members, we employ the term n. This is the ratio of the modulus of elasticity of steel to that of concrete, or $n = E_s/E_c$. E_s is taken as 29,000 ksi [200,000 MPa].

Consider a concrete for which f_c' is 4000 psi and w is 145 lb/ft^3. Then $E_c = 57,000\sqrt{f_c'} = 57,000\sqrt{4000} = 3,600,000$ psi and $n = E_s/E_c = 29,000/3,600 = 8.055$. The values for n for four different strengths of concrete are given in Table 13-1. As is the usual practice, the values for n are rounded off to those given in the table.

13-9 Flexural Design Formulas for Rectangular Beams

This discussion refers to a rectangular concrete section, tension reinforcing only.

Referring to Fig. 13-1, the following are defined:

b = width of the concrete compression zone,

d = effective depth of the section for stress analysis; from the centroid of the steel to the edge of the compression zone,

A_s = cross-sectional area of the reinforcing,

p = percentage of reinforcing, defined as

$$p = \frac{A_s}{bd}$$

n = elastic ratio = $\dfrac{E \text{ of the steel reinforcing}}{E \text{ of the concrete}}$

kd = height of the compression stress zone; used to locate the neutral axis of the stressed section; expressed as a percentage (k) of d,

jd = internal moment arm, between the net tension force and the net compression force; expressed as a percentage (j) of d,

f_c = maximum compressive stress in the concrete,

f_s = tensile stress in the reinforcing.

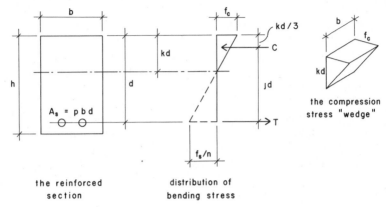

the reinforced section

distribution of bending stress

the compression stress "wedge"

FIGURE 13-1. Moment resistance of a rectangular concrete section with tension reinforcing.

The compression force, C, may be expressed as the volume of the compression stress "wedge," as shown in the figure.

$$C = \tfrac{1}{2}(kd)(b)(f_c) = \tfrac{1}{2} kf_c bd$$

Using the compression force, we may express the moment resistance of the section as

$$M = Cjd = (\tfrac{1}{2} kf_c bd)(jd) = \tfrac{1}{2} kjf_c bd^2 \quad \text{(Formula 13-9-1)}$$

This may be used to derive an expression for the concrete stress:

$$f_c = \frac{2M}{kjbd^2} \quad \text{(Formula 13-9-2)}$$

The resisting moment may also be expressed in terms of the steel and the steel stress as

$$M = Tjd = (A_s)(f_s)(jd)$$

This may be used for determination of the steel stress or for finding the required area of steel:

$$f_s = \frac{M}{A_s jd} \quad \text{(Formula 13-9-3)}$$

$$A_s = \frac{M}{f_s jd} \quad \text{(Formula 13-9-4)}$$

A useful reference is the so-called balanced section, which occurs when use of the exact amount of reinforcing results in the simultaneous limiting stresses in the concrete and steel. The properties which establish this relationship may be expressed as follows:

$$\text{balanced } k = \frac{1}{1 + f_s/nf_c} \qquad \text{(Formula 13-9-5)}$$

$$j = 1 - \frac{k}{3} \qquad \text{(Formula 13-9-6)}$$

$$p = \frac{f_c k}{2f_s} \qquad \text{(Formula 13-9-7)}$$

$$M = Rbd^2 \qquad \text{(Formula 13-9-8)}$$

in which

$$R = \tfrac{1}{2} kjf_c \qquad \text{(Formula 13-9-9)}$$

derived from formula (1). If the limiting compression stress in the concrete ($f_c = 0.45 f_c'$) and the limiting stress in the steel are entered in Formula (5), the balanced section value for k may be found. Then the corresponding values for j, p, and R may be found. The balanced p may be used to determine the maximum amount of tensile reinforcing that may be used in a section without the addition of compressive reinforcing. If less tensile reinforcing is used, the moment will be limited by the steel stress, the maximum stress in the concrete will be below the limit of $0.45 f_c'$, the value of k will be slightly lower than the balanced value, and the value of j will be slightly higher than the balanced value. These relationships are useful in design for the determination of approximate requirements for cross sections.

Table 13-3 gives the balanced section properties for various combinations of concrete strength and limiting steel stress. The values of n, k, j, and p are all without units. However, R must be expressed in particular units; the unit used in the table is kip-inches (k-in).

When the area of steel used is less than the balanced p, the true value of k may be determined by the following formula:

$$k = \sqrt{2np - (np)^2} - np \qquad \text{(Formula 13-9-10)}$$

TABLE 13-3.   Balanced Section Properties for Rectangular Concrete Sections with Tension Reinforcing Only

f_s		f'_c		n	k	j	p	R	
ksi	MPa	ksi	MPa					k-in	kN-m
16	110	2.0	13.79	11.3	0.389	0.870	0.0109	0.152	1045
		2.5	17.24	10.1	0.415	0.862	0.0146	0.201	1382
		3.0	20.68	9.2	0.437	0.854	0.0184	0.252	1733
		4.0	27.58	8.0	0.474	0.842	0.0266	0.359	2468
20	138	2.0	13.79	11.3	0.337	0.888	0.0076	0.135	928
		2.5	17.24	10.1	0.362	0.879	0.0102	0.179	1231
		3.0	20.68	9.2	0.383	0.872	0.0129	0.226	1554
		4.0	27.58	8.0	0.419	0.860	0.0188	0.324	2228
24	165	2.0	13.79	11.3	0.298	0.901	0.0056	0.121	832
		2.5	17.24	10.1	0.321	0.893	0.0075	0.161	1107
		3.0	20.68	9.2	0.341	0.886	0.0096	0.204	1403
		4.0	27.58	8.0	0.375	0.875	0.0141	0.295	2028

Figure 13-2 may be used to find approximate k values for various combinations of p and n.

In the design of concrete beams, there are two situations that commonly occur. The first occurs when the beam is entirely undetermined; that is, the concrete dimensions and the reinforcing are unknown. The second occurs when the concrete dimensions are given, and the required reinforcing for a specific bending moment must be determined. The following examples illustrate the use of the formulas just developed for each of these problems.

Example 1.   A rectangular concrete beam of concrete with f'_c of 3000 psi [20.7 MPa] and steel reinforcing with $f_s = 20$ ksi [138 MPa] must sustain a bending moment of 200 kip-ft [271 kN-m]. Select the beam dimensions and the reinforcing for a section with tension reinforcing only.

Solution: (1) With tension reinforcing only, the minimum size beam will be a balanced section, since a smaller beam would have

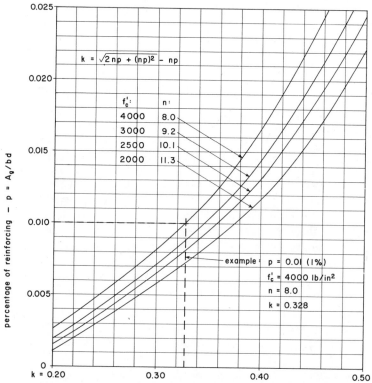

FIGURE 13-2. k factors for rectangular concrete sections with tension reinforcing—as a function of p and n.

to be stressed beyond the capacity of the concrete to develop the required moment. Using Formula (8),

$$M = Rbd^2 = 200 \text{ kip-ft } [271 \text{ kN-m}]$$

Then from Table 13-3, for f'_c of 3000 psi and f_s of 20 ksi,

$$R = 0.226 \text{ (in units of kip-in) } [1554 \text{ in units of kN-m}]$$

Therefore

$$M = 200 \times 12 = 0.226(bd^2), \text{ and } bd^2 = 10,619$$

$$[M = 271 = 1554(bd^2), \text{ and } bd^2 = 0.1744]$$

(2) Various combinations of b and d may be found; for example,

$$b = 10 \text{ in.}, \quad d = \sqrt{\frac{10,619}{10}}$$

$$= 32.6 \text{ in. } [b = 0.254 \ m, \ d = 0.829 \ m]$$

$$b = 15 \text{ in.}, \quad d = \sqrt{\frac{10,619}{15}}$$

$$= 26.6 \text{ in. } [b = 0.381 \ m, \ d = 0.677 \ m]$$

Although they are not given in this example, there are often some considerations other than flexural behavior alone that influence the choice of specific dimensions for a beam. These situations are discussed in Chapters 14 and 15. If the beam is of the ordinary form shown in Fig. 13-3, the specified dimension is usually that given as h. Assuming the use of a No. 3 U-stirrup, a cover of 1.5 in. [38 mm], and an average-size reinforcing bar of 1-in. [25-mm] diameter (No. 8 bar), the design dimension d will be less than h by 2.375 in. [60 mm]. Lacking other considerations, we will assume a b of 15 in. [380 mm] and an h of 29 in. [740 mm], with the resulting d of $29 - 2.375 = 26.625$ in. [680 mm].

(3) We next use the specific value for d with Formula (4) to find the required area of steel A_s. Since our selection is very close to the balanced section, we may use the value of j from Table 13-3.

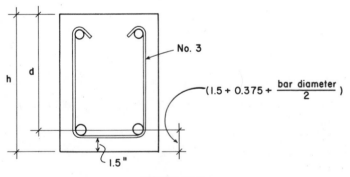

FIGURE 13-3.

Thus

$$A_s = \frac{M}{f_s jd} = \frac{200 \times 12}{20 \times 0.872 \times 26.625} = 5.17 \text{ in}^2$$

$$\left[A_s = \frac{271,000}{0.138 \times 0.872 \times 680} = 3312 \text{ mm}^2 \right]$$

Or using the formula for the definition of p and the balanced p value from Table 13-3,

$$A_s = pbd = 0.0129(15 \times 26.625) = 5.15 \text{ in}^2$$

$$[A_s = 0.0129(380 \times 680) = 3333 \text{ mm}^2]$$

(4) We next select a set of reinforcing bars to obtain this area. As with the beam dimensions, there are other concerns, as discussed in Chapters 14 and 15. For the purpose of our example, if we select bars all of a single size (see Table 13-2), the number required will be:

$$\text{for No. 6 bars, } \frac{5.17}{0.44} = 11.75, \text{ or } 12 \quad \left[\frac{3312}{284} = 11.66 \right]$$

$$\text{for No. 7 bars, } \frac{5.17}{0.60} = 8.62, \text{ or } 9 \quad \left[\frac{3312}{387} = 8.56 \right]$$

$$\text{for No. 8 bars, } \frac{5.17}{0.79} = 6.54, \text{ or } 7 \quad \left[\frac{3312}{510} = 6.49 \right]$$

$$\text{for No. 9 bars, } \frac{5.17}{1.00} = 5.17, \text{ or } 6 \quad \left[\frac{3312}{645} = 5.13 \right]$$

$$\text{for No. 10 bars, } \frac{5.17}{1.27} = 4.07, \text{ or } 5 \quad \left[\frac{3312}{819} = 4.04 \right]$$

$$\text{for No. 11 bars, } \frac{5.17}{1.56} = 3.31, \text{ or } 4 \quad \left[\frac{3312}{1006} = 3.29 \right]$$

For all except the No. 11 bars, the requirements for bar spacing (as discussed in Chapter 14) would result in the need to place the bars in stacked layers in the 15-in.-wide beam. While this is possible, it would require some increase in the dimension h in order to maintain the effective depth of approximately 26.6 in., since the centroid of the steel bar areas would move farther away from the edge of the concrete.

Example 2. A rectangular concrete beam of concrete with f'_c of 3000 psi [20.7 MPa] and steel with f_s of 20 ksi [138 MPa] has dimensions of $b = 15$ in. [380 mm] and $h = 36$ in. [910 mm]. Find the area required for the steel reinforcing for a moment of 200 kip-ft [271 kN-m].

Solution: The first step in this case is to determine the balanced moment capacity of the beam with the given dimensions. If we assume the section to be as shown in Fig. 13-3, we may assume an approximate value for d to be h minus 2.5 in. [64 mm], or 33.5 in. [851 mm]. Then with the value for R from Table 13-3,

$$M = Rbd^2 = 0.226 \times 15 \times (33.5)^2 = 3804 \text{ kip-in}$$

$$\text{or } M = \frac{3804}{12} = 317 \text{ kip-ft}$$

$$[M = 1554 \times 0.380 \times (0.850)^2 = 427 \text{ kN-m}]$$

Since this value is considerably larger than the required moment, it is thus established that the given section is larger than that required for a balanced stress condition. As a result, the concrete flexural stress will be lower than the limit of $0.45 f'_c$, and the section is qualified as being under-reinforced; which is to say that the reinforcing required will be less than that required to produce a balanced section (with moment capacity of 317 kip-ft). In order to find the required area of steel, we use Formula (4) just as we did in the preceding example. However, the true value for j in the formula will be something greater than that for the balanced section (0.872 from Table 13-3).

As the amount of reinforcing in the section decreases below the full amount required for a balanced section, the value of k decreases and the value of j increases. However, the range for j is small: from 0.872 up to something less than 1.0. A reasonable procedure is to assume a value for j, find the corresponding required area, and then perform an investigation to verify the assumed value for j, as follows.

Assume $j = 0.90$

Then

$$A_s = \frac{M}{f_s jd} = \frac{200 \times 12}{20 \times 0.90 \times 33.5} = 3.98 \text{ in}^2$$

and

$$p = \frac{A_s}{bd} = \frac{3.98}{15 \times 33.5} = 0.00792$$

$$\left[A_s = \frac{271,000}{0.138 \times 0.90 \times 850} = 2567 \text{ mm}^2 \right.$$

$$\left. p = \frac{2567}{380 \times 850} = 0.00795 \right]$$

Using this value for p in Fig. 13-2, we find $k = 0.313$. Using Formula (6) we then determine j to be

$$j = 1 - \frac{k}{3} = 1 - \frac{0.313}{3} = 0.896$$

which is reasonably close to our assumption, so the computed area is adequate for design.

Problem 13-9-A. A rectangular concrete beam has concrete with $f'_c = 3000$ psi [20.7 MPa] and steel reinforcing with $f_s = 20$ ksi [138 MPa]. Select the beam dimensions and reinforcing for a balanced section if the beam sustains a bending moment of 240 k-ft [325 kN-m].

Problem 13-9-B. Find the area of steel reinforcing required and select the bars for the beam in Problem 13-9-A if the section dimensions are $b = 16$ in. and $d = 32$ in.

13-10 Shear

Shear occurs in a variety of situations in reinforced concrete structures. This discussion is limited to two common situations, beam shear and peripheral shear, as shown in Fig. 13-4.

Critical shear conditions in beams typically occur near the ends of the beam spans. The stress of real concern is actually not shear but the diagonal tension that accompanies shear stress. Since the unit value of the diagonal tension stress is the same as that of the shear stress, computation of the shear stress will ac-

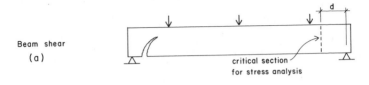

Beam shear

(a)

critical section
for stress analysis

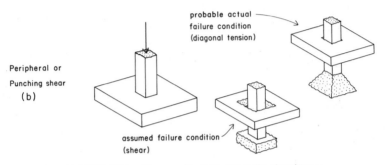

probable actual
failure condition
(diagonal tension)

Peripheral or
Punching shear

(b)

assumed failure condition
(shear)

FIGURE 13-4.   Shear actions in concrete structures.

count for the tension effect. For design use, the code permits the use of a critical shear stress located at a d (effective beam depth) distance from the beam support. The concrete alone is permitted to take a stress of 1.1 $\sqrt{f'_c}$, with any additional stress required to be developed by shear reinforcing, commonly in the form of U-shaped stirrups (see Fig. 13-3).

Peripheral shear, sometimes called punching shear, occurs where a concentrated force is applied to a surface element of concrete, such as a wall, floor slab, or column footing. The situation of the column footing is shown in Fig. 13-4b. Diagonal tension failure in this case results in the punching out of a pyramidal or conical portion of the footing beneath the column. For simulation of this failure, the code permits computation of a design shear stress at a circumscribed periphery around the column a distance of $d/2$ from the column face. The column load is assumed to develop a uniform shear stress on this peripheral section of the footing, for which the stress is limited to 2 $\sqrt{f'_c}$.

In summary, these two shear conditions are treated as follows:

1. Beam shear

   $v_c = \dfrac{V}{bd}$ where V is taken at a d distance from supports and maximum stress on concrete $= 1.1 \sqrt{f'_c} [0.09 \sqrt{f'_c}$ for SI]

2. Peripheral shear

   $v_c = \dfrac{V}{\Sigma w \times d}$ where Σw is the peripheral circumference and maximum stress on concrete $= 2 \sqrt{f'_c} [0.17 \sqrt{f'_c}$ for SI]

13-11 Shear Reinforcement for Beams

The shear diagram for a simple span beam with a uniformly distributed load consists of two triangles, as shown in Fig. 3-18b. This diagram is repeated in Fig. 13-5a. If the unit shear stress in a concrete beam exceeds the limit for the concrete alone, the situation is as represented in Fig. 13-5b, for which we note that

V = maximum shear force at the end of the beam,

V_c = shear force capability of the concrete alone, as limited by the maximum permitted stress v_c,

V' = shear force required to be developed by shear reinforcing (stirrups).

The shaded portion of the diagram in Fig. 13-5b indicates the shear force range over which reinforcing must be provided. This range theoretically ends at a distance a from the support, at which location the shear force is low enough for the beam to resist it without reinforcing. Although the maximum shear force occurs at the end of the span, the code permits the consideration of a maximum shear stress for design at a distance from the support equal to the effective depth of the beam, d.

Figure 13-5c is a modification of Fig. 13-5b, incorporating the adjustment for use of the critical shear stress at d distance from

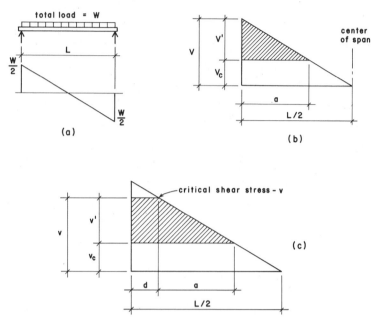

FIGURE 13-5. Design considerations for shear reinforcing.

the support. Shear values for Fig. 13-5c are shown in *stress* units, rather than *force* units, to indicate more clearly the relation to stress limitations used for design. The shear *stress* diagram is identical in form to the shear *force* diagram, with its values obtained by simply dividing the shear force values by the quantity bd, using the formula for shear stress, $v = V/bd$.

13-12 Design of Shear Reinforcement for Beams

Figure 13-6a shows one-half of a uniformly loaded beam, for which the form of the shear diagram is as shown in Fig. 13-6b. Shear reinforcing, consisting of U-shape stirrups, is provided over the distance R from the support. We note the following:

1. Critical design v is determined at the distance d from the support, as discussed in Section 13-11.

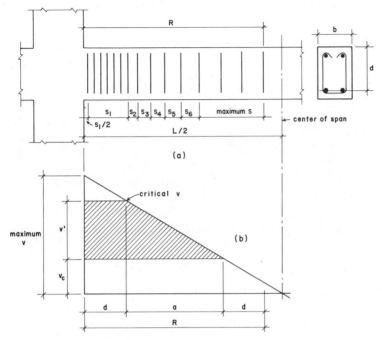

FIGURE 13-6.

2. The stirrups must be designed for the excess unit shear stress v', which extends over the shaded portion of the diagram.

Stirrups in a single beam usually will be all of a single type and bar size, so the stirrup design consists of determining the center-to-center spacing of the stirrups (the values for s, as shown in the figure). For determination of the required spacings, the following considerations are made:

1. The limiting value for v_c (stress limit for the concrete) is $1.1 \sqrt{f'_c}$ [$0.09 \sqrt{f'_c}$ for SI].
2. Maximum spacing is $d/2$ except for values of v' over $2 \sqrt{f'_c}$, for which maximum spacing is $d/4$ [$0.17 \sqrt{f'_c}$ for SI].

3. The absolute maximum value for v' is $4.4 \sqrt{f_c'}$ [$0.37 \sqrt{f_c'}$ for SI].

4. Stirrups required at the critical shear stress location (d distance from the support) are provided at that spacing over the distance between the support and the location of critical stress. The first stirrup is placed at one-half this required spacing from the support.

5. Stirrups must be provided at the maximum permitted spacing ($d/2$) for a distance d beyond the computed distance a. (See Fig. 13-5c.)

With regard to the last point, it may be observed that the range distance R, over which stirrups are required, is established as $R = a + 2d$.

The required spacing for stirrups at any location is determined from the formula

$$s = \frac{A_v f_s}{v' b}$$

in which A_v = total cross-sectional area of steel bars in a single stirrup (2×0.11 for a No. 3 U-stirrup),

f_s = permitted unit tensile stress in the stirrup,

v' = computed unit shear stress in excess of v_c at the point being considered,

b = width of the beam web.

Although stirrup design is most practically accomplished by using various design aids from handbooks, the following example illustrates the process for a simple case.

Example.   The concrete beam in Fig. 13-7 is made from concrete with f_c' of 3000 psi [20.7 MPa]. Determine the layout for a set of No. 3 U-stirrups for one end of the beam, using f_s = 20 ksi [138 MPa] for the stirrups.

Solution: With the total load of 45 kips [200 kN], the end shear on the beam will be one-half the load, or 22.5 kips [100 kN]. The maximum shear stress at the support is thus

$$v = \frac{V}{bd} = \frac{22,500}{10 \times 20} = 112.5 \text{ psi} \left[\frac{0.100}{0.254 \times 0.508} = 0.775 \text{ MPa} \right]$$

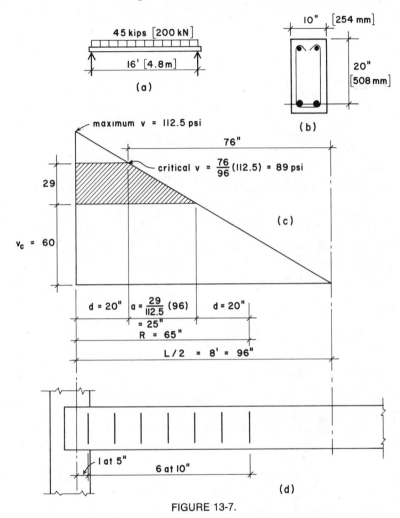

FIGURE 13-7.

Referring to Fig. 13-7c, we can compute the value for the beam shear stress at the critical location (*d* distance from the support) by observing the proportionality of the similar triangles, thus

$$v = \frac{76}{96}(112.5) = 89 \text{ psi } [0.614 \text{ MPa}]$$

For the limiting concrete stress

$$v_c = 1.1 \sqrt{f'_c} = 1.1 \sqrt{3000} = 60 \text{ psi}$$

$$[v_c = 0.09 \sqrt{f'_c} = 0.09 \sqrt{20.7} = 0.409 \text{ MPa}]$$

The values for v', a, and R are now established on the figure, as shown. The required stirrup spacing at the maximum v' value is

$$s = \frac{A_v f_s}{v'b} = \frac{(2 \times 0.11) \times 20,000}{29 \times 10} = 15.2 \text{ in.}$$

$$\left[s = \frac{(2 \times 71 \times 10^{-6}) \times (138)}{0.20 \times 0.254} = 0.385 \text{ m} \right]$$

Since this value is greater than the maximum spacing limit of $d/2 = 10$ in. [0.254 m], the calculated s values are not critical for this example, and spacing will be determined on the basis of the various limits, as shown in the figure (See Fig. 13-7d.)

Problem 13-12-A. A concrete beam similar to that shown in Fig. 13-7 sustains a total load of 60 kips [267 kN] on a span of 24 ft [7.32 m]. Determine the layout for a set of No. 3 U-stirrups using $f_s = 20$ ksi [138 MPa] and $f'_c = 3000$ psi [20.7 MPa]. The section dimensions are $b = 12$ in. [305 mm] and $d = 26$ in. [660 mm].

13-13 Bond Stress

Bond stress is the essential interactive relationship between a steel reinforcing bar and the concrete mass of the structural element in which it is embedded. Bond stresses are developed on the surfaces of all reinforcing bars whenever some structural action requires the steel and concrete to share load. The basic concept of bond stress development may be illustrated by the simple example shown in Fig. 13-8, in which a steel bar is embedded in a block of concrete and is required to resist a pull-out tension force.

Figure 13-8b shows the static equilibrium relationship for the steel bar, with the pull-out force developed as the product of a tensile stress times the area of the bar cross section $\left(f_s \times \frac{\pi}{4} D^2 \right)$ and the resisting force developed by a bond stress (u) operating on the surface of the bar ($u \times \pi D \times L$). By equating these two

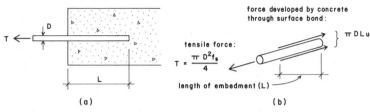

FIGURE 13-8. Development of bond stress.

forces, we can derive an expression either for the unit bond stress or the required embedment length for a limiting bond stress.

Bond stress development is affected by a number of considerations; some of the major ones are the following:

1. Grade of Steel. As the f_y of the steel is increased, the allowable f_s value will also increase, requiring the development of higher bond stresses or the need for greater embedment lengths.

2. Strength of Concrete. In general, as f'_c is increased, the capability for development of bond stress is also increased.

3. Bar Size. Consideration of the expression for the tension force in the bar in Fig. 13-8b will indicate that the force capability of the bar increases with the square of the diameter. On the other hand, the resistance developed by bond stress increases only linearly with increase of the bar diameter. Thus bond stresses tend to be more critical on bars of large diameter.

4. Concrete Encasement. The bonding force must be developed in the concrete mass around each bar. This development is limited when this mass is constrained due to closely spaced groups of bars or where bars are placed close to the edge of the concrete member.

5. Location of Bars. When concrete is poured into forms and cured into its hardened state, the concrete near the bottom of the member tends to develop slightly higher quality than that near the top. The weight of the concrete mass above produces a denser material in the lower con-

crete, and the exposed top surface tends to dry more rapidly, resulting in less well-cured concrete near the top. This difference in quality affects the potential for bond resistance, so some adjustment is made for bars placed near the top (such as reinforcement for negative moment in beams).

In times past, working stress procedures included the establishment of allowable stresses for bond and the computation of bond stresses for various situations. At present, however, the codes deal with this problem as one of development length, as discussed in the next section.

13-14   Development of Reinforcement

The 1977 ACI Code defines *development length* as the length of embedment required to develop the design strength of the reinforcing at a critical section. For beams, critical sections occur at points of maximum stress and at points within the span where some of the reinforcement terminates or is bent up or down. For a uniformly loaded simple span beam, one critical section is at midspan, where the bending moment is a maximum. The tensile reinforcing required for flexure at this point must extend on both sides a sufficient distance to develop the stress in the bars; however, except for very short spans with large bars, the bar lengths will ordinarily be more than sufficient.

In the simple beam, the bottom reinforcing required for the maximum moment at midspan is not entirely required as the moment decreases toward the end of the span. It is thus sometimes the practice to make only part of the midspan reinforcing continuous for the whole beam length. In this case it may be necessary to assure that the bars that are of partial length are extended sufficiently from the midspan point and that the bars remaining beyond the cutoff point can develop the stress required at that point.

When beams are continuous through the supports, top reinforcing is required for the negative moments at the supports. These top bars must be investigated for the development lengths in terms of the distance they extend from the supports.

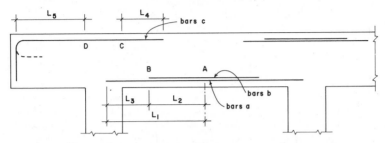

FIGURE 13-9. Development lengths for beam reinforcing.

Some of the situations just described are illustrated in Fig. 13-9. Referring to the figure, we note the following:

1. Bars a, extending the full span, must have development length L_1 sufficient for the maximum moment at midspan point A.

2. Partial length bars b must have development length L_2 sufficient for the stress at point A.

3. Full length bars a must extend beyond the cutoff point for bars b a distance of L_3 that is sufficient to develop the stress in flexure at point B (the cutoff point).

4. Top bars c must extend into the interior span the distance L_4 of sufficient magnitude to develop the stress in flexure at the face of the support—point C.

5. Bars c are also extended to provide for the cantilever moment at point D. Thus length L_5 must be sufficient to develop the flexural stress in bars c due to the cantilever moment. If L_5 is not sufficient, it is common practice to use a hooked end on the bars to assist the development. This may be achieved by simply bending the bar at right angles or by bending it into a complete 180° hook, as shown by the dotted line in the figure.

Table 13-4 yields values for development lengths based on the requirements of Chapter 12 of the 1977 ACI Code. The table incorporates some of the variables discussed in Section 13-13. Note that these lengths apply only to tensile reinforcement.

TABLE 13-4.   Minimum Development Length for Tensile Reinforcement (inches)

| Bar Size (No.) | $f_y = 40$ ksi [276 MPa] | | | | $f_y = 60$ ksi [414 MPa] | | | |
| | $f'_c = 3$ ksi [20.7 MPa] | | $f'_c = 4$ ksi [27.6 MPa] | | $f'_c = 3$ ksi [20.7 MPa] | | $f'_c = 4$ ksi [27.6 MPa] | |
	Top Bars[a]	Other Bars	Top Bars[a]	Other Bars	Top Bars[a]	Other Bars	Top Bars[a]	Other Bars
3	12	12	12	12	13	12	13	12
4	12	12	12	12	17	12	17	12
5	14	12	12	12	21	15	21	15
6	18	13	16	12	27	19	15	18
7	25	18	21	15	37	26	32	23
8	32	23	28	20	48	35	42	30
9	41	29	36	25	61	44	53	38
10	52	37	45	32	78	56	68	48
11	64	46	55	40	96	68	83	59
14	87	62	75	54	130	93	113	81
18	113	80	98	70	169	120	146	104

Note: Lengths are based on requirements of the 1977 ACI Code.

[a] Horizontal bars so placed that more than 12 in. [305 mm] of concrete is cast in the member below the reinforcement.

Steel reinforcement is also used to develop compression in some situations, most notably in concrete columns and bearing walls. Negative moment resistance in beams is often assisted by compression reinforcement, usually achieved by extension of the bottom bars that already exist for positive moment. Development lengths for this situation must also be considered, although in general the variables are fewer. A common situation in the design of column footings for concrete structures is the concern for a thickness of the footing that is adequate for the required development length for the column reinforcement; this is discussed in Chapter 17.

14

Reinforced Concrete Beams

||||||||||||||||||||||||||||||||||||||||||||||||||||||||||||||

14-1 Typical Beams

In the preceding chapter we dealt exclusively with simply supported beams as the basis for our discussion of flexure, shear, and
diagonal tension, and development length of reinforcement. In
practice, however, reinforced concrete floor systems are poured
simultaneously for several adjacent spans, and if the reinforcement is arranged properly, continuity is developed between the
spans and we have *continuous* or *restrained* beams (Section 2-4).

In the design of reinforced concrete beams *with uniformly distributed loads,* three span conditions occur repeatedly. First,
there is the simple beam, a *single span* with no restraint at the
ends. Next, there is the *end span* of a continuous beam, with no
restraint at the noncontinuous end. Finally, there is the *interior
span* of a fully continuous beam. These three beams are shown
diagrammatically in Fig. 14-1a, b, and c, respectively. These figures contain much valuable information. The total uniformly distributed load is represented by W in each instance. For each of
the three beams, the magnitude as well as the position of M, the
maximum bending moment, is given. For each beam the value of
V, the maximum vertical shear, is shown; note that different val-

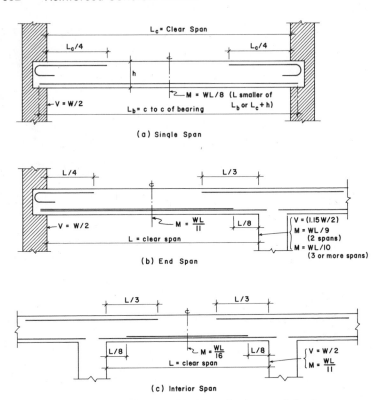

FIGURE 14-1.   Design considerations for beam reinforcing.

ues are given for the two ends of the end span. In order to have a sufficient area of contact between the tensile reinforcement and the concrete, there must be an ample length of embedment; these lengths are shown in the figure as fractions of L. Hooks on tensile reinforcement are used only at beam terminations where there is insufficient concrete to provide an adequate length of bar.

14-2   Length of Span

In computations for reinforced concrete, the span length of freely supported beams (simple beams) is generally taken as the distance between centers of supports or bearing areas; it should not exceed the clear span plus the depth of beam or slab. The span length for continuous or restrained beams is taken as the clear distance between faces of supports. These lengths are shown in Fig. 14-1.

The single-span condition illustrated in Fig. 14-1*a* is similar to that shown in Fig. 13-5. Although the bending moment is theoretically zero at this simple support, some restraint may be developed by the weight of the wall above the beam, thereby creating a negative moment in the top of the beam at the face of the wall when the member is loaded. For this reason it is common practice to place some tensile reinforcement in the top at a support of this nature by either bending up some of the bottom steel or using short, straight bars.

14-3 Bending Moments

For a simple beam, that is, a single span having no restraint at the supports, the maximum bending moment for a uniformly distributed load is at the center of the span, and its magnitude is $M = WL/8$ (Section 3-11). The moment is zero at the supports and is positive over the entire span length. In continuous beams, however, negative bending moments (Section 3-7) are developed at the supports and positive moments at or near midspan. This may be readily observed from the exaggerated deformation curve of Fig. 14-2*a*. The exact values of the bending moments depend on several factors, but in the case of approximately equal spans supporting uniform loads, when the live load does not exceed three times the dead load, the bending moment values given in Fig. 14-2 may be used for design.

The values given in Fig. 14-2 are in general agreement with the recommendations of Chapter 8 of the 1977 ACI Code. These values have been adjusted to account for partial live loading of multiple-span beams. Note that these values apply only to uniformly loaded beams. Approximate designs for other loadings may be achieved by using the equivalent tabular load factors given in Fig. 3-19. Chapter 8 of the 1977 ACI Code also gives some factors for end support conditions other than the simple supports shown in Fig. 14-2.

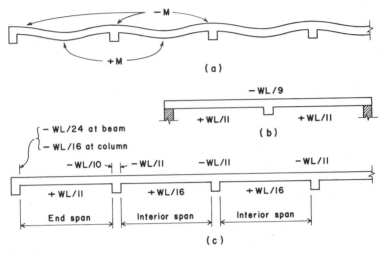

FIGURE 14-2.   Design moments for continuous beams.

14-4 T-Beams

When a floor slab and its supporting beams are poured at the same time, the result is a monolithic construction in which a portion of the slab on each side of the beam serves as the flange of a T-beam. The part of the beam that projects below the slab is called the *web* or *stem* of the T-beam. This type of beam affords an economical form of construction and is commonly used; it is shown in Fig. 14-3a. For a simple beam the flange is in the compression zone, and there is ample concrete to resist the compressive stresses, as shown in Fig. 14-3b. However, in a continuous beam there are negative bending moments over the supports, the flange here is in the tension zone, and the compressive stresses are in the web. (See Fig. 14-3c.) It is important to remember that only the area formed by the width of the web b' and the effective depth d is to be considered in computing the resistance to shear and to bending stresses over the supports. This is the hatched area $b'd$ shown in Fig. 14-3d. It is customary, when conditions permit, to have b', the width of the web, one-half to one-third of d, the effective depth.

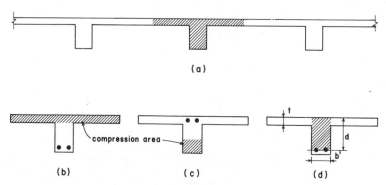

FIGURE 14-3.   Beam actions in monolithic beam-slab construction.

The effective flange width to be used in the design of symmetrical T-beams shall not exceed one-quarter the span length of the beam, and its overhanging width on either side of the web shall not exceed eight times the thickness of the slab nor one-half the clear distance to the next beam.

For average conditions the flange area is invariably large enough to resist the compression stresses that result from the positive bending moment at midspan. To convince yourself that this is true, consider the following discussion. It is not a required step in the design of a T-beam.

Both Fig. 14-4a and b represent T-beam sections; the neutral axis is indicated for each section. In Fig. 14-4a the neutral axis lies in the slab, the flange area, whereas in Fig. 14-4b it falls below the slab. The areas above the neutral axis, the hatched areas, resist compressive stresses for positive moments. In Fig. 14-4b the area of the web below the slab subjected to compressive stresses is ignored in computations. If the dimensions of the parts of a T-beam are known, the position of the neutral surface may be accurately computed.

The design of a T-beam is commonly limited to the determination of the required steel reinforcement. The dimension t is usually established by design of the slab, and the dimensions of the concrete stem (b' and d in Fig. 14-4) are usually established by design for negative moments at the supports, except for the case of simple span beams. To find A_s for the positive bending moment

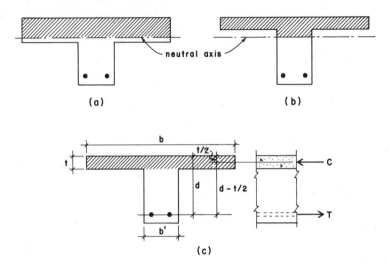

FIGURE 14-4. T-beam actions and design considerations.

between the supports, we use the formula

$$A_s = \frac{M}{f_s[d - (t/2)]}$$

in which A_s = total area of the positive moment reinforcement for the T-beam,

M = maximum positive bending moment,

f_s = allowable unit tensile stress in the steel,

d = effective depth of the T-beam,

t = thickness of the flange of the T-beam.

This is a slightly conservative formula based on the assumption that the compressive stress in the concrete is uniform over the entire area of the flange of the T-beam.

14-5 Design of a Continuous Concrete Beam

Figure 14-5 shows an interior span of a typical continuous concrete beam, existing as part of a beam-and-slab system, where the beams and slabs are poured at the same time. At midspan points,

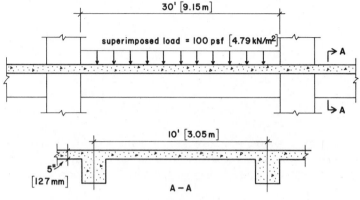

FIGURE 14-5.

the beam will perform as a T-section, as discussed in Section 14-4, while at supports the beam will perform as a rectangular section for the development of negative moment (tension stress in the top of the beam). We will use this example to illustrate the general procedure for design of continuous beams.

Various factors may affect the choice of the slab thickness and the dimensions of the beam stem. We will assume the use of a 5-in. [127 mm] thick slab, as shown, and will freely choose the stem dimensions for this example. The stem size is primarily affected by considerations for the maximum negative moment and the maximum shear, both of which occur at the supports, where the T-effect does not apply. Using the approximation value from Fig. 14-2, we find the maximum negative moment as

$$M = \frac{1}{10} \, wL \quad \text{or} \quad \frac{1}{10} \, wL^2$$

For w we consider the following:

superimposed load = 100 psf × 10-ft beam spacing = 1000 lb/ft,

$$\text{slab weight} = 5 \text{ in.} \times \frac{150 \text{ pcf}}{12} = 62.5 \text{ psf},$$

slab load on beam = 62.5 × 10 = 625 lb/ft

assumed stem weight (a guess) = 300 lb/ft.

The total load on the beam is thus

$$w = 1000 + 625 + 300 = 1925 \text{ lb/ft } [28.09 \text{ kN/m}]$$

and the maximum moment is

$$M = \frac{1}{11} wL^2 = \frac{1}{11} (1.925)(30)^2 = 157.5 \text{ kip-ft } [214 \text{ kN-m}]$$

Assuming a concrete strength of 3000 psi [20.7 MPa] and a maximum steel flexural stress of 20 ksi [138 MPa], we establish the limiting balanced section as described in Section 13-9.

From Table 13-3, $R = 0.226$

and

$$M = Rbd^2 = (0.226)bd^2 = 157.5$$

thus

$$\text{required } bd^2 = \frac{M}{R} = \frac{157.5 \times 12}{0.226} = 8363$$

For a first try we assume $b = 16$ in. Then

$$\text{required } d = \sqrt{\frac{8363}{16}} = \sqrt{523} = 22.9 \text{ in.}$$

$$[b = 406 \text{ mm}, d = 582 \text{ mm}]$$

At this point a decision must be made as to whether this section may be developed using some compressive reinforcement as well as the tensile reinforcement. Since we have not developed the procedures for the design of such a section, we will not do this in this example, although the practice is quite common. This compressive reinforcement is often provided by simply extending the bottom reinforcing bars (for midspan positive moment) through the support points and into the adjacent span. If compressive reinforcing is used, it is possible to consider the use of a depth less than that required for the balanced section.

Without compressive reinforcing, we must select a dimension for the overall beam stem height that will result in an effective depth not less than that of the computed value of 22.9 in. Assuming a concrete cover of 1.5 in., the use of No. 3 U-stirrups, and a single layer of reinforcing bars, the minimum overall height of the beam (stem plus slab) should be approximately 22.9 + 2.5 = 25.4 in. We will choose an even inch dimension of 26 in. for the height, resulting in an approximate effective depth of 23.5 in. [597 mm]. Note that the exact value for the effective depth will depend on the diameter of the reinforcing bars, which have not yet been chosen.

Before proceeding with the flexural design, we should investigate the shear stress condition for our selected beam size, to assure that excessive shear reinforcement will not be required. Referring to the discussion in Section 13-10, we find

$$\text{maximum end shear force} = \frac{wL}{2} = \frac{1.925 \times 30}{2} = 28.875 \text{ kips}$$

Critical shear force at d distance from the support is

$$V = 28.875 - \frac{23.5}{12} (1.925) = 25.1 \text{ kips}$$

$$\text{Critical sense} \quad v = \frac{V}{bd} = \frac{25,100}{16 \times 23.5} = 66.8 \text{ psi [0.461 MPa]}$$

Since this value is only slightly over the limit for the concrete with no reinforcing ($1.1 \sqrt{f'_c} = 60$ psi [0.414 MPa]), it is evident that shear design will not be critical.

14-6 Flexural Reinforcement for the Continuous Beam

Primary consideration for flexural reinforcement is the provision of an adequate area of bars at the two critical moment locations. At the support, the maximum negative moment as previously computed is 157.5 kip-ft [214 kN-m]. Since our choice for the section is very close to the balanced section for this moment, we may use the balanced section value for j from Table 13-3. Thus

$$A_s = \frac{M}{f_s jd} = \frac{157.5 \times 12}{(20)(0.872)(23.5)} = 4.61 \text{ in}^2 \text{ [2974 mm}^2\text{]}$$

From Fig. 14-2 the maximum moment at midspan is

$$M = \frac{wL^2}{16} = \frac{1.925(30)^2}{16} = 108.3 \text{ kip-ft } [146.9 \text{ kN-m}]$$

and for the T-section, as discussed in Section 14-4,

$$A_s = \frac{M}{f_s(d - t/2)} = \frac{108.3 \times 12}{(20)(21.0)} = 3.09 \text{ in}^2 \ [1994\text{mm}^2]$$

Choice of the beam reinforcing must be made with several considerations, which include:

1. The area requirements at the two critical points.
2. Coordinated selection of bent bars (alternate layout shown in Fig. 14-6). These bars are used at both points.
3. Placement of the bars in a properly spaced single layer. Clear space between bars must be not less than the bar diameter or 1 in., whichever is greater.

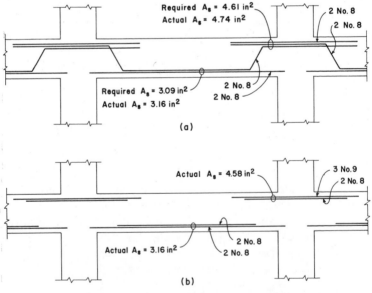

FIGURE 14-6. Reinforcing choices for the beam: (a) with a combination of bent and straight bars, (b) with all straight bars.

4. Mixed combinations of bar sizes, including bars not greater than two sizes away from each other. That is, No. 8 plus No. 6 but not No. 8 plus No. 5.
5. Consideration of development lengths.

The two general alternatives for bar layouts are as shown in Fig. 14-6. Use of bent bars reduces the number of separate bars that must be handled and supported during pouring. However, the greater cost of fabricating bent bars and the difficulty in placing them in the forms have made the use of all straight bars more popular in recent years. Use of straight bars also eliminates the additional complexity of coordinated bar selection for the top and bottom reinforcing.

Bars can be placed in more than one layer if necessary, but this results in some loss of efficiency, as the centroid of the reinforcing areas moves farther away from the edge of the concrete, and the effective d is thus reduced. Installation is also somewhat more difficult with multiple layers.

If the beam section is as shown in Fig. 14-7, the spacing requirements for the bottom bars (four No. 8 bars) dictates the need for a clear inside distance between the stirrup legs of at least 7 in. For the 16-in.-wide beam with cover of 1.5 in. and No. 3 U-stirrups, this dimension is 12.25 in., so spacing is not a problem for the bottom bars. For the top bars, if the bars are all placed

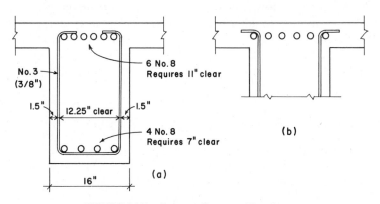

FIGURE 14-7.   Bar spacing considerations.

inside the stirrups as shown in Fig. 14-7a, the required total width for the six bars is 11 in. This is still less than that provided, but the bars will be quite close and will offer some difficulty during pouring of the concrete. In order to gain more space, or simply to spread the bars to make placing of the concrete easier, it is common practice to place some bars outside the stirrup, as shown in Fig. 14-7b. If the slab is quite thin, the provision of proper cover for these bars should be investigated.

14-7 Development Length for the Continuous Beam

Development lengths are usually most critical for large-diameter bars and for heavily loaded beams of short span. For uniformly loaded beams, the cutoff points for top bars and bend points for bent bars are usually established by standard detailing practices, which generally assure that conditions are not critical for development length. If partial length bars such as those shown in Fig. 14-6b are used, their cutoff points should be checked for proper development length, using the values from Table 13-4.

14-8 Shear Reinforcement for the Continuous Beam

The maximum shear stress condition for this beam was investigated previously for the purpose of establishing the required stem size. Since the critical shear stress was previously computed as 66.8 psi, we may now construct the shear stress diagram shown in Fig. 14-8 in order to determine the required stirrup spacings. For the diagram we have rounded off the critical shear stress to 67 psi and the effective beam depth to 24 in. The distance a on the diagram may be determined by observing the similar triangles as follows:

$$\frac{7}{a} = \frac{67}{156}, \quad a = \frac{156}{67} \, (7) = 16.3 \text{ in., or approximately 16 in.}$$

The total length over which shear reinforcing must be provided is the distance R shown on the diagram, which is determined as (see Section 13-12)

$$R = 24 + 16 + 24 = 64 \text{ in.}$$

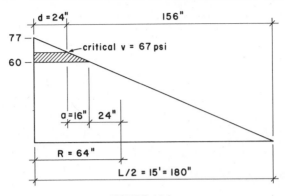

FIGURE 14-8.

At the point of critical maximum shear stress, the excess stress required to be taken by the reinforcing is 7 psi, and the spacing required is determined as

$$s = \frac{A_v f_v}{v'b} = \frac{(2 \times 0.11)(20)(10^3)}{(7)(16)} = 39.3 \text{ in.}$$

Since this value exceeds the maximum spacing of $d/2 = 12$ in., stirrups are provided at the maximum allowable spacing.

In theory, if the computed shear stress at the critical location is less than the limit for the concrete alone, no reinforcement is required. However, the present code (ACI 318-77) has a provision in Appendix B (Article B-7-5-5) that requires a minimum reinforcement for beams even when the computed shear stress is as low as one-half of the limit for the concrete. In addition to the usual restrictions for stirrups, the maximum spacing for this reinforcing is

$$s = \frac{A_v f_y}{50b} \left[s = \frac{A_v f_y}{0.34b} \right]$$

in which f_y is the yield stress of the reinforcing.

Ordinary reinforcing bars are available in grades with yield stress of 40, 50, and 60 ksi. If we assume the lower value of 40 ksi [276 MPa] for a conservative design, the required spacing of minimum reinforcement for this beam is

$$s = \frac{A_v f_y}{50b} = \frac{(2 \times 0.11)(40)(10^3)}{50 \times 16} = 11 \text{ in. } [279 \text{ mm}]$$

There is no code recommendation as to the number of minimum stirrups or the range over which they must be provided. One recommendation is for a minimum range of $1.5\ d$ or one-eighth of the clear span, whichever is greater. On this basis, for the beam with $d = 23.5$ in. and a span of 30 ft,

$$R = (1.5)(23.5) = 35 \text{ in.}$$

or $R = \dfrac{30 \times 12}{8} = 45$ in., which is the greater value.

If the computed critical shear stress were less than 60 psi, a possible choice for minimum shear reinforcing would be stirrups with the following spacing:

$$1 \text{ at } 6 \text{ in., } 4 \text{ at } 11 \text{ in. } = 50 \text{ in. total}$$

To satisfy the data in Fig. 14-8, a possible choice would be

$$1 \text{ at } 6 \text{ in., } 5 \text{ at } 12 \text{ in. } = 66 \text{ in. total}$$

Thus the computed reinforcing is only slightly greater, with approximately the same spacing, as the required minimum reinforcing.

14-9 Shear Reinforcement for Girders

Our consideration of reinforced concrete beam design thus far has been limited to beams carrying uniformly distributed loads. The principal loads on girders, however, are concentrated loads, with uniformly distributed loads limited to the girder weight and the floor loading directly above the girder. For beams or girders unsymmetrically loaded or for beams subjected to both concentrated and distributed loads, it is advisable to draw the shear and bending moment diagrams to determine the positions and magnitudes of the shear and moments. This procedure is explained in

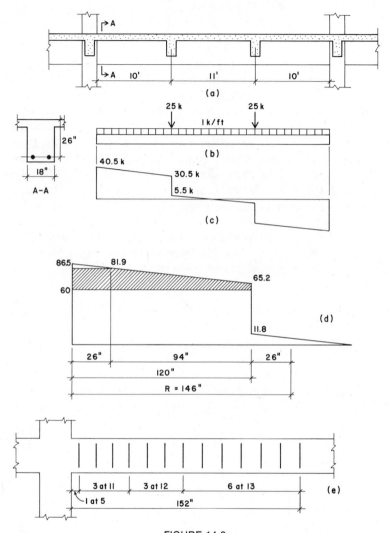

FIGURE 14-9.

365

Chapter 3. The following example illustrates the design of shear reinforcement for a typical girder.

A structural system is shown in Fig. 14-9a, consisting of continuous girders that are supported by columns and that in turn provide support for a series of beams at 11-ft centers. The beams deliver their end reactions to the girders in the form of concentrated loads. The girder also carries some uniformly distributed load consisting of its own dead weight and a portion of the floor directly over the girder. This combined load is shown in Fig. 14-9b, and the resulting shear diagram is shown in Fig. 14-9c. Because of the symmetry the shear diagram is the same as that for a simply supported beam, even though the girder is continuous through its supports.

For the girder section shown, we assume the following: $b = 18$ in., $d = 26$ in., $f'_c = 3$ ksi, $f_s = 20$ ksi (for stirrups), No. 3 U-stirrups. For the stirrup design the following computations are made.

Maximum shear stress at support (used only to construct the complete shear stress diagram),

$$v = \frac{V}{bd} = \frac{40,500}{18 \times 26} = 86.5 \text{ psi}$$

Critical shear stress at d distance (26 in.) from support,

$$v = \frac{38,333}{18 \times 26} = 81.9 \text{ psi}$$

Shear stresses at the concentrated load point,

$$v = \frac{30,500}{18 \times 26} = 65.2 \text{ psi}$$

and

$$v = \frac{5,500}{18 \times 26} = 11.8 \text{ psi}$$

Allowable shear on the concrete alone,

$$v_c = 1.1\sqrt{f'_c} = 1.1\sqrt{3,000} = 60 \text{ psi}$$

These values are displayed on the shear stress diagram in Fig. 14-9d together with the girder span dimensions that affect the stirrup design. For the stirrup spacing determination, we do the following computations.

$$\text{Maximum stirrup spacing, } s = \frac{d}{2} = \frac{26}{2} = 13 \text{ in.}$$

Maximum v' value at d distance from support,

$$v' = 81.9 - 60 = 21.9 \text{ psi}$$

Required stirrup spacing for the maximum v' value,

$$s = \frac{A_v f_s}{v' b} = \frac{(2 \times 0.11)(20,000)}{21.9 \times 18} = 11.16 \text{ in.}$$

Since this value is less than the maximum permitted spacing, it wil be used as the limit for the stirrups at the end of the span. As a guide to determination of the required stirrup layout, we may compute some additional shear stress values at 1-ft increments of the span beyond the critical d distance from the support. Referring to the shear diagram, we observe that the shear values will decrease by 1.0 kip (the uniformly distributed load) for each foot of span. Thus

at $d + 1$ ft, $V = 38.33 - 1 = 37.33$ kips

$$v = \frac{V}{bd} = \frac{37,333}{18 \times 26} = 79.8 \text{ psi}$$

$$v' = 79.8 - 60 = 19.8 \text{ psi}$$

$$s = \frac{A_v f_s}{v' b} = \frac{4400}{19.8 \times 18} = 12.35 \text{ in.}$$

At $d + 2$ ft, $V = 38.33 - 2 = 36.33$ kips

$$v = \frac{36,333}{18 \times 26} = 77.6 \text{ psi}$$

$$v' = 77.6 - 60 = 17.6 \text{ psi}$$

$$s = \frac{4400}{17.6 \times 18} = 13.89 \text{ in.}$$

We may thus observe that the spacing may be increased to 12 in. at a little less than 1 ft past the d distance and may be increased to the maximum of 13 in. at a little over 1 ft beyond the d distance.

Since an excess of shear stress (above the limit of 60 psi for the concrete) exists for the entire distance from the girder support to the beam location, this distance comprises the required distance a as developed for the uniformly loaded beam in the preceding examples. Thus stirrups must be provided for an additional d distance beyond this point, or a total of 146 in. from the end of the span.

On the basis of these computations and the various requirements previously discussed, a possible choice for the stirrup spacing layout is that shown in Fig. 14-9e.

15

Reinforced Concrete Floor Systems

|||||||||||||||||||||||||||||||||||||||||||||||||||||||||||

15-1 Introduction

There are many different reinforced concrete floor systems, both cast in place and precast. The cast-in-place systems are generally of one of the following types:

1. One-way solid slab and beam.
2. Two-way solid slab and beam.
3. One-way concrete joist construction.
4. Two-way flat slab or flat plate without beams.
5. Two-way joist construction, called waffle construction.

Each system has its distinct advantages and limitations, depending on the spacing of supports, magnitude of loads, required fire rating, and cost of construction. The floor plan of the building and the purpose for which the building is to be used determine loading conditions and the layout of supports. Whenever possible, columns should be aligned in rows and spaced at regular intervals in order to simplify and lower the cost of the building construction.

15-2   One-Way Solid Slabs

One of the most commonly used concrete floor systems consists of a solid slab that is continuous over parallel supports. The supports may consist of bearing walls of masonry or concrete but most often consist of sets of evenly spaced concrete beams. The beams are usually supported by girders, which in turn are supported by columns. In this type of slab the principal reinforcement runs in one direction, parallel to the slab span and perpendicular to the supports. For this reason it is called a *one-way solid slab*. The number and spacing of supporting beams depends on their span, the column spacing, and the magnitude of the loads. Most often the beams are spaced uniformly and frame into the girders at the center, third, or quarter points. The form work for this type of floor is readily constructed, and the one-way slab is most economical for medium and heavy floor loads for relatively short spans, 6–12 ft. For long spans the slab thickness must be increased, resulting in considerable dead weight of the construction, which increases the cost of the slab and its reinforcing as well as the cost of supporting beams, girders, columns, and foundations. An example of this type of construction is shown in Fig. 15-1.

To design a one-way slab, we consider a strip 12 in. wide. This strip is designed as a beam whose width is 12 in. and on which is a uniformly distributed load. As with any rectangular beam, the effective depth and tensile reinforcement are computed as explained in Section 14-5. A minimum slab thickness is often determined on the basis of the fire rating requirements of the applicable building code. A minimum thickness is also required to prevent excessive deflection. Based on deflection limitations, slab thicknesses should not be less than those given in Table 15-1.

15-3   Shrinkage and Temperature Reinforcement

While flexural reinforcement is required in only one direction in the one-way slab, reinforcement at right angles to the flexural reinforcement is also provided for stresses due to shrinkage of the concrete and temperature fluctuations. The amount of this reinforcement is specified as a percentage, p, of the gross cross-

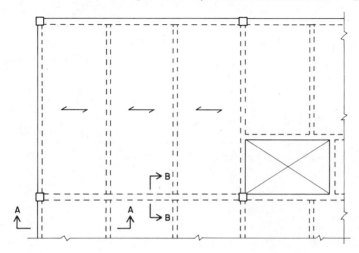

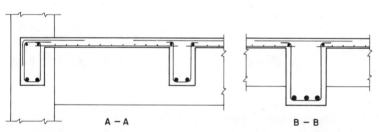

FIGURE 15-1.   Typical slab-beam-girder construction with one-way slabs.

sectional area of the concrete, as follows:

for slabs reinforced with Grade 40 or Grade 50 deformed bars,

$$p = \frac{A_s}{b \times t} = 0.0020$$

and for slabs reinforced with Grade 60 deformed bars,

$$p = 0.0018$$

Center-to-center bar spacing must not be greater than five times the slab thickness nor 18 in.

TABLE 15-1. Minimum Thickness of One-Way Slabs or Beams Unless Deflections Are Computed

Type of Member	End Conditions	Minimum Thickness of Slab or Height of Beam	
		$f_y = 40$ ksi [276 MPa]	$f_y = 60$ ksi [414 MPa]
Solid one-way slabs[a]	Simple support	$L/25$	$L/20$
	One end continuous	$L/30$	$L/24$
	Both ends continuous	$L/35$	$L/28$
	Cantilever	$L/12.5$	$L/10$
Beams or joists	Simple support	$L/20$	$L/16$
	One end continuous	$L/23$	$L/18.5$
	Both ends continuous	$L/26$	$L/21$
	Cantilever	$L/10$	$L/8$

Source: Data adapted from *Building Code Requirements for Reinforced Concrete* (ACI 318-77), 1977 ed., with permission of the publishers, American Concrete Institute.

[a] Valid only for members not supporting or attached to partitions or other construction likely to be damaged by large deflections.

15-4 Design of a One-Way Solid Slab

As discussed in Section 15-2, the one-way slab is designed by assuming the slab to consist of a series of 12-in.-[305-mm-] wide segments. The bending moment for this 12-in.-wide rectangular element is determined, and the required effective depth and area of tensile reinforcement, A_s, is computed using the procedures for a rectangular beam as discussed in Section 14-5. The A_s thus determined is the average amount of steel per 1 ft width of slab that is required. The maximum spacing for this reinforcement is three times the slab thickness or a maximum of 18 in. [457 mm]. The size and spacing of bars may be selected by use of Table 15-2.

It is not practicable to use shear reinforcement in one-way slabs, and consequently the maximum unit shear stress must be

TABLE 15-2. Areas of Bars in Reinforced Concrete Slabs
per Foot of Width

Spacing (in.)	Areas of Bars (in square inches)									
	#2ᵃ	#3	#4	#5	#6	#7	#8	#9	#10	#11
3	0.20	0.44	0.79	1.23	1.77	2.41	3.14	4.00		
3 1/2	0.17	0.38	0.67	1.05	1.51	2.06	2.69	3.43	4.36	
4	0.15	0.33	0.59	0.92	1.33	1.80	2.36	3.00	3.81	4.68
4 1/2	0.13	0.29	0.52	0.82	1.18	1.60	2.09	2.67	3.39	4.16
5	0.12	0.26	0.47	0.74	1.06	1.44	1.88	2.40	3.05	3.74
5 1/2	0.11	0.24	0.43	0.67	0.96	1.31	1.71	2.18	2.77	3.40
6	0.10	0.22	0.39	0.61	0.88	1.20	1.57	2.00	2.54	3.12
6 1/2	0.09	0.20	0.36	0.57	0.82	1.11	1.45	1.85	2.35	2.88
7	0.08	0.19	0.34	0.53	0.76	1.03	1.35	1.71	2.18	2.67
7 1/2	0.08	0.18	0.31	0.49	0.71	0.96	1.26	1.60	2.03	2.50
8	0.07	0.17	0.29	0.46	0.66	0.90	1.18	1.50	1.91	2.34
8 1/2	0.07	0.16	0.28	0.43	0.62	0.85	1.11	1.41	1.79	2.20
9	0.07	0.15	0.26	0.41	0.59	0.80	1.05	1.33	1.69	2.08
9 1/2	0.06	0.14	0.25	0.39	0.56	0.76	0.99	1.26	1.60	1.97
10	0.06	0.13	0.24	0.37	0.53	0.72	0.94	1.20	1.52	1.87
11	0.05	0.12	0.21	0.33	0.48	0.66	0.86	1.09	1.39	1.70
12	0.05	0.11	0.20	0.31	0.44	0.60	0.79	1.00	1.27	1.56

ᵃ This bar (0.25-in. diameter) is a plain bar; all others listed are deformed.

kept within the limit for the concrete alone. The usual procedure is to check the shear stress with the effective depth determined for bending before proceeding to find A_s. Except for very short span slabs with excessively heavy loadings, shear stress is seldom critical.

Although simply supported single span slabs are sometimes encountered, the majority of slabs used in building construction are continuous through multiple spans. In our discussion of the design of one-way slabs, we will use the approximate bending moment and shear factors given in Fig. 14-1 in Section 14-3, which are adequate for average conditions.

The following example illustrates the procedure for the design of a continuous solid one-way slab.

Example.   A solid one-way slab is to be used for a framing system similar to that shown in Fig. 15-1. Column spacing is 30 ft, with evenly spaced beams occurring at 10 ft center to center. Superimposed loads on the structure (floor live load plus other construction dead load) are a total of 160 psf. Use $f'_c = 3$ ksi [20.7 MPa] and Grade 40 reinforcement with $f_y = 40$ ksi [276 MPa] and $f_s = 20$ ksi [138 MPa]. Determine the thickness for the slab, and pick its reinforcement.

Solution: To find the slab thickness, we consider three factors: the minimum thickness for deflection, the minimum effective depth for the maximum moment, and the minimum effective depth for the maximum shear. For all of these we must first determine the span of the slab. For design purpose this is taken as the clear span, which is the dimension from face to face of the supporting beams, as shown in Fig. 15-2. With the beams at 10 ft centers, this dimension is 10 ft, less the width of one beam. Since the beams are not given, we will assume a dimension for them. In practice we would proceed from the slab design to the beam design, after which the assumed dimension could be verified. For this example we will assume a beam width of 12 in., yielding a clear span of 9 ft.

We consider first the minimum thickness required for deflection. If the slabs in all spans have the same thickness (which is the most common practice), the critical slab is the end span, since there is no continuity of the slab beyond the end beam. While the beam will offer some restraint, it is best to consider this as a simple support; thus we use the factor of $L/30$ from Table 15-1.

$$\text{Minimum } t = \frac{L}{30} = \frac{9 \times 12}{30} = 3.6 \text{ in.}$$

We will try a 4-in.-thick slab, for which the dead weight of the slab will be

$$w = \frac{4}{12} \times 150 = 50 \text{ lb/ft}^2$$

and the total design loading will thus be $50 + 160 = 210 \text{ lb/ft}^2$.

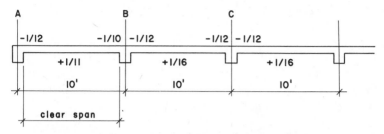

FIGURE 15-2.   Design moments for continuous one-way slabs.

We next consider the maximum bending moment. Inspection of the moment values given for various locations in Fig. 15-2 shows the maximum value to be $\frac{1}{10}\,wL^2$. These values have been adapted from those given in Fig. 14-2. With the span and loading as determined, the maximum moment is thus

$$M = \frac{1}{10}\,wL^2 = \frac{1}{10}\,(210)(9)^2 = 1701 \text{ ft-lb}$$

This moment should now be compared to the balanced moment capacity for the design section, using the relationships as discussed for rectangular beams in Section 13-9. For this computation we must assume an effective depth for the design section. This dimension will be the slab thickness minus the concrete cover and one-half the bar diameter. With the reinforcing not yet determined, we will assume an effective depth equal to the slab thickness minus 1.125 inch, which will be exactly true with the usual cover of $\frac{3}{4}$ in. and a $\frac{3}{4}$-in.-diameter (No. 6) bar. Then using the balanced R factor from Table 13-3, the maximum resisting moment for the 12-in.-wide design section is

$$M_R = Rbd^2 = 0.226(12)(2.875)^2 = 22.416 \text{ kip-in}$$

or,

$$M_R = 22.416 \times \frac{1000}{12} = 1868 \text{ ft-lb}$$

As this value is in excess of the required maximum moment, the slab will be adequate for flexural stress.

Finally, before proceeding with the design of the reinforcing, we should verify our slab thickness for shear stress. For an interior span, the maximum shear will be $wL/2$, but for the end span it is the usual practice to consider some unbalanced condition for the shear due to the discontinuous end. We therefore use a maximum shear of $1.15\ wL/2$, or an increase of 15% over the simple beam shear value. Thus

$$\text{maximum shear} = V = 1.15\ \frac{wL}{2} = 1.15 \times \frac{210 \times 9}{2} = 1087\ \text{lb}$$

$$\text{and maximum shear stress} = v = \frac{V}{bd} = \frac{1087}{12 \times 2.875} = 31.5\ \text{psi}$$

This is considerably less than the limit for the concrete alone ($v_c = 1.1\ \sqrt{f'_c} = 60$ psi), so the assumed slab thickness is not critical for shear stress.

Having thus verified our choice for the slab thickness, we may now proceed with the design of the reinforcing. For a balanced section, Table 13-3 yields a value of 0.872 for the j factor. However, since all of our reinforced sections will be classified as under-reinforced (actual moment less than the balanced limit), we will use a slightly higher value, say 0.90, for j in the design of the reinforcing.

Referring to Fig. 15-3, we note that there are five critical locations for which a moment must be determined and the required steel area computed. Reinforcing in the top of the slab must be computed for the negative moments at the end support, at the first interior beam, and at the typical interior beam. Reinforcing in the bottom of the slab must be computed for the positive moments at midspan in the first span and the typical interior spans. The design for these conditions is summarized in Fig. 15-3. For the data displayed in the figure we note the following:

$$\text{maximum spacing of reinforcing} = 3 \times t = 3 \times 4 = 12\ \text{in.}$$

$$\begin{aligned}
\text{design moment} = M &= (\text{moment factor } F) \times wL^2 \\
&= F \times (210)(9)^2 \times 12 \\
&= F \times 204{,}120 \qquad \text{(in in-lb units)}
\end{aligned}$$

$$\text{required } A_s \quad = \frac{M}{f_s jd} = \frac{F \times 204{,}120}{(20{,}000)(0.9)(2.875)} = F \times 3.944$$

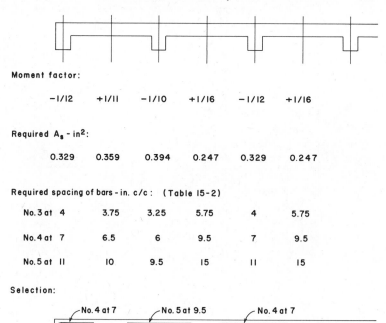

Moment factor:

$-1/12$ $+1/11$ $-1/10$ $+1/16$ $-1/12$ $+1/16$

Required A_s - in^2:

0.329   0.359   0.394   0.247   0.329   0.247

Required spacing of bars - in. c/c : (Table 15-2)

No. 3 at	4	3.75	3.25	5.75	4	5.75
No. 4 at	7	6.5	6	9.5	7	9.5
No. 5 at	11	10	9.5	15	11	15

Selection:

FIGURE 15-3. Design of the slab reinforcing.

Using data from Table 15-2, Fig. 15-3 shows required spacings for No. 3, 4, and 5 bars. A possible choice for the slab reinforcing, using all straight bars, is shown at the bottom of the figure.

Problem 15-4-A. A solid one-way slab is to be used for a framing system similar to that shown in Fig. 15-1. Column spacing is 36 ft, with regularly spaced beams occurring at 12 ft center to center. Superimposed loads on the structure are a total of 180 psf [8.62 kN/m^2]. Use $f'_c = 3$ ksi [20.7 MPa] and Grade 40 reinforcing with $f_y = 40$ ksi [276 MPa] and $f_s = 20$ ksi [138 MPa]. Determine the thickness for the slab, and select the size and spacing for the bars.

15-5 Concrete One-Way Joist Construction

Figure 15-4 shows a partial framing plan and some details for a type of construction that utilizes a series of very closely spaced

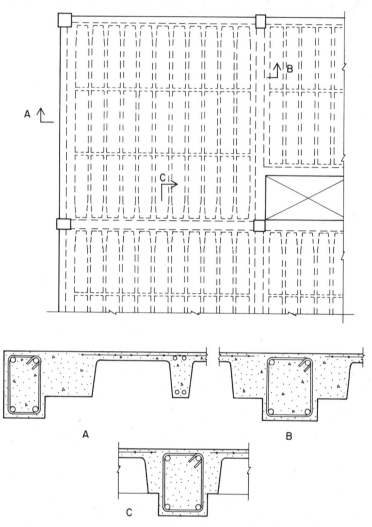

FIGURE 15-4. Typical concrete one-way joist construction.

beams and a relatively thin solid slab. Because of its resemblance to ordinary wood joist construction, this is called concrete joist construction. It is generally the lightest (in dead weight) of any type of poured-in-place concrete construction for flat spans. It is best suited for the situation of relatively long spans and light loadings.

Slabs as thin as 2 in. and joists as narrow as 4 in. are used with this construction. Because of the thinness of the parts and the small amount of cover provided for reinforcement (typically only $\frac{3}{4}$–1 in.), the construction has very low resistance to fire in exposed condition. It is therefore necessary to provide protection, as for steel construction, or to restrict its use to situations where high fire ratings are not required.

Joist construction is most often produced by using metal pans that are used to form the void spaces between joists. Once the concrete is hardened, the pans are pried off and reused. Other systems of forming are also available, using elements of fiber glass reinforced plastic, plastic coated cardboard, or precast lightweight concrete blocks.

The relatively thin short span slabs are typically reinforced with welded wire mesh rather than ordinary deformed bars. Joists are typically tapered at their ends, as shown in the framing plan in Fig. 15-4. This is done to provide additional resistance to the shear and negative bending moments at the ends of the spans. Shear reinforcement in the form of single vertical bars may be provided but is not frequently used.

This was a highly popular form of construction in earlier times, but it has been largely replaced by various forms of precast concrete construction at present.

15-6   Concrete Waffle Construction

Waffle construction consists of two-way spanning joists that are formed in a manner similar to that for one-way spanning joists, using forming units of metal, plastic, or cardboard to produce the void spaces between the joists. The most widely used type of waffle construction is the waffle flat slab, in which solid portions around column supports are produced by omitting the void-making forms. An example of a portion of such a system is shown in Fig. 15-5. This type of system is analogous to the solid flat slab which will be discussed in Section 15-7. At points of discontinuity in the plan—such as at large openings or at edges of the building—it is usually necessary to form beams. These beams may be produced as projections below the waffle, as shown in Fig. 15-5, or may be created within the waffle depth by omitting a row of the void-making forms, as shown in Fig. 15-6.

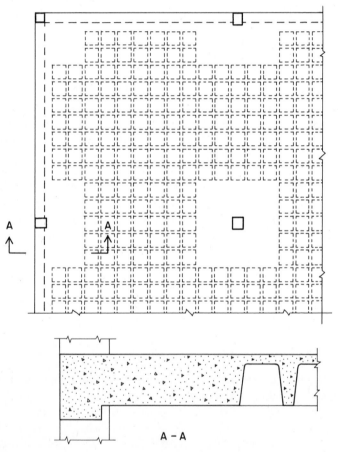

FIGURE 15-5.   Typical concrete waffle construction with edge beams only.

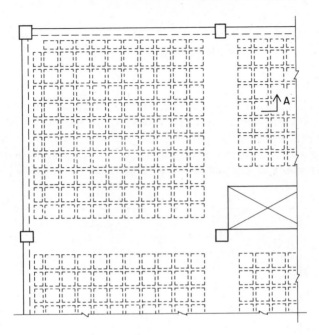

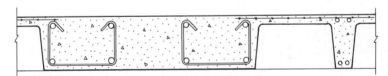

FIGURE 15-6.   Typical concrete waffle construction with column-line beams within the waffle depth.

If beams are provided on all of the column lines, as shown in Fig. 15-6, the construction is analogous to the two-way solid slab with edge supports, as discussed in Section 15-7. With this system the solid portions around the column are not required, since the waffle itself does not achieve the transfer of high shear or development of the high negative moments at the columns.

As with the one-way joist construction, fire ratings are low for ordinary waffle construction. The system is best suited for situations involving relatively light loads, medium to long spans, approximately square column bays, and a reasonable number of multiple bays in each direction.

15-7   Two-Way-Spanning Solid Slab Construction

If reinforced in both directions, the solid concrete slab may span two ways as well as one. The widest use of such a slab is in flat slab or flat plate construction. In flat slab construction, beams are used only at points of discontinuity, with the typical system consisting only of the slab and the strengthening elements used at

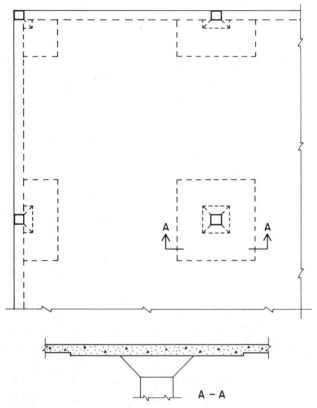

FIGURE 15-7.   Typical concrete flat slab construction with drop panels and column caps.

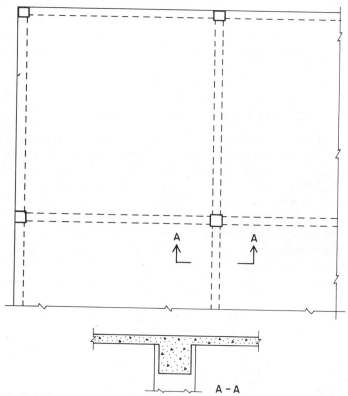

FIGURE 15-8. Typical two-way-spanning concrete slab with edge
supports.

column supports. Typical details for a flat slab system are shown
in Fig. 15-7. Drop panels consisting of thickened portions square
in plan are used to give additional resistance to the high shear and
negative moment that develops at the column supports. Enlarged
portions are also sometimes provided at the tops of the columns
(called column capitals), to further reduce the stresses in the slab.

Two-way slab construction consists of multiple bays of solid
two-way-spanning slabs with edge supports consisting of bearing
walls of concrete or masonry or of column-line beams formed in
the usual manner. Typical details for such a system are shown in
Fig. 15-8.

Two-way solid slab construction is generally favored over waffle construction where higher fire rating is required for the unprotected structure or where spans are short and loadings high. As with all types of two-way spanning systems, they function most efficiently where the spans in each direction are approximately the same.

15-8   Slab Design Using Handbooks

Design of both one-way and two-way slab construction, as well as one-way joist construction and waffle systems, may be achieved using data from tables in handbooks. One widely used source for such work is the *CRSI Handbook,* published by the Concrete Reinforcing Steel Institute. Material for this purpose is also published by the American Concrete Institute (publishers of the ACI Code) and by the Portland Cement Association. Even where loading conditions or span dimensions may not be in exact agreement with those given in the tables, the handbooks can still be used to establish approximate data for preliminary design.

16

Reinforced Concrete Columns

||||||||||||||||||||||||||||||||||||||||||||||||||||||||||||

16-1   Introduction

The practicing structural designer customarily uses tables or a
computer-aided procedure to determine the dimensions and reinforcing for concrete columns. The complexity of analytical formulas and the large number of variables make it impractical to
perform design for a large number of columns solely by hand
computation. The provisions relating to the design of columns in
the 1977 ACI Code are quite different from those of the working
stress design method in the 1963 Code. The current code does not
permit design of columns by the working stress method, but it
rather requires that the service load capacity of columns be determined as 40% of that computed by strength design procedures.
The strength design procedures are discussed briefly in Chapter
18. The discussions in this chapter are limited to the working
stress design procedures, as provided for in the 1963 ACI Code.

   Most concrete columns in building construction are relatively
stout. Although the code provides for reduction of axial compression on the basis of slenderness, the reductions do not become
significant until the ratio of the column height to its least lateral
dimension exceeds about 12. For slenderness beyond this ratio,
the code reductions should be considered.

16-2   Columns with Axial Load plus Bending Moment

Due to the nature of most concrete structures, current design practices generally do not consider the possibility of a concrete column with axial compression alone. That is to say, the existence of some bending moment is always considered together with the axial force. Figure 16-1 illustrates the nature of the so-called *interaction response* for a concrete column, with a range of combinations of axial load plus bending moment. In general, there are three basic ranges of this behavior, as follows:

1. Large Axial Force, Minor Moment.   For this case the moment has little effect, and the resistance to pure axial force is only negligibly reduced.
2. Significant Values for Both Axial Force and Moment. For this case the analysis for design must include the full combined force effects, that is, the interaction of the axial force and the bending moment.
3. Large Bending Moment, Minor Axial Force.   For this case the column behaves essentially as a doubly reinforced (tension and compression reinforced) member, with its capacity for moment resistance affected only slightly by the axial force.

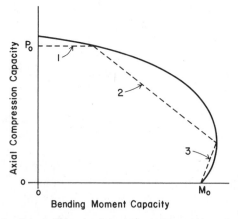

FIGURE 16-1.   Interaction of axial compression and bending moment for a reinforced concrete column.

In Fig. 16-1 the solid line on the graph represents the true response of the column—a form of behavior verified by many load tests on laboratory specimens. The dashed line figure on the graph represents the generalization of the three types of response just described.

The terminal points of the interaction response—pure axial compression or pure bending moment—may be reasonably easily determined. The interaction responses between these two limits require complex analyses, beyond the scope of this book.

16-3   Types of Reinforced Concrete Columns

Reinforced concrete columns for buildings generally fall into one of the following categories:

1. Square tied columns.
2. Round spiral columns.
3. Rectangular tied columns.
4. Columns of other geometries (hexagonal, L-shape, T-shape, etc.) with either ties or spirals.

In tied columns the longitudinal reinforcing is held in place by loop ties made of small-diameter reinforcing bars, commonly No. 2–No. 4. Such a column is represented by the square section shown in Fig. 16-2a. This type of reinforcing can quite readily accommodate other geometries as well as the square. The design of such a column is discussed in Section 16-4.

Spiral columns are those in which the longitudinal reinforcing is placed in a circle, with the whole group of bars enclosed by a continuous cyclindrical spiral made from steel rod or large-diameter steel wire. Although this reinforcing system obviously works best with a round column section, it can be used also with other geometries. A round column of this type is shown in Fig. 16-2b.

Experience has shown the spiral column to be slightly stronger than an equivalent tied column with the same amount of concrete and reinforcing. For this reason code provisions allow slightly more load on spiral columns. Spiral reinforcing tends to be expensive, however, and the round bar pattern does not always mesh well with other construction details in buildings. Thus tied

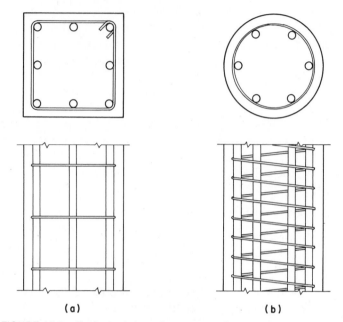

(a)   (b)

FIGURE 16-2.   Typical reinforced concrete columns: (a) with loop ties, (b) with a spiral wrap.

columns are often favored where restrictions on the outer dimensions of the sections are not severe.

16-4   General Requirements for Reinforced Concrete Columns

Code provisions and practical construction considerations place a number of restrictions on column dimensions and choice of reinforcing.

Column Size.   Rectangular tied columns are limited to a minimum area of 96 in^2 and a side dimension of 10 in. if square and 8 in. if oblong rectangular. Spiral columns are limited to a minimum size of 12 in. if either round or square.

Reinforcing.   Minimum bar size is No. 5. The minimum number of bars is four for tied columns, six for spiral columns. The minimum amount of area of steel is 1% of the gross column area. A maximum area of steel of 8% of the gross area is permitted, but bar spacing limitations makes this difficult to achieve; 4% is a more practical limit.

Ties. Ties shall be at least No. 3 for bars No. 10 and smaller. No. 4 ties should be used for bars that are No. 11 and larger. Vertical spacing of ties shall be not more than 16 times the bar diameter, 48 times the tie diameter, or the least dimension of the column. Ties shall be arranged so that every corner and alternate longitudinal bar is held by the corner of a tie with an included angle of not greater than 135°, and no bar shall be farther than 6 in. clear from such a supported bar. Complete circular ties may be used for bars placed in a circular pattern.

Concrete Cover. A minimum of 1.5 in. is needed when the column surface is not exposed to weather or in contact with the ground; 2 in. should be used for formed surfaces exposed to the weather or in contact with ground; 3 in. are necessary if the concrete is cast against earth.

Spacing of Bars. Clear distance between bars shall not be less than 1.5 times the bar diameter, 1.33 times the maximum specified size for the coarse aggregate, or 1.5 in.

16-5   Design of Tied Columns

The 1963 ACI Code limits the axial compression load on a tied column to

$$P = 0.85[A_g(0.25f_c' + f_s p_g)]$$

in which P = maximum permissible axial load,

A_g = gross area of the column,

f_c' = ultimate compressive strength of the concrete,

f_s = allowable compressive stress in the reinforcing, taken as 40% of the yield stress but not to exceed 30,000 psi,

p_g = percent of steel = A_s/A_g,

A_s = cross-sectional area of the reinforcing.

The following example illustrates the use of this formula for the determination of the allowable load on a given column.

Example 1. A 16-in. square tied column is reinforced with four No. 10 bars; $f'_c = 4000$ psi and $f_s = 20,000$ psi. Find the safe load for the column.

Solution: For use in the formula, we determine

$$A_g = 16 \times 16 = 256 \text{ in}^2$$

$$p_g = \frac{A_s}{A_g} = \frac{(4 \times 1.27)}{256} = 0.0198$$

Then

$$P = 0.85[256(0.25 \times 4 + 20 \times 0.0198)] = 303.8 \text{ kips}$$

In most building structures, concrete columns will sustain some computed bending moment in addition to the axial compression load. (See Figure 16-3). Even when a computed moment is not present, however, it is well to consider some amount of accidental eccentricity or other source of moment. It is recommended, therefore, that the maximum safe load be limited to that given for a minimum eccentricity of 10% of the column dimension.

Figure 16-4 gives safe loads for a selected number of sizes of square tied columns. Loads are given for various degrees of eccentricity, which is a means for expressing axial load and bending moment combinations. The computed moment on the column is translated into an equivalent eccentric loading, as shown in Fig. 16-3. Data for the curves was computed by using 40% of the load determined by strength design methods, as required by the 1977 ACI Code.

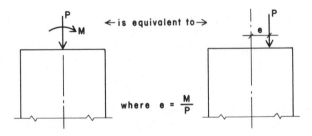

FIGURE 16-3.

For the column in Example 1, it may be noted that Fig. 16-4 yields a maximum value of approximately 260 kips, which is only about 80% of the value previously determined by the working stress formula. The discrepancy occurs because the code requires the use of a minimum eccentricity for all columns. Thus the curves in Fig. 16-4 do not begin at zero eccentricity.

The following examples illustrate the use of Fig. 16-4 for the design of tied columns.

Example 2. A column with $f'_c = 4$ ksi and steel with $f_y = 60$ ksi sustains an axial compression load of 400 kips. Find the minimum practical column size if reinforcing is a maximum of 4% and the maximum size if reinforcing is a minimum of 1%.

Solution: Using Fig. 16-4a, we find from the sizes given:

Minimum column is 20 in. square with 8 No. 9 (Curve No. 14).

Maximum capacity is 410 kips, $p_g = 2.0\%$.

Maximum size is 24 in. square with 4 No. 11 (Curve No. 17).

Maximum capacity is 510 kips, $p_g = 1.08\%$.

It should be apparent that it is possible to use an 18-in. or 19-in. column as the minimum size and to use a 22-in. or 23-in. column as the maximum size. Since these sizes are not given in the figure, we cannot verify them for certain without using strength design procedures.

Example 3. A square tied column with $f'_c = 4$ ksi and steel with $f_y = 60$ ksi sustains an axial load of 400 kips and a bending moment of 200 kip-ft. Determine the minimum size column and its reinforcing.

Solution: We first determine the equivalent eccentricity, as shown in Fig. 16-3. Thus

$$e = \frac{M}{P} = \frac{200 \times 12}{400} = 6 \text{ in.}$$

Then, from Fig. 16-4 we find:

Minimum size is 24 in. square with 16 No. 10 bars.

Capacity at 6 in. eccentricity is 410 kips.

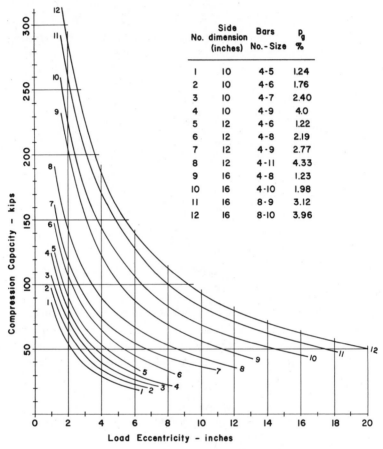

No.	Side dimension (inches)	Bars No.-Size	p_g %
1	10	4-5	1.24
2	10	4-6	1.76
3	10	4-7	2.40
4	10	4-9	4.0
5	12	4-6	1.22
6	12	4-8	2.19
7	12	4-9	2.77
8	12	4-11	4.33
9	16	4-8	1.23
10	16	4-10	1.98
11	16	8-9	3.12
12	16	8-10	3.96

FIGURE 16-4.   Safe service loads for square tied columns with f'_c = 4 ksi and f_y = 60 ksi.

There is usually a number of possible combinations of reinforcing bars that may be assembled to satisfy the steel area requirement for a given column. Aside from providing for the area, the number of bars must also work reasonably in the layout of the column. Figure 16-5 shows a number of tied columns with various number of bars. When a column is small, the preferred choice is

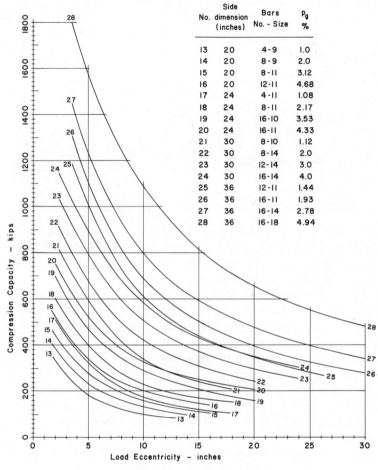

No.	Side dimension (inches)	Bars No. - Size	P_g %
13	20	4 - 9	1.0
14	20	8 - 9	2.0
15	20	8 - 11	3.12
16	20	12 - 11	4.68
17	24	4 - 11	1.08
18	24	8 - 11	2.17
19	24	16 - 10	3.53
20	24	16 - 11	4.33
21	30	8 - 10	1.12
22	30	8 - 14	2.0
23	30	12 - 14	3.0
24	30	16 - 14	4.0
25	36	12 - 11	1.44
26	36	16 - 11	1.93
27	36	16 - 14	2.78
28	36	16 - 18	4.94

FIGURE 16-4. (*Continued*)

usually that of the simple four-bar layout, with one bar in each corner and a single peripheral tie. As the column gets larger, the distance between the corner bars gets larger, and it is best to use more bars so that the reinforcing is spread out around the column periphery. For a symmetrical layout and the simplest of tie layouts, the best choice is for numbers that are multiples of four, as

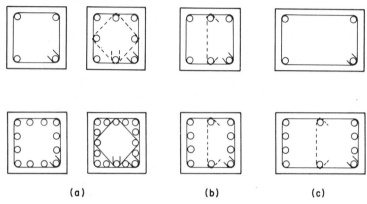

(a)   **(b)**   **(c)**

FIGURE 16-5. Typical bar placement and tie patterns for tied columns.

shown in Fig. 16-5a. The number of additional ties required for
these layouts depends on the size of the column and the considerations discussed in Section 16-4.

An unsymmetrical bar arrangement is not necessarily bad,
even though the column and its construction details are otherwise
not oriented differently on the two axes. In situations where moments may be greater on one axis, the unsymmetrical layout is
actually preferred; in fact, the column shape will also be more
effective if it is unsymmetrical, as shown for the oblong shapes in
Fig. 16-5c.

Problems 16-5-A*-B-C*-D-E.   Using Fig. 16-5, pick the minimum size
square tied column and its reinforcing for the following combinations of
axial load and bending moment.

	Axial Compressive Load in kips	Bending Moment in kip-ft
A	100	25
B	100	50
C	150	75
D	200	100
E	300	150

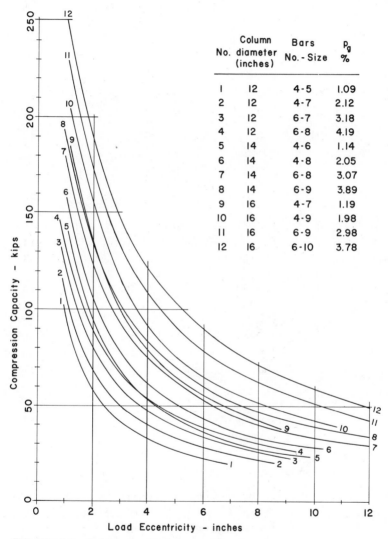

No.	Column diameter (inches)	Bars No.-Size	P_g %
1	12	4-5	1.09
2	12	4-7	2.12
3	12	6-7	3.18
4	12	6-8	4.19
5	14	4-6	1.14
6	14	4-8	2.05
7	14	6-8	3.07
8	14	6-9	3.89
9	16	4-7	1.19
10	16	4-9	1.98
11	16	6-9	2.98
12	16	6-10	3.78

FIGURE 16-6. Safe service loads for round tied columns with f'_c = 4 ksi and f_y = 60 ksi.

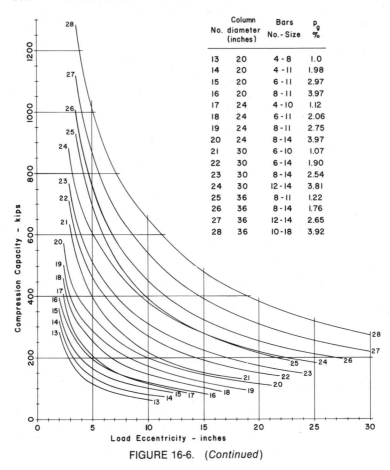

No.	Column diameter (inches)	Bars No.-Size	p_g %
13	20	4 - 8	1.0
14	20	4 - 11	1.98
15	20	6 - 11	2.97
16	20	8 - 11	3.97
17	24	4 - 10	1.12
18	24	6 - 11	2.06
19	24	8 - 11	2.75
20	24	8 - 14	3.97
21	30	6 - 10	1.07
22	30	6 - 14	1.90
23	30	8 - 14	2.54
24	30	12 - 14	3.81
25	36	8 - 11	1.22
26	36	8 - 14	1.76
27	36	12 - 14	2.65
28	36	10 - 18	3.92

FIGURE 16-6. (*Continued*)

16-6 Design of Round Columns

Round columns may be designed and built as spiral columns, as described in Section 16-3, or they may be developed as tied columns with the bars placed in a circle and held by a series of round circumferential ties. Because of the cost of spirals, it is usually more economical to use the tied column, so they are often used unless the additional strength or other behavioral characteristics of the spiral column are required.

It is also possible to use rectangular bar layouts and tie patterns as shown in Fig. 16-5 inside a round column form. In such cases, the column is usually designed as a square column using the square shape that can be included within the round form. It is thus possible to use a four-bar column for small-diameter, round column forms.

Figure 16-6 gives safe loads for round columns that are designed as tied columns. Load values have been adapted from values determined by strength design methods. The curves in Fig. 16-6 are similar to those for the square columns in Fig. 16-4, and their use is similar to that demonstrated in Examples 2 and 3.

Problems 16-6-A*-B-C*-D-E.   Using Fig. 16-6, pick the minimum size round column and its reinforcing for the load and moment combinations in Problem 16-5.

17

Footings

||||||||||||||||||||||||||||||||||||||||||||||||||||||||||||||||

17-1 Introduction

The primary purpose of a footing is to spread the loads so that the allowable bearing capacity of the foundation material is not exceeded. In cities where experience and tests have established the allowable strengths of various foundation soils, local building codes may be consulted to determine the bearing capacities used in design. In the absence of such information, borings or load tests should be made. For sizable structures, borings at the site should always be made and their results interpreted by a qualified soils engineer.

Footings may be classified as wall footings and column footings. In the former type the load is brought to the foundation as a uniform load per linear ft of wall; in the latter it is concentrated at the base of the column. Column loads are sometimes combined on a single footing, especially where columns are very closely spaced. However, the independent footing supporting a single column is the most common type and the one to be considered in this chapter.

17-2 Independent Column Footings

The great majority of independent or isolated column footings are square in plan, with reinforcing consisting of two sets of bars at right angles to each other. This is known as two-way reinforcement. The column may be placed directly on the footing block, or it may be supported by a pedestal. A pedestal, or pier, is a short, wide compression block that serves to reduce the punching effect on the footing. For steel columns a pier may also serve to raise the bottom of the steel column above ground level.

The design of a column footing is usually based on the following considerations:

1. Maximum Soil Pressure. The sum of the superimposed load on the footing and the weight of the footing must not exceed the limit for bearing pressure on the supporting material. The required total plan area of the footing is determined on this basis.

2. Control of Settlement. Where buildings rest on highly compressible soil, it may be necessary to select footing areas that assure a uniform settlement of all the building columns rather than to strive for a maximum use of the allowable soil pressure.

3. Size of the Column. The larger the column, the less will be the shear, flexural, and bond stresses in the footing, since these are developed by the cantilever effect of the footing projection beyond the edges of the column.

4. Shear Stress Limit for the Concrete. For square-plan footings this is usually the only critical stress condition for the concrete. In order to reduce the required amount of reinforcing, the footing depth is usually established well above that required by the flexural stress limit for the concrete.

5. Flexural Stress and Development Length Limits for the Bars. These are considered on the basis of the moment developed in the cantilevered footing at the face of the column.

6. Footing Thickness for Development of Column Reinforcing. When a footing supports a reinforced concrete column, the compressive force in the column bars must be transferred to the footing by bond stress—called *doweling*

of the bars. The thickness of the footing must be sufficient for the necessary development length of the column bars.

The following example illustrates the design process for a simple, square column footing.

Example. A 16-in. [406-mm] square concrete column exerts a load of 240 kips [1068 kN] on a square column footing. Determine the footing dimensions and the necessary reinforcing using the following data: $f'_c = 3$ ksi [20.7 MPa], Grade 40 bars with $f_y = 40$ ksi [276 MPa] and $f_s = 20$ ksi [138 MPa], maximum permissible soil pressure = 4000 psf [192 kN/m²].

Solution: The first decision to be made is that of the height, or thickness, of the footing. This has to be a raw first guess unless the dimensions of similar footings are known. In practice this knowledge is generally available from previous design work or from handbook tables. In lieu of this, a reasonable guess is made, the design work is performed, and an adjustment is made if the assumed thickness proves inadequate. We will assume a footing thickness of 20 in. [508 mm] for a first try for this example.

The footing thickness establishes the weight of the footing on a per-square-ft basis. This weight is then subtracted from the maximum permissible soil pressure, and the net value is then usable for the superimposed load on the footing. Thus

$$\text{footing weight} = \frac{20}{12} (150 \text{ psf}) = 250 \text{ psf } [12 \text{ kN/m}^2]$$

net usable soil pressure = 4000 − 250 = 3750 psf [180 kN/m²]

$$\text{required footing plan area} = \frac{240,000}{3750} = 64 \text{ ft}^2 \left[\frac{1068}{180} = 5.93 \text{ m}^2 \right]$$

and the length of the side of the square footing,

$$L = \sqrt{64} = 8 \text{ ft} \quad [\sqrt{5.93} = 2.44 \text{ m}]$$

Two shear stress situations must be considered for the concrete. The first occurs as ordinary beam shear in the cantilevered portion and is computed at a critical section at a distance d (effective depth of the beam) from the face of the column as shown in

Fig. 17-1a. The shear stress at this section is computed in the same manner as for a beam, as discussed in Section 13-10, and the stress limit is $v_c = 1.1\sqrt{f'_c}$. The second shear stress condition is that of peripheral shear, or so-called "punching" shear, and is investigated at a circumferential section around the column at a distance of $d/2$ from the column face as shown in Fig. 17-1b. For this condition the allowable stress is $v_c = 2.0\sqrt{f'_c}$.

With two-way reinforcing, it is necessary to place the bars in one direction on top of the bars in the other direction. Thus, although the footing is supposed to be the same in both directions, there are actually two different d distances—one for each layer of bars. It is common practice to use the average of these two distances for the design value of d; that is, d = the footing thickness less the sum of the concrete cover and the bar diameter. With the bar diameter as yet undetermined, we will assume an approximate d of the footing thickness less 4 in. [102 mm] (a concrete cover of 3 in. plus a No. 8 bar). For the example this becomes

$$d = t - 4 = 20 - 4 = 16 \text{ in. } [406 \text{ mm}]$$

It should be noted that it is the *net* soil pressure that causes stresses in the footing, since there will be no bending or shear in the footing when it rests along on the soil. We thus use the net soil pressure of 3750 psf [180 kN/m²] to determine the shear and bending effects for the footing.

For the beam shear investigation, we determine the shear force generated by the net soil pressure acting on the shaded portion of the footing plan area shown in Fig. 17-1a. Thus

$$V = 3750 \times 8 \times \frac{24}{12} = 60,000 \text{ lb } [180 \times 2.44 \times 0.609 = 267.5 \text{ kN}]$$

and, using the formula for shear stress in a beam (Section 13-10),

$$v = \frac{V}{bd} = \frac{60,000}{96 \times 16} = 39.1 \text{ psi } \left[\frac{0.2675}{2.44 \times 0.406} = 0.270 \text{ MPa}\right]$$

which is compared to the allowable stress of

$$v_c = 1.1\sqrt{f'_c} = 1.1\sqrt{3000} = 60 \text{ psi } [0.414 \text{ MPa}]$$

indicating that this condition is not critical.

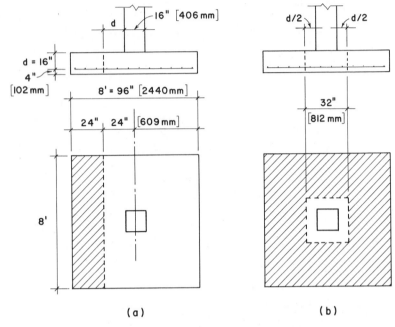

FIGURE 17-1.   Shear considerations for the footing.

For the peripheral shear investigation, we determine the shear force generated by the net soil pressure acting on the shaded portion of the footing area shown in Fig. 17-1*b*. Thus

$$V = 3750 \left[(8)^2 - \left(\frac{32}{12}\right)^2 \right] = 213{,}333 \text{ lb}$$

$$[V = 180(2.44^2 - 0.812^2) = 953 \text{ kN}]$$

Shear stress for this case is determined with the same formula as for beam shear, with the dimension *b* being the total peripheral circumference. Thus

$$v = \frac{V}{bd} = \frac{213{,}333}{(4 \times 32) \times 16} = 104.2 \text{ psi}$$

$$\left[v = \frac{0.953}{(4 \times 0.812) \times 0.406} = 0.723 \text{ MPa} \right]$$

which is compared to the allowable stress of

$$v_c = 2\sqrt{f'_c} = 2\sqrt{3000} = 109.5 \text{ psi } [0.755 \text{ MPa}]$$

This computation indicates that the peripheral shear stress is not critical, but since the actual stress is quite close to the limit, the assumed thickness of 20 in. is probably the least full-inch value that can be used. Flexural stress in the concrete should also be considered, although it is seldom critical for a square footing. One way to verify this is to compute the balanced moment capacity of the section with $b = 96$ in. and $d = 16$ in. Using the factor for a balanced section from Table 13-3, we find

$$M_R = Rbd^2 = 0.226(96)(16)^2 = 5554 \text{ kip-in or } 463 \text{ kip-ft}$$

which may be compared with the actual moment computed in the next step.

For the reinforcing we consider the stresses developed at a section at the edge of the column as shown in Fig. 17-2. The cantilever moment for the 40-in. [1016-mm] projection of the foot-

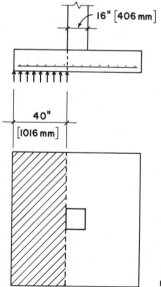

FIGURE 17-2. Bending and development length considerations for the footing.

ing beyond the column is

$$M = 3750 \times 8 \times \frac{40}{12} \times \frac{1}{2}\left(\frac{40}{12}\right) = 166,667 \text{ lb-ft}$$

$$\left[M = 180 \times 2.44 \times 1.016 \times \frac{1.016}{2} = 227 \text{ kN-m}\right]$$

Using the formula for required steel area in a beam, with a conservative guess of 0.9 for j, we find (see Section 13-9)

$$A_s = \frac{M}{f_s jd} = \frac{166,667 \times 12}{20 \times 0.9 \times 16 \times 10^3} = 6.95 \text{ in}^2 \text{ [4502 mm}^2\text{]}$$

This requirement may be met by various combinations of bars, such as those in Table 17-1. Data for consideration of the development length and the center-to-center bar spacing is also given in the table. The flexural stress in the bars must be developed by the embedment length equal to the projection of the bars beyond the column edge, as discussed in Section 13-14. With a minimum of 2 in. [51 mm] of concrete cover at the edge of the footing, this length is 38 in. [965 mm]. The required development lengths indicated in the table are taken from Table 13-4; it may be noted that all of the combinations in the table are adequate in this regard.

TABLE 17-1. Reinforcing Alternatives for the Column Footing

Number and Size of Bars	Area of Steel Provided		Required Development Length[a]		Center-to-Center Spacing	
	(in²)	(mm²)	(in.)	(mm)	in.	(mm)
12 No. 7	7.20	4645	18	457	8.2	208
9 No. 8	7.11	4687	23	584	11.3	286
7 No. 9	7.00	4516	29	737	15	381
6 No. 10	7.62	4916	37	940	18	458

[a] From Table 13-4; values for "other bars," $f_y = 40$ ksi, $f'_c = 3$ ksi.

If the distance from the edge of the footing to the first bar at each side is approximately 3 in. [76 mm], the center-to-center distance for the two outside bars will be 96 − 2(3) = 90 in. [2286 mm], and with the rest of the bars evenly spaced, the spacing will be 90 divided by the number of total bars less one. This value is shown in the table for each set of bars. The maximum permitted spacing is 18 in. [457 mm], and the minimum should be a distance that is adequate to permit good flow of the wet concrete between the two-way grid of bars—say 4 in. [102 mm] or more.

All of the bar combinations in Table 17-1 are adequate for the footing. Many designers prefer to use the largest possible bar, as this reduces the number of bars that must be handled and supported during construction. On this basis, the footing will be the following:

8 ft square by 20 in. thick with six No. 10 bars each way.

Problem 17-2-A.   Design a square footing for a 14-in. [356-mm] square concrete column with a load of 219 kips [974 kN]. The maximum permissible soil pressure is 3000 psf [144 kN/m²]. Use concrete with $f'_c = 3$ ksi [20.7 MPa] and reinforcing of Grade 40 bars with $f_y = 40$ ksi [276 MPa] and $f_s = 20$ ksi [138 MPa].

17-3   Load Tables for Column Footings

For ordinary situations we often design square column footings by using data from tables in various references. Even where special circumstances make it necessary to perform the type of design illustrated in Section 17-3, such tables will assist in making a first guess for the footing dimensions.

Table 17-2 gives the allowable superimposed load for a range of footings and soil pressures. This material has been adapted from a more extensive table in *Simplified Design of Building Foundations* by James Ambrose (New York, Wiley, 1982). Designs are given for concrete strengths of 2000 and 3000 psi. The low strength of 2000 psi is sometimes used for small buildings, since many building codes permit the omission of testing of the concrete if this value is used for design.

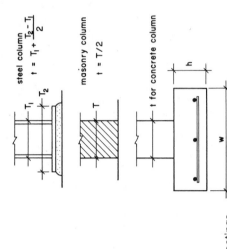

steel column
$t = T_1 + \dfrac{T_2 - T_1}{2}$

masonry column
$t = T/2$

t for concrete column

TABLE 17-2. Allowable Loads on Square Column Footings

Maximum Soil Pressure (lb/ft²)	Minimum Column Width t (in.)	$f'_c = 2000$ psi				$f'_c = 3000$ psi			
		Allowable Load on Footing (k)	Footing Dimensions h (in.)	w (ft)	Reinforcing Each Way	Allowable Load on Footing (k)	Footing Dimensions h (in.)	w (ft)	Reinforcing Each Way
1000	8	7.9	10	3.0	2 No. 3	7.9	10	3.0	2 No. 3
	8	10.7	10	3.5	3 No. 3	10.7	10	3.5	3 No. 3
	8	14.0	10	4.0	3 No. 4	14.0	10	4.0	3 No. 4

8	4 No. 4	4.5	10	17.7	4 No. 4	4.5	10	17.7
8	4 No. 5	5.0	10	22	4 No. 5	5.0	10	22
8	5 No. 6	6.0	10	31	5 No. 6	6.0	10	31
8	7 No. 6	7.0	11	42	6 No. 6	7.0	12	42
8	3 No. 3	3.0	10	12.4	3 No. 3	3.0	10	12.4
8	3 No. 4	3.5	10	16.8	3 No. 4	3.5	10	16.8
8	4 No. 4	4.0	10	22	4 No. 4	4.0	10	22
8	4 No. 5	4.5	10	28	4 No. 5	4.5	10	28
8	6 No. 5	5.0	10	34	5 No. 5	5.0	11	34
8	6 No. 6	6.0	11	49	6 No. 6	6.0	12	48
8	6 No. 7	7.0	13	65	7 No. 6	7.0	14	65
8	7 No. 7	8.0	15	84	7 No. 7	8.0	16	83
8	10 No. 7	9.0	16	105	8 No. 7	9.0	18	103
8	4 No. 3	3.0	10	17	4 No. 3	3.0	10	17
8	4 No. 4	3.5	10	23	4 No. 4	3.5	10	23
8	6 No. 4	4.0	10	30	6 No. 4	4.0	10	30
8	6 No. 5	4.5	10	38	5 No. 5	4.5	11	37
8	5 No. 6	5.0	11	46	6 No. 5	5.0	12	46
8	7 No. 6	6.0	13	66	6 No. 6	6.0	14	65
8	7 No. 7	7.0	15	89	8 No. 6	7.0	16	88
8	9 No. 7	8.0	17	114	8 No. 7	8.0	18	113
8	8 No. 8	9.0	19	143	8 No. 8	9.0	20	142
10	10 No. 8	10.0	20	175	9 No. 8	10.0	21	174

1500

2000

TABLE 17-2. (Continued)

Maximum Soil Pressure (lb/ft²)	Minimum Column Width t (in.)	f'_c = 2000 psi				f'_c = 3000 psi			
		Allowable Load on Footing (k)	Footing Dimensions h (in.)	w (ft)	Reinforcing Each Way	Allowable Load on Footing (k)	Footing Dimensions h (in.)	w (ft)	Reinforcing Each Way
3000	8	26	10	3.0	3 No. 4	26	10	3.0	3 No. 4
	8	35	10	3.5	4 No. 5	35	10	3.5	4 No. 5
	8	45	12	4.0	4 No. 5	46	11	4.0	5 No. 5
	8	57	13	4.5	6 No. 5	57	12	4.5	6 No. 5
	8	70	14	5.0	5 No. 6	71	13	5.0	6 No. 6
	8	100	17	6.0	7 No. 6	101	15	6.0	8 No. 6
	10	135	19	7.0	7 No. 7	136	18	7.0	8 No. 7
	10	175	21	8.0	10 No. 7	177	19	8.0	8 No. 8
	12	219	23	9.0	9 No. 8	221	21	9.0	10 No. 8
	12	269	25	10.0	11 No. 8	271	23	10.0	10 No. 9
	12	320	28	11.0	11 No. 9	323	26	11.0	12 No. 9
	14	378	30	12.0	12 No. 9	381	28	12.0	11 No. 10
4000	8	35	10	3.0	4 No. 4	35	10	3.0	4 No. 4
	8	47	12	3.5	4 No. 5	47	11	3.5	4 No. 5

8	61	13	5 No. 5	4.0	61	12	4.0	6 No. 5
8	77	15	5 No. 6	4.5	77	13	4.5	6 No. 6
8	95	16	6 No. 6	5.0	95	15	5.0	6 No. 6
8	135	19	8 No. 6	6.0	136	18	6.0	7 No. 7
10	182	22	8 No. 7	7.0	184	20	7.0	9 No. 7
10	237	24	9 No. 8	8.0	238	22	8.0	9 No. 8
12	297	26	10 No. 8	9.0	299	24	9.0	9 No. 9
12	364	29	13 No. 8	10.0	366	27	10.0	11 No. 9
14	435	32	12 No. 9	11.0	440	29	11.0	11 No. 10
14	515	34	14 No. 9	12.0	520	31	12.0	13 No. 10
16	600	36	17 No. 9	13.0	606	33	13.0	15 No. 10
16	688	39	15 No. 10	14.0	696	36	14.0	14 No. 11
18	784	41	17 No. 10	15.0	793	38	15.0	16 No. 11

Note: Allowable loads do not include the weight of the footing, which has been deducted from the total bearing capacity. Criteria: $f_s = 20$ ksi, $v_c = 1.1\sqrt{f_c'}$ for beam shear, $v_c = 2\sqrt{f_c'}$ for peripheral shear.

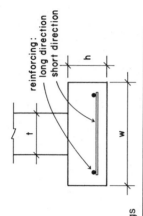

reinforcing:
long direction
short direction

TABLE 17-3. Allowable Loads on Wall Footings

| Maximum Soil Pressure (lb/ft²) | Minimum Wall Thickness | | Allowable Load on Footing (lb/ft) | Footing Dimensions | | Reinforcing | |
	Concrete t (in.)	Masonry t (in.)		h (in.)	w (in.)	Long Direction	Short Direction
1000	4	8	2,625	10	36	3 No. 4	No. 3 at 16
	4	8	3,062	10	42	2 No. 5	No. 3 at 12
	6	12	3,500	10	48	4 No. 4	No. 4 at 16
	6	12	3,938	10	54	3 No. 5	No. 4 at 13
	6	12	4,375	10	60	3 No. 5	No. 4 at 10
	6	12	5,250	10	72	4 No. 5	No. 5 at 11
1500	4	8	4,125	10	36	3 No. 4	No. 3 at 10

f'_c							
	4	8	4,812	10	42	2 No. 5	No. 4 at 13
	6	12	5,500	10	48	4 No. 4	No. 4 at 11
	6	12	6,131	11	54	3 No. 5	No. 5 at 15
	6	12	6,812	11	60	5 No. 4	No. 5 at 12
	8	16	8,100	12	72	5 No. 5	No. 5 at 10
2000	4	8	5,625	10	36	3 No. 4	No. 4 at 14
	6	12	6,562	10	42	2 No. 5	No. 4 at 11
	6	12	7,500	10	48	4 No. 4	No. 5 at 12
	6	12	8,381	11	54	3 No. 5	No. 5 at 11
	6	12	9,250	12	60	4 No. 5	No. 5 at 10
	8	16	10,875	15	72	6 No. 5	No. 5 at 9
3000	6	12	8,625	10	36	3 No. 4	No. 4 at 10
	6	12	10,019	11	42	4 No. 4	No. 5 at 13
	6	12	11,400	12	48	3 No. 5	No. 5 at 10
	6	12	12,712	14	54	6 No. 4	No. 5 at 10
	8	16	14,062	15	60	5 No. 5	No. 5 at 9
	8	16	16,725	17	72	6 No. 5	No. 6 at 10

Note: Allowable loads do not include the weight of the footing, which has been deducted from the total bearing capacity. Criteria: $f'_c = 2000$ psi, $f_s = 20$ ksi, $v_c = 1.1\sqrt{f'_c}$.

17-4   Load Tables for Wall Footings

Wall footings may be designed by the same process that was illustrated in Section 17-3, omitting the investigation for peripheral shear. The principal reinforcing in wall footings is placed in only one direction, perpendicular to the plane of the wall. Minimum reinforcing is commonly placed in the long direction (parallel to the wall) to provide for temperature and shrinkage stresses. Table 17-3 gives allowable loads for a range of wall footings and soil pressures. This material is adapted from a more extensive table in the reference cited in Section 17-3.

18

Ultimate-Strength Design

IIIIIIIIIIIIIIIIIIIIIIIIIIIIIIIIIIIIIIIIIIIIIIIIIIIIIIIIIIIIII

18-1 Introduction

The straight-line distribution of compressive stress (Fig. 13-1) is valid at working stress levels because the stresses developed under load vary approximately with the distance from the neutral axis, in accordance with elastic theory. However, shrinkage and cracking of the concrete, together with the phenomenon of creep under sustained loading, complicate the stress distribution. Over time, stresses computed in reinforced concrete members on the basis of elastic theory are not realistic. Generally speaking, the serviceability of the working stress design method is maintained by the differentials provided between the allowable compressive stress f_c and the specified compressive strength of the concrete f'_c and between the allowable tensile stress f_s and the yield strength of the steel reinforcement f_y. These differentials are, in effect, factors of safety (Section 1-12).

Laboratory investigations over several years have revealed that stresses in both concrete and steel at *ultimate load* can be determined with greater precision than at working or *service load*. This condition has led to the development of ultimate-strength design which has become the predominant design method for important building structures. As noted earlier (Section 13-2), the 1977 ACI Code is built primarily around ultimate-strength design, referred to in the code as *strength design.* Extended consideration of this method is beyond the scope of this book, but the following brief discussion should serve as an introduction to the considerations on which it is based.

18-2 Loads and Load Factors

The live and dead loads we have dealt with thus far in the book are called *service loads*; taken together, they represent a best estimate of the actual load a structural member may be called upon to support. When the strength design method is used, the service loads must be increased by a specified *load factor* in order to provide a factor of safety. These load factors are different for live load, dead load, and wind or earthquake loading. Section 9-3 of ACI 318-77 specifies several combinations of load factors to be considered, but we will limit our attention to the basic relationship of Code equation (9-1):

$$U = 1.4D + 1.7L$$

in which U represents the *design* load, *design* shear, or *design* moment, as the case may be; D represents the service dead load; and L represents the service live load.

18-3 Capacity Reduction Factors

In addition to the use of load factors, the theoretical capacity of a structural member is reduced by a capacity reduction factor, called the ϕ (phi) factor. The ϕ factors provide for variations in materials, care and skill of workers, construction dimensions, and so on, and also take into account the structural importance of a member and the adequacy of the theory on which its strength calculations are based. The basic equations for strength in bending, shear, and column action are multiplied by the appropriate ϕ factors to give magnitudes less than the theoretical values. Capacity reduction factors prescribed by the 1977 Code include the following:

bending calculations = 0.90

spiral columns = 0.75

tied columns = 0.70

shear = 0.85

bearing on concrete = 0.70

bending in plain concrete = 0.65

It will be noted that columns have lower factors than beams.

18-4 Bending Stresses in Rectangular Beams

Figure 18-1a is an abridgment of Fig. 13-1 showing the straight-line compressive stress distribution assumed in working stress design; f_c represents the maximum allowable working value of the extreme fiber stress. Figure 18-1b illustrates an assumed parabolic stress distribution when the value of the extreme fiber stress has reached f_c', the specified compressive strength of the concrete (i.e., the "ultimate strength"). Figure 18-1c shows the equivalent rectangular concrete stress distribution permitted by the ACI Code for use in the strength design method. The rectangular "stress block" is based on the assumption that a concrete stress of $0.85f_c'$ is uniformly distributed over the compression zone, which has dimensions equal to the beam width b and the distance a which locates a line parallel to and above the neutral axis. The value of a is determined from the expression $a = \beta_1 \times c$, where β_1 (beta one) is a factor that varies with the compressive strength of the concrete, and c is the distance from the extreme fiber to the neutral axis. For concrete having f_c' equal to or less than 4000 psi [27.6 MPa], the code gives a = 0.85 c.

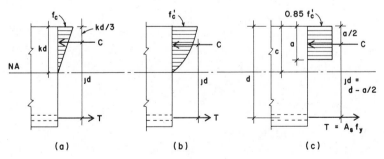

FIGURE 18-1.

With the rectangular stress block, the magnitude of the compressive force in the concrete is expressed as

$$C = 0.85f'_c \times b \times a$$

and it acts at a distance of $a/2$ from the top of the beam. The arm of the resisting force couple then becomes $d - (a/2)$, and the developed resisting moment as governed by the concrete is

$$M_t = C\left(d - \frac{a}{2}\right) = 0.85f'_c ba \times \left(d - \frac{a}{2}\right)$$

With T expressed as $A_s \times f_y$, the developed moment as governed by the reinforcing is

$$M_t = T\left(d - \frac{a}{2}\right) = A_s f_y \left(d - \frac{a}{2}\right)$$

A formula for the dimension a of the stress block can be derived by equating the compression and tension forces; thus

$$0.85f'_c ba = A_s f_y \quad \text{and} \quad a = \frac{A_s f_y}{0.85f'_c b}$$

Expressing the area of steel in terms of a percentage, ρ, the formula for a may be modified as follows:

$$\rho = \frac{A_s}{bd} \quad \text{or} \quad A_s = \rho bd$$

$$a = \frac{(\rho bd)f_y}{0.85f'_c b} = \frac{\rho df_y}{0.85f'_c}$$

The balanced section for strength design is visualized in terms of strain rather than stress. The limit for a balanced section is expressed in the form of the percentage of steel required to produce balanced conditions. The formula for this percentage is

$$\rho_b = \frac{0.85f'_c}{f_y} \times \frac{87}{87 + f_y}$$

in which f'_c and f_y are in units of ksi.

The ACI Code limits the percentage of steel to 75% of this balanced value in beams with tension reinforcing only.

Returning to the formula for the developed resisting moment, as expressed in terms of the steel, we see that a useful formula may be derived as follows:

$$M_t = A_s f_y \left(d - \frac{a}{2} \right) = (\rho b d)(f_y) \left(d - \frac{a}{2} \right)$$

$$= (\rho b d)(f_y)(d) \left(1 - \frac{1}{2} \frac{a}{d} \right)$$

$$= (b d^2) \left[\rho f_y \left(1 - \frac{1}{2} \frac{a}{d} \right) \right]$$

$$= R b d^2$$

where $R = \rho f_y \left(1 - \frac{1}{2} \frac{a}{d} \right)$

With the reduction factor applied, as discussed in Section 18-3, the design moment for a section is limited to nine-tenths of the theoretical resisting moment.

Values for the balanced section factors—ρ, R, and a/d—are given in Table 18-1 for various combinations of f_c' and f_y. The use of these data for a simple design problem is illustrated in the example in the next section.

18-5   Design of a Beam with Tension Reinforcing Only

The balanced section, as discussed in the preceding section, is not necessarily a practical one for design. In most cases economy

TABLE 18-1.   Balanced Section Factors—Strength Design

f_c' (psi)	$f_y = 40$ ksi			$f_y = 50$ ksi			$f_y = 60$ ksi		
	a/d	ρ	R (lb-in)	a/d	ρ	R (lb-in)	a/d	ρ	R (lb-in)
2500	0.437	0.0232	726	0.405	0.0172	686	0.377	0.0134	652
3000	0.437	0.0279	872	0.405	0.0207	824	0.377	0.0160	779
4000	0.437	0.0371	1160	0.405	0.0275	1097	0.377	0.0214	1042
5000	0.411	0.0437	1389	0.381	0.0324	1311	0.355	0.0251	1239
6000	0.385	0.0491	1586	0.357	0.0364	1496	0.333	0.0283	1416

will be achieved by using less than the balanced reinforcing for a given concrete section. In special circumstances it may also be possible, or even desirable, to use compressive reinforcing in addition to tension reinforcing. Nevertheless, just as in the working stress method, the balanced section is often a useful reference when design is performed. The following example illustrates a procedure for the design of a simple rectangular beam section with tension reinforcing only.

Example.   The service load bending moments on a beam are 58 kip-ft [78.6 kN-m] for dead load and 38 kip-ft [51.5 kN-m] for live load. The beam is 10 in. [254 mm] wide, f'_c is 4000 psi [27.6 MPa], and f_y is 60 ksi [414 MPa]. Determine the depth of the beam and the tensile reinforcing required.

Solution: (1) The first step is to determine the design moment, using the load factors, as discussed in Section 18-2. Thus

$$U = 1.4D + 1.7L$$

$$M_u = 1.4(M_{DL}) + 1.7(M_{LL})$$

$$= 1.4(58) + 1.7(38) = 145.8 \text{ kip-ft [197.7 kN-m]}$$

(2) With the capacity reduction factor of 0.90 applied, as discussed in Section 18-3, the required moment capacity of the section is determined as

$$M_t = \frac{M_u}{0.90} = \frac{145.8}{0.90} = 162 \text{ k-ft} \quad \text{or} \quad 1944 \text{ kip-in [220 kN-m]}$$

(3) From Table 18-1 we obtain the following factors for a balanced section: $a/d = 0.377$, $\rho = 0.0214$, $R = 1042$. To find the minimum required effective depth d, we use the formula for resisting moment,

$$M = Rbd^2 = (1042)(10)d^2 = 1,944,000$$

from which

$$d^2 = 186.6, \quad d = \sqrt{186.6} = 13.7 \text{ in.}$$

If the limiting depth is used, the required steel area may be found using the balanced percentage factor, thus

$$A_s = \rho(bd) = 0.0214(10 \times 13.7) = 2.93 \text{ in}^2$$

If a greater depth than that required for the balanced section is used, the balanced factors from Table 18-1 for a/d and ρ will not apply. Design aids are available from various handbooks to assist in the design of such a section. However, a relatively simple approach is to assume some value for a and to use it in a simple two-step procedure. To estimate a value for a, we first determine the balanced value for a as follows:

$$a/d = 0.377, \quad a = 0.377(d) = 0.377(13.7) = 5.16 \text{ in.}$$

If a greater value is used for d, the value of a will be less than this, since the compression force is reduced as the internal moment arm $(d - a/2)$ increases. Let us assume a d of 20 in. and, for a first trial, assume a value of 4 in. for a. Then from the formula for the resisting moment based on the steel, we find A_s as

$$A_s = \frac{M_t}{f_y(d - a/2)} = \frac{1{,}944{,}000}{60{,}000(20 - 2)} = 1.80 \text{ in}^2$$

With this steel area the actual percentage of steel is

$$\frac{A_s}{bd} = \frac{1.80}{10 \times 20} = 0.0090$$

and with the formula derived for a in Section 18.4,

$$a = \frac{df_y}{0.85f_c'} = \frac{(0.0090)(20)(60)}{(0.85)(4)} = 3.18 \text{ in.}$$

This indicates that our first guess for a was high and that the internal moment arm is even larger. Thus the required steel area is actually slightly less and will result in an a value less than 3.18. We therefore try a second time with an a value of 3.0 in. and determine the new value for A_s to be

$$\frac{1{,}944{,}000}{60{,}000(20 - 1.5)} = 1.75 \text{ in}^2$$

Since this results in a change of only about 3% in the steel area, further correction is of no practical value, and the steel bars may be selected for the revised area of 1.75 in^2.

The ACI Code stipulates various other restrictions for beams, including a requirement for a minimum amount of reinforcing. Note that the procedure in the preceding example dealt only with consideration of bending moment. In a real design situation, consideration must be given also to shear and the required development length for the bars.

Problem 18-5-A-B-C. Using f'_c = 3 ksi [20.7 MPa] and f_y = 50 ksi [345 MPa], find the minimum depth required for a balanced section for the given data. Also find the area of reinforcement required if the depth chosen is 1.5 times that required for the balanced section. Use strength design methods.

Moment Due To

	Dead Load		Live Load		Beam Width	
	(kip-ft)	(kN-m)	(kip-ft)	(kN-m)	(in.)	(mm)
A	40	54.2	20	27.1	12	305
B	80	108.5	40	54.2	15	381
C	100	135.6	50	67.8	18	457

18-6   Scope of Ultimate-Strength Design

The strength design method has applications far beyond beam design. It is particularly significant in the design of columns and in rigid frames where all members are subjected to combined bending and direct force. The reader who wishes to study strength design and its applications further is referred to the following publications: *Reinforced Concrete Fundamentals,* 4th ed., by Phil M. Ferguson (New York, Wiley, 1979) and *Structural Design Guide to the ACI Building Code,* 2d ed., by P. F. Rice and E. S. Hoffman (New York, Van Nostrand Reinhold, 1979).

V

ROOF TRUSSES

|||||||||||||||||||||||||||||||||||||||||||||||||||||||||||||||||||||

19

Forces in Trusses

||||||||||||||||||||||||||||||||||||||||||||||||||||||||

19-1   Introduction

A truss is a framed structure composed of straight members so arranged and connected at their ends that the forces in the members are either tension or compression. Basically a truss consists of a number of triangles framed together. (See Fig. 19-1.)

The terminology used for the typical elements of a truss is illustrated in Fig. 19-1. A highly significant relationship for a truss is the ratio of the span to the rise or height; this is a critical factor affecting the relative efficiency of the truss. While this ratio may be quite high for beams or slabs, it must be relatively low for a truss if it is to achieve a significant structural efficiency.

Trusses are ordinarily used in sets or series, in the manner of beams or girders in ordinary framing systems. Fig. 19-2 shows the elements of a typical roof truss system utilizing intermediate spanning elements and a surfacing deck. While the sloping-top or gable-form truss is a common one for roofs, trusses with flat top and bottom chords are used for floor structures and for flat roofs.

19-2   Loads on Trusses

The first step in the design of a truss consists of computing the loads the truss will be required to support. These comprise both dead and live loads. The former includes the weight of all con-

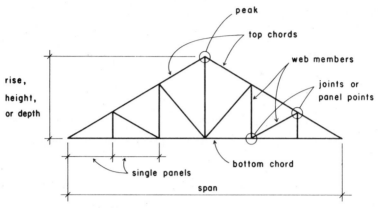

FIGURE 19-1.   Elements of a truss.

struction materials supported by the truss, and the latter includes loads resulting from snow, wind, occupancy of the building, construction activity, and possible ponding of water when drainage is impaired on flat roofs.

Precise values for dead loads cannot be established until the truss is completely designed and all other details of the construc-

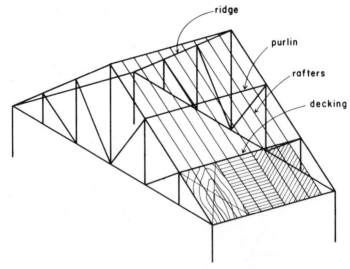

FIGURE 19-2.   Elements of truss systems.

tion determined. Since this information is usually not completely available at early stages of design work, reasonable estimates must be used for the dead loads; these must then be confirmed when better information is obtained.

Minimum design live loads are specified by local building codes. Where snow accumulation is a potential problem, the live load for roofs is usually based on this factor with data from local weather histories. Otherwise the specified roof load essentially is intended to provide some capacity for sustaining loads experienced during construction and maintenance of the roof. The basic required roof load can usually be reduced by some percent when the roof slope is of some significant angle or when a large area of roof surface is supported by a single truss. Magnitudes of design wind pressures and various other requirements for wind design are specified by local building codes. The code in force for a specific building location should be used for any design work.

19-3   Analysis for Internal Forces in Truss Members

For simple planar trusses, the analysis performed for design generally proceeds as follows:

1. Determination of the truss loadings. This consists of the separate computation of the dead loads, live (gravity) loads, wind loads, and—where required—earthquake loads. The truss is then analyzed separately for each loading.

2. Determination of the forces (reactions) generated at the truss supports by each of the loadings.

3. Determination of the internal forces in the truss members caused by each loading.

4. Determination of the critical combination of internal forces for each individual truss member.

For simple truss action the loads are assumed to be applied to the truss at its joints. For each loading an individual truss member will typically sustain some magnitude of either axial tension or axial compression. For different combinations of the loads, some members may sustain a reversal of the character of the internal force, that is, a net tension with one combination and a net compression with another. Where this reversal effect occurs, it is necessary to design the member both as a tension element and as a compression element. Otherwise, members are designed as either one or the other.

For statically determinate trusses, the analysis for support reactions and internal forces can be performed by employing the ordinary procedures for investigation of static equilibrium. This work may be executed either by algebraic methods or by special graphic techniques. For a complete discussion and illustration of these procedures, the reader is referred to *Simplified Design of Building Trusses,* 3d ed., by James Ambrose (New York, Wiley, 1982). An example of the application of these procedures is given in the next section.

19-4   Analysis and Design of a Roof Truss

The following example illustrates the principal considerations in the design of a simple roof truss. The loading conditions are limited to those of dead and live gravity loads. Although wind must be considered in the design of any roof truss, it is often the case that the wind load is not critical where local wind conditions are not severe. This is due to the fact that codes generally permit an increase in allowable stresses when wind is included in loading combinations; thus the wind load is not of significance unless it is of considerable magnitude.

The general form of the truss and the two loading conditions are shown in Fig. 19-3. Also shown in the figure are the reaction forces at the supports. Although the truss is not symmetrical, the loadings are; thus the reaction forces are equal at both supports.

It may be noted that the loadings are similar in distribution on the truss and differ only in magnitude. This situation permits the performance of a single analysis for internal forces, since the values for one loading may be simply determined by proportion from the analysis for the other loading. In fact, the analysis may be performed using a hypothetical unit loading, with the values for both real loadings determined by proportion.

Figure 19-4 illustrates an algebraic analysis by the *method of joints.* In Fig. 19-4*a* each joint of the truss is displayed with all of

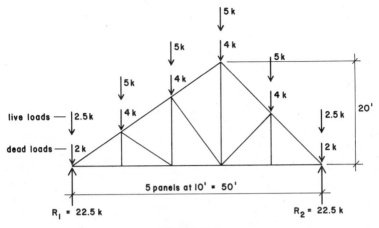

FIGURE 19-3.

the forces that are possible at each joint, consisting of the loads, the support reactions, and the internal forces exerted by the truss members. Each joint constitutes an individual set of coplanar, concurrent forces for which static equilibrium can be established by satisfying only two conditions: $\Sigma F_v = 0$ and $\Sigma F_h = 0$. That is, the sum of the vertical components of force is zero, and the sum of the horizontal components of force is zero.

In Fig. 19-4a the forces due to the loads and reactions are completely represented, while the forces due to the members are indicated in direction only. Application of the conditions of statics permits the determination of the unknown internal forces at any joint where not more than two unknown forces exist. On this basis the analysis of forces at either of the supports can be achieved initially. Where more than two unknown internal forces occur at a joint, the results of analysis of other joints must first be obtained to reduce the number of unknowns to that limited by the number of conditions of equilibrium.

It may be noted in Fig. 19-4b that vertical and horizontal members are represented by a single force while sloping members are represented by two components—vertical and horizontal. This is done to simplify the algebraic analysis, although the number of forces is actually increased. At such joints additional algebraic

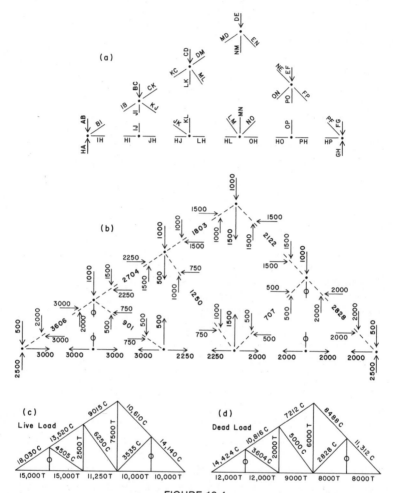

FIGURE 19-4.

conditions can be used, consisting of the known value of proportionality between the components for a sloping member. (For example, the components are equal if the slope is 45°, etc.)

The complete results of the algebraic analysis are displayed in Fig. 19-4b. Both the component forces and the actual net axial force are shown for each sloping member. The former may be used to verify the conditions of equilibrium at the joints, while the latter is used for actual design of the member.

The results of the analysis for internal forces are displayed in Fig. 19-4c (for live load) and in Fig. 19-4d (for dead load). These have been directly proportioned from the results displayed for the unit loading in Fig. 19-4b. A net tension force is indicated as T and a compression force as C. Some so-called *zero stress members* occur due to the manner in which the loading is assumed to be applied. These members usually serve some useful purpose but may truly be quite insignificantly stressed by the truss action.

For this example it is not really necessary to perform separate analyses for the dead and live loadings. For design purposes the results of the two loadings will simply be added. However, when wind loads or other loading situations also occur, it is often necessary to consider various combinations, all of which will include dead load but not necessarily live load. Thus in practice it is usually necessary to consider the loads separately.

The results of the analysis in this example will be used in the next chapter for the example of the design of a steel truss.

19-5 Graphical Analysis of Trusses

For most planar trusses it is possible to perform an approximate analysis for internal forces by means of a relatively simple graphic construction. Figure 19-5 shows the elements that are utilized for such a construction, the first of which is a scaled layout of the truss and its loads, as shown in Fig. 19-5a. On this figure—called a space diagram—letters are placed between each of the external forces and in each of the interior triangular spaces of the truss. These letters may be used to designate each internal and external force, using a two-letter identification. Thus the left reaction is designated as *HA*, the top chord member at the left end as *BI*, and so on.

The Maxwell diagram, shown in Fig. 19-5b, is constructed to find the values of the internal forces. The process consists of constructing individual force polygons for each of the truss joints. The graphic method therefore has a direct relationship to the algebraic method of joints, and just as with that method, a joint cannot be solved if there are more than two unknown forces at

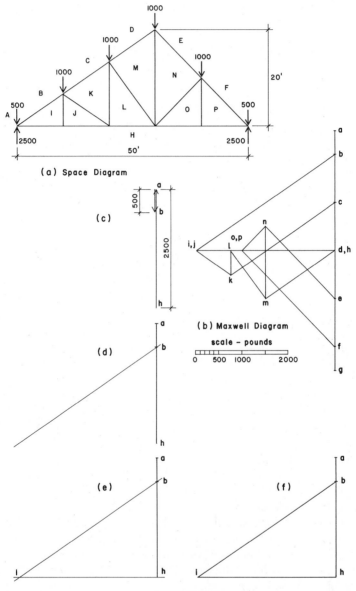

(a) Space Diagram

(c)

(b) Maxwell Diagram

scale - pounds

0   500 1000   2000

(d)

(e)

(f)

FIGURE 19-5.

the joint. Thus the sequence in which the polygons are constructed is restricted in the same manner as for the method of joints.

Let us first demonstrate the procedure for the construction of the individual joint force polygons. We will do this by constructing the polygon for the joint at the left support. This joint is chosen because there are, initially, only two unknown forces at the joint. We note that there are four forces at the joint and that the value of two (the reaction and the load) are known. The construction is therefore begun by laying out the two known forces, as shown in Fig. 19-5c. On the force polygon we use lowercase letters for the ends of the forces, corresponding to the two-letter designations established on the space diagram. Thus on the polygon, *ha* designates the vector for the reaction and *ab* the vector for the load.

It now remains to add the vectors for *bi* (the top chord) and *ih* (the bottom chord) on the polygon. This amounts to locating point *i* on the polygon, since points *b* and *h* are already determined. To do this we note that the vector for *bi* will be in a direction parallel to the top chord and will pass through point *b*. A line is therefore drawn through point *b* in the proper direction, as shown in Fig. 19-5d. We then proceed in a similar manner to establish the vector for *ih* by drawing a horizontal line through *h*, as shown in Fig. 19-5e. The intersection of these two lines locates the unknown point *i*, and the completed force polygon is shown in Fig. 19-5f.

The two-letter designations for the forces could be written in either sequence, that is, either *ha* or *ah* for the reaction. However, for the construction of the Maxwell diagram—which is a composite of the force polygons for all the truss joints—it is necessary to follow a consistent procedure for the two-letter designations. This procedure consists of always reading around the joint in a clockwise manner to establish the two-letter designations.

When the Maxwell diagram is completed, the internal force values may be determined by measurement of the appropriate vectors for the members, using the same scale that was used to construct the vectors for the known forces—the loads and reactions. The character (tension or compression) of the forces can be

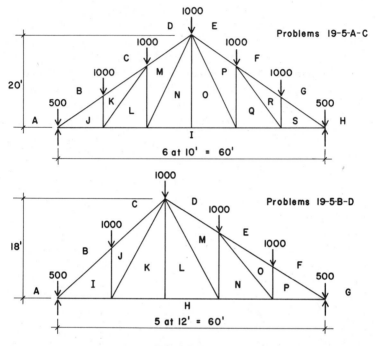

FIGURE 19-6.

determined by reading the individual joint polygons in the clock-wise sequence noted for the joint. Thus we observe that reading from b and i on the force polygon we move downward to the left. At the joint this direction of movement indicates that the force acts toward the joint; that is, the force HI is a compression force.

Using the scale shown in the figure, we can verify that the values for the internal force vectors correspond with reasonable accuracy to the values shown on Fig. 19-5b, as obtained from the algebraic analysis.

Problem 19-5-A-B. Determine the character and magnitude of the internal forces in all of the members of the trusses shown in Fig. 19-6, using the algebraic method of joints.

Problem 19-5-C-D. Determine the character and magnitude of the internal forces in all of the members of the trusses shown in Fig. 19-6, using graphic methods.

20

Design of a Steel Truss
||||||||||||||||||||||||||||||||||||||||||||||||||||||||||

20-1 Introduction

The example in this chapter illustrates the process of design for a light steel roof truss. The form and loading for the truss is that shown in Fig. 19-3, and the truss will be designed for the internal forces computed in the example in the preceding chapter. Although there are various options for the construction of such a truss, the example will employ the use of double-angle members and joints with high-strength bolts and steel gusset plates. Explanations of the design of such elements, and data necessary for the work, are provided in the chapters in Part 2.

20-2 Selection of Truss Members

The truss web members will of necessity each be individual elements. However, it is common to make chords of a single element continuous through several panels, limited only by the available lengths of the steel shapes from suppliers. These length limits must be determined by inquiry with local suppliers but are usually of a general nature. For this truss it would in theory be possible to make each of the sloping top chords and the horizontal bottom chord of a single-piece element. This is least likely for the bottom chord, since the element would be approximately 50 ft long; thus at least one splice should be anticipated for the bottom chord. For

433

the long top chord, the total length required is approximately 36 ft, which is questionable but not completely unimaginable.

Based on a reasonable speculation of available lengths, we will assume the truss member layout shown in Fig. 20-1, which indicates 11 different elements for the truss. Each of these elements must be individually designed, and the data for the design is summarized in Table 20-1. The selections indicated in the table are based on the use of A36 steel for the angles and $\frac{3}{4}$-in. A325-type bolts for the connections. The reader is referred to the appropriate sections in Part 2 for the design of axially loaded tension and compression members and bolted joints.

The following are some of the general considerations that would be made in selecting the truss members and developing the details for the truss construction:

1. Selection of the type and size of connector. This should be done with consideration of current steel fabricating practices. High-strength bolts are favored, since they will produce truss joints with little deformation. The size of the bolts must be matched to the sizes of the truss members; for this truss we will use all $\frac{3}{4}$-in. bolts.

2. Selection of thickness for the gusset plate. This must be appropriate for the size of the members and size and capacity of the bolts. Principal concerns for the gusset plate are its buckling in compression and its tearing in tension (Section 7-5). A thickness commonly used for trusses of this size is $\frac{3}{8}$ in., although careful analysis may determine that a thinner plate is possible.

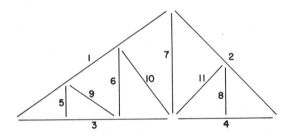

FIGURE 20-1.   Configuration of the truss members.

TABLE 20-1. Design Data for the Truss Members

Truss Member (see Fig. 20-1)	Axial Force[a] (kips)	Design Length (ft)	Minimum Required Radius of Gyration[b] r (in.)	Required Net Area[c] A (in²)	Selection from Table 6-5 2 Angles, Long Legs Back-to-Back	Actual Capacity of Member[d] (kips)
1	32.45C	12	0.72	—	4 × 3 × 5/16	47.1C
2	25.45C	14.14	0.85	—	4 × 3 × 5/16	34.8C
3	27.0T	10	0.50	0.931	3 × 2 × 1/4	31.7T
4	18.0T	10	0.50	0.621	2 1/2 × 2 × 1/4	24.5T
5	0	6.67	0.33	—	2 1/2 × 2 × 1/4	—
6	4.5T	13.33	0.67	0.155	2 1/2 × 2 × 1/4	24.5T
7	13.5T	20	1.0	0.466	2 1/2 × 2 × 1/4	24.5T
8	0	10	0.50	—	2 1/2 × 2 × 1/4	—
9	8.1C	12	0.72	—	2 1/2 × 2 × 1/4	9.4C
10	11.25C	16.7	1.0	—	3 1/2 × 2 1/2 × 1/4	12.8C
11	6.36C	14.14	0.85	—	3 × 2 × 1/4	9.9C

[a] From investigation performed in Section 19-4. Sum of values for dead load and live load, Fig. 19-4. Sum of values for dead load and live load, Fig. 19-4*c* and *d*.
[b] $L/240$ for tension member; $L/200$ for compression member; allowable $F_t = 29$ ksi.
[c] Based on stress at bolt holes; allowable $F_t = 29$ ksi.
[d] For tension: actual net area of connected leg only at 29 ksi. For compression: actual gross area times allowable stress for axial compression based on actual L/r.

435

3. Selection of the type of angle (equal or unequal legs), the angle arrangement (long legs back-to-back or short legs back-to-back, if unequal legs), and the minimum-size member to be used. There are various considerations to be made for this selection. For the example, we will use unequal-leg angles with long legs back-to-back and will limit our choices to the sizes given in Table 6-5. Based on the $\frac{3}{4}$-in. bolt, normal angle-gage lines (Table 7-4), and minimum edge distance (Table 7-2), the smallest angle leg that can be bolted is 2.5 in. Our minimum member from Table 6-5 will thus be the $2\frac{1}{2} \times 2 \times \frac{1}{4}$ angle.

4. Support conditions and details. Trusses may be supported in a number of ways: bearing on a masonry wall, sitting on top of a steel column, framing into the face of a column, and so on. Details of the truss joint at the support will be different for different types of support.

5. Framing of the supported roof or floor. When trusses are widely spaced, some type of framing will generally be used to span between trusses, and provision must be made for the connection of these elements to the trusses. For closely spaced trusses, structural decks may be directly supported by the chords of the trusses.

6. Support of miscellaneous elements. In addition to supporting roofs and floors, trusses may also support other items such as ceilings, lighting fixtures, ductwork, catwalks, and so on. Requirements for attachment of these items and the effects on the chord members of the truss will require consideration.

7. Lateral bracing of the trusses and effective design lengths of truss members. Most trusses require some form of lateral bracing. When every joint of the truss is laterally braced, the effective design length (for lateral buckling) of all truss members will be their joint-to-joint length. When the lateral bracing occurs less frequently, the truss chords may have unbraced lengths that are longer.

8. Combined functions of chord members. When loadings such as those discussed in items (5) and (6) occur, the truss

chords often must be designed for beam-spanning effects in addition to the axial tension or compression forces due to the truss action.

For this example we have assumed that the effective length of all truss members is the joint-to-joint length and that the members sustain only the forces due to the truss action.

20-3 Design of the Truss Joints

Figure 20-2 shows a possible layout for one of the bottom chord joints of the truss. In general, it is desirable to make the joint as compact as possible, as this reduces the required size of the gusset plate. For prevention of twisting in the joint, the action lines of the members should intersect at a common point. While these action lines should in theory be along the centroidal axes of the members, it is more common to use the gage lines for the bolts, as shown in the illustration.

It would be possible to make the joint slightly more compact by trimming the ends of some members or by cutting them at an angle. However, the cost of such special cutting would probably more than offset the resulting savings due to the slightly smaller gusset plate.

The number of bolts used in the end of each member will normally be based on the axial force in the member. The code requires a minimum of two bolts for each member, and for this

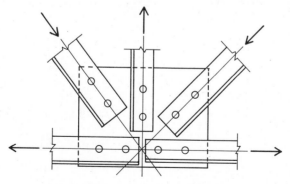

FIGURE 20-2.   Layout of the truss joint.

joint it happens that the capacity of two of the chosen bolts exceeds all of the member forces. Thus only two bolts are required for all the members at this joint.

The layout of the joint is subject to the various considerations of bolt spacing, edge distance, tearing, and so on, as discussed in Chapter 7.

Problem 20-3-A. Select all of the truss members and design the layout for one bottom chord joint for the truss in Problem 19-5-A. Use $\frac{3}{8}$-in. gusset plates, A36 steel for the members, and A325 bolts for the joints.

Problem 20-3-B. Select all of the truss members and design the layout for one bottom chord joint for the truss in Problem 19-5-B. Use $\frac{3}{8}$-in. gusset plates, A36 steel for the members, and A325 bolts for the joints.

Answers to Selected Problems

||||||||||||||||||||||||||||||||||||||||||||||||||||||||||||||||

The answers given below are for those problems marked with asterisks (*) in the text. In general, numeric answers are carried to three significant figures except in cases where additional accuracy seemed desirable as an aid in interpreting the result.

Chapter 1

1-7-A.   3.33 in^2 [2150 mm]

1-7-C.   Required area = 0.60 in^2 [387 mm^2]; diameter = 0.874 in. ($\frac{7}{8}$ in.) [22.2 mm]

1-7-F.   Area = 21.8 in^2 [14,065 mm^2]; load = 196 kips [872 kN]

1-15-A. 19,333 lb [86.0 kN]

1-15-C. E = 29,550 ksi [204 GPa]

Chapter 2

2-6-B. R_1 = 3 kips [13.34 kN]; R_2 = 3 kips [13.34 kN]

2-6-E. R_1 = 14.4 kips [64.1 kN]; R_2 = 9.6 kips [42.7 kN]

2-6-G. R_1 = 10,100 lb [44.925 kN]; R_2 = 11,900 lb [52.931 kN]

2-7-B. R_1 = 4.38 kips [19.5 kN]; R_2 = 5.62 kips [25.0 kN]

2-7-D. R_1 = 4430 lb [19.7 kN]; R_2 = 7570 lb [33.7 kN]

2-7-H. R_1 = 5625 kips [25.02 kN]; R_2 = 4375 lb [19.46 kN]

2-8-C. R_1 = 6092 lb [27.1 kN]; R_2 = 2608 lb [11.6 kN]

2-8-D. R_1 = 2.46 kips [10.94 kN]; R_2 = 3.14 kips [13.97 kN]

2-8-I. R_1 = 6750 lb [30.0 kN]; R_2 = 5250 lb [23.35 kN]

Chapter 3

3-3-B. Max V = 1827 lb [8.10 kN]; V = 0 at x = 9 ft [2.7 m]

3-3-C. Max V = 9050 lb [40 kN]; V = 0 at x = 11.22 ft [3.42 m]

3-3-G. Max V = 4333 lb [19.1 kN]; V = 0 at x = 9 ft [2.7 m]

3-6-A. Max V = 5250 lb [23.35 kN]; max M = 18,375 ft-lb [24.92 kN-m]

3-6-B. Max V = 1114 lb [4.96 kN]; max M = 4286 ft-lb [5.81 kN-m]

3-6-D. Max V = 13.67 kips [60.8 kN]; max M = 65.4 k-ft [86.7 kN-m]

3-6-J. Max V = 6700 lb [29.8 kN]; max M = 28,300 ft-lb [38.4 kN-m]

3-7-A. Max M = 44 k-ft [59.7 kN-m]

3-7-C. Max M = 22,300 ft-lb [30.2 kN-m]

3-7-G. Max M = 15 k-ft [20.34 kN-m]

3-8-B. Max V = 1500 lb [6.7 kN]; max M = 9500 ft-lb [12.9 kN-m]

3-8-D. Max V = 2700 lb [12.01 kN]; max M = 12,750 ft-lb [17.29 kN-m]

3-12-A. Max V = 1.5 P; max M = $PL/2$

3-12-B. Max V = 2 P; max M = 3 $PL/5$

Chapter 4

4-4-A. c_y = 2.6 in. [66 mm]

4-4-D. c_x = 1.293 in. [32.8 mm]; c_y = 3.42 in. [86.9 mm]

4-5-B. 205.3 in^4 [85.4 × 10^6 mm^4]

4-5-F. 682.3 in^4 [284 × 10^6 mm^4]

4-8-A. 399.8 in^4 [166 × 10^6 mm^4]

4-8-B. 420 in ^{4} [175 × 10^6 mm^4]

4-9-C. r = 5.31 in. [134.9 mm]

Chapter 5

5-10-B.   W 12 × 22
5-10-D.   W 24 × 55
5-10-G.   W 12 × 19
5-10-J.   W 16 × 26
5-11-A.   80.39 kips [358 kN]
5-12-A.   0.833 in. [21.2 mm]
5-13-A.   W 10 × 12, W 5 × 19
5-13-E.   W 18 × 35, W 12 × 53
5-15-A.   W 14 × 34 is adequate
5-15-B.   W 16 × 40 or W 18 × 40
5-24-A.   17 × 9 × $\frac{11}{16}$ in. [432 mm × 229 mm × 17 mm]
5-25-A.   24H8
5-25-C.   22H6

Chapter 6

6-3-A. $L/r = 95$
6-6-A.   430 kips [1913 kN]
6-7-A.   W 8 × 31
6-9-B.   TS 4 × 4 × $\frac{1}{4}$
6-10-A.   78 kips [347 kN]
6-12-D.   W 14 × 61
6-12-H.   (approximately) PL 1.25 × 15 × 16 in. [32 × 380 × 410 mm]

Chapter 7

7-6-A.   6 bolts; outer plate $\frac{7}{8}$ in. [22 mm], inner plate 1 in. [25 mm]

Chapter 8

8-5-A. $L_1 = 11$ in. [279 mm]; $L_2 = 5$ in. [127 mm]

Chapter 10

10-4-A. 3×12

10-5-A. Maximum shear stress is 83.1 psi; allowable stress is 85 psi; beam is adequate

10-6-A. Allowable deflection is 0.8 in. [20 mm]; actual deflection is 0.33 in. [8.4 mm]; beam is adequate

10-6-D. Allowable deflection is 0.8 in. [20 mm]; actual deflection is 0.30 in. [7.6 mm]; beam is adequate

10-9-A. 2×10

10-10-A. 2×6

Chapter 11

11-3-A. 7.82 kips [35 kN]

11-3-E. 8×8

Chapter 12

12-2-A. 18.5 kips [82 kN]

Chapter 16

16-5-A. 12 in. square; four No. 11 bars

16-5-C. 20 in. square; four No. 9 bars

16-6-A. 14 in. round; six No. 9 bars

16-6-C. 24 in. round; eight No. 11 bars

Index

||||||||||||||||||||||||||||||||||||||||||||||||||||||

443